Learn Aspen Plus in 24 Hours

ABOUT THE AUTHOR

Thomas A. Adams II is an associate professor and associate chair in the Department of Chemical Engineering at McMaster University in Hamilton, Ontario, Canada. He received dual bachelor's degrees from Michigan State University in 2003, one in chemical engineering and the other in computer science. He received his Ph.D. in 2008 from the University of Pennsylvania under the supervision of Prof. Warren D. Seider and completed a postdoctoral appointment under Prof. Paul Barton at the Massachusetts Institute of Technology. He is also a licensed professional engineer, and is an associate editor with the *Canadian Journal of Chemical Engineering* and with *Frontiers in Energy Research.*

Professor Adams' research focuses on the design and simulation of sustainable energy conversion systems, including areas such as synthetic fuels, alternative fuels, biofuels, fuel cells, waste-to-energy, integrated community energy, polygeneration, and process intensification. The primary goal of his research is to create new chemical process systems and devices which will lead to worldwide global change in the way we make and use energy, following the principles of the triple-bottom-line of sustainability. He has over 20 years of experience using Aspen Plus, as well as other related software, for research and problem solving.

Professor Adams has received numerous awards for his research and service, including the *Canadian Journal of Chemical Engineering*'s Lectureship Award, the Canadian Society for Chemical Engineering's Emerging Leader of Chemical Engineering Award, membership in *Industrial & Engineering Chemistry Research*'s 2018 Class of Influential Researchers, an Ontario Early Researcher Award, a Joseph Ip Distinguished Engineering Fellowship, the President's Award for Excellence in Graduate Supervision, and the American Institute of Chemical Engineers' David Himmelblau Award for Innovations in Computer-Based Chemical Engineering Education, and he is now honored as a University Scholar at his home institution of McMaster University. He has published over 75 research articles in peer-reviewed journals which used Aspen Plus or related products as a key component in the research methodology. His research has been featured in the popular media such as Bloomberg TV, *Wired, Scientific American,* the Discovery Channel, and various radio programs in the United States and Canada. But he is much more proud of the accomplishments of his graduate students, who include a Vanier Scholar, an Ontario Trillium Scholar, and a Governor-General's Medal recipient, and who are active researchers and engineers all over the world.

Learn Aspen Plus in 24 Hours

THOMAS A. ADAMS II

Second Edition

CACHE
Computer Aids
for Chemical Engineering

New York   Chicago   San Francisco   Athens   London
Madrid   Mexico City   Milan   New Delhi
Singapore   Sydney   Toronto

Library of Congress Cataloging-in-Publication Data

Names: Adams, Thomas A., II, author.
Title: Learn Aspen Plus in 24 hours / Thomas A. Adams, II.
Description: Second edition. | New York : McGraw Hill, 2021. | Includes
    bibliographical references and index.
Identifiers: LCCN 2021044703 | ISBN 9781264266654 | ISBN 9781264266661 (ebook)
Subjects: LCSH: Aspen plus. | Chemical processes—Computer simulation. |
    Chemical process control—Computer programs.
Classification: LCC TP155.7 .A28 2021 | DDC 660/.280028553—dc23/eng/20211108
LC record available at https://lccn.loc.gov/2021044703

McGraw Hill books are available at special quantity discounts to use as premiums and sales promotions or for use in corporate training programs. To contact a representative, please visit the Contact Us page at www.mhprofessional.com.

Learn Aspen Plus in 24 Hours, Second Edition

References and screen images from Aspen Plus®, Aspen Energy Analyzer®, Aspen Capital Cost Estimator®, and Aspen Simulation Workbook® are reprinted with permission from Aspen Technology, Inc. AspenTech®, Aspen Plus®, Aspen Energy Analyzer®, Aspen Capital Cost Estimator®, Aspen Simulation Workbook®, aspenONE®, and the AspenTech leaf logo are trademarks of Aspen Technology, Inc. All rights reserved.

1 2 3 4 5 6 7 8 9   CCD   26 25 24 23 22 21

ISBN 978-1-264-26665-4
MHID 1-264-26665-0

Sponsoring Editor
Robin Najar

Editorial Supervisor
Stephen M. Smith

Production Supervisor
Lynn M. Messina

Acquisitions Coordinator
Elizabeth M. Houde

Project Managers
Adwiti Pradhan and Anamika Singh, MPS Limited

Copy Editor
Bindu Singh, MPS Limited

Proofreader
Subashree Baskaran, MPS Limited

Indexer
Edwin Durbin

Art Director, Cover
Jeff Weeks

Composition
MPS Limited

CONTENTS

ABOUT THIS BOOK

The material for this book (starting with the first edition) was developed at McMaster University over a 10-year period for an undergraduate course focused on problem solving strategies that use chemical process flowsheeting software. The primary objective of the course is to teach students how to solve problems relating to chemical processes and conceptual process design. The text in this book is intended for students and practitioners to teach themselves how to use the software in a series of twelve 2-hour guided tutorials in our undergraduate computing labs. Plus, for this second edition of the book, we added several bonus tutorials, based on reader feedback and suggestions.

This is not a user guide to Aspen Plus! If you are looking for specific details on a specific model or feature, you should consult the user guide or help files included with the program. Instead, this book will help you teach yourself how to solve problems using the software. It will provide readers with the ability to select and use the appropriate tools for solving many kinds of chemical engineering problems related to chemical processes, separations, reactions, mass transfer, heat transfer, and thermodynamics. It is geared toward the undergraduate level and informed by a large amount of undergraduate student feedback, but graduate students and professionals will also find this book very helpful in getting up and running quickly.

This book uses Aspen Plus V12, the latest version available at the time of writing. If you are using older or newer versions, the book will still be very useful, since most features and problem solving principles remain essentially unchanged from version to version. Over time, new editions will be released with pertinent updates as they come. The Computer Aids for Chemical Engineering Corporation (or CAChE Corp.), which has provided funding in support of this book, is hoping to develop a larger body of materials which encompass many different computer-aided process engineering tools ("CAPE tools" as they are commonly called) and software far beyond the scope of Aspen Plus. If you would like to find out about other software, tools, or methods, or would like to contribute your own chapters and modules to future editions, you are encouraged to visit our website at:

http://PSEcommunity.org/

For solution files and source code, and a handy list of hyperlinks to all of the linked material in the book (helpful for print-edition readers), go to

http://PSEcommunity.org/books/lap24

This book was made possible by the support of many teaching assistants, reviewers, and contributors who have contributed in a number of ways over the years since the first edition. A huge thank you to Tia Ghantous (McMaster) for extensive testing, upgrading, fact checking, challenging, and editing with regard to updating this book for the second edition, and finding (hopefully!) all my mistakes. Thank you especially to Dr. Jaffer Ghouse (U.S. Department of Energy), Dr. Vida Meidanshahi (McMaster), Prof. Jake Nease (McMaster), Prof. Yaser Khojestah Salkuyeh (Concordia University), and Trevor West (McMaster) for making edits, additions, suggestions, and other contributions. A big thank you to Dr. Chinedu Okoli (US DOE) for the original development of Tutorials 4 and 11. Thank you to Prof. Scott Guelcher (Vanderbilt), Prof. Russel Dunn (Vanderbilt), Prof. Fernando Lima (West Virginia), Prof. Fengqi You (Cornell), Prof. Debangsu Bhattacharyya (West Virginia), and Prof. Mario Eden (Auburn) for peer-reviewing the material or providing helpful critical feedback and ideas across the different editions of the book. A special thank you as well to the CAChE Corporation for providing financial support for editing and development costs. Finally, a most special thank you to my wife, Ariane, who is basically a superhero raising five young kids during the pandemic with wisdom, grace, and patience not found in mere mortals.

INTRODUCTION

Aspen Plus is a computer-aided process engineering (CAPE) tool which has been in continual development for several decades. Its primary use is to aid in the rapid computer simulation of chemical plants that operate at steady state, although it can do some basic batch simulations as well. Aspen Plus contains a collection of mathematical models for different kinds of chemical process equipment such as heat exchangers, pumps, compressors, turbines, distillation columns, absorbers, strippers, and chemical reactors. A mathematical model is essentially a collection of equations which describe the important parts of the equipment and how it works. Users can select from different pre-made models, enter in key information about how it is used (such as the chemicals involved, temperatures, pressures, flow rates, sizes, and dimensions), and then use the model to compute unknown pieces of information (such as reaction conversions, efficiencies, performance criteria, output conditions, energy usage, and costs). Although some of the models might be simple enough to use "by hand" individually, the real power of the software is the ability to link together hundreds of models into a process system, thus constructing a large model for an entire chemical plant containing potentially millions of equations. The user can then run a simulation using the model, which essentially means to solve the equations in order to find the important unknowns about the process. To do this, Aspen Plus contains a variety of time-tested algorithms which are useful and often very effective in solving the system of equations quickly and accurately.

The models in Aspen Plus are quite generic. This means that they can be used for many different kinds of applications with many different chemicals. For example, a heat exchanger model can be used to compute how much energy it takes to heat a certain chemical from one temperature or another. To do this, the models need information about the chemicals involved. The heat exchanger model, for example, needs to know not just what chemical is being heated, but the heat capacity of the chemical involved (which usually changes with temperature), and perhaps other information such as the boiling point and heat of vaporization of the chemical if it goes through a phase change. If there is a mixture of chemicals inside the heat exchanger, then it needs to know this information for all of the individual chemicals as well as how it should handle the effects of mixing.

Finding this information in the literature can be quite tedious, time consuming, and even expensive, especially for simulations with many chemicals. Fortunately, Aspen Plus contains a massive database (known as Aspen Properties) containing physical property information on literally hundreds of thousands of chemicals. This includes correlations for heat capacities, thermal conductivities, viscosity, surface tension, molecular weights, densities, and critical properties. It contains parameters for equations of state models that connect temperature, pressure, enthalpy, entropy, molar volume, and fugacity, such as Peng-Robinson, Soave-Redlich-Kwong, Chao-Seader, PC-SAFT, Non-Random Two-Liquid (NRTL), UNIQUAC, and many others. It even has the capability of using certain theoretical methods (such as UNIFAC) to predict parameters where information is missing or for chemicals which are not in the database at all, just from the structure of the molecule itself. However, it is very important to recognize that these are just models, and that they can have varying degrees of accuracy depending on the temperature, pressure, and mixture conditions in which they are used. Fortunately, the database also contains a very large amount of experimental measurements for physical properties, so you can quickly determine how well the models you have chosen to use match relevant experimental data. Overall, the physical property models and data included with the software is one of its most useful features.

On top of this, Aspen Plus includes a direct connection to Aspen Capital Cost Estimator, which can be used to estimate the capital costs of a piece of equipment with remarkable levels of detail. Based on plant construction data and very detailed cost models (e.g., literally including a line item called nuts and bolts, or counting the number of coats of paint), these estimates are routinely updated and provide a much more accurate and rigorous estimation of the capital costs of a piece of equipment than general correlations found in many process design textbooks. This is combined with recently added features that make it easy to compute the costs of utilities and even the global warming potential of using those utilities (e.g., the carbon dioxide emissions associated with burning natural gas for heat), with suggested values provided based on recent scientific studies. Moreover, these models are updated with each new version of the software, maintaining a timely relevance. Altogether, the software is extremely useful for quickly constructing rigorous chemical process models for the purposes of process simulation, rapid prototyping, or chemical engineering problem solving.

How It Works

SEQUENTIAL MODULAR FLOWSHEETING

At its core, Aspen Plus is a collection of *modules*, where each module contains both a mathematical model of a chemical process unit operation and a computer algorithm written specifically to solve it. For example, consider one such module for a flash drum called FLASH2 (the "2" means that it considers two phases—vapor and liquid). You can put an instance of a FLASH2 module on a flowsheet (I named it MYFLASH). In order to use this drum, you have to connect a material stream leading into it (FEED), and provide two streams for the outlet ports, one for the VAPOR and one for the LIQUID, as shown in Figure 0.1.

In order to simulate the flash drum, Aspen Plus will execute the computer algorithm associated with it. The algorithm operates essentially like any function in a programming language, such as C/C++, Matlab, Python, or FORTRAN, where the program expects (and requires) certain inputs to the program in order to run. Once given the inputs, it executes the program and computes the outputs. In the case of Aspen Plus, the inputs always consist of two things: the degrees of freedom that defines the flash drum (often referred to as the model parameters) and the contents of the feed streams. For example, in Figure 0.1 I specified my feed stream as being 50 mol% water and 50% methanol at 50°C, 1.2 bar, and 100 kmol/hr, which are the model inputs, and I specified MYFLASH as operating at 80°C and with a 0.2 bar pressure drop, which are the model parameters.

The model equations for the flash drum include relationships such as mass balances, energy balances, fugacity balances, and physical property correlations. Since the input stream and the model parameters are

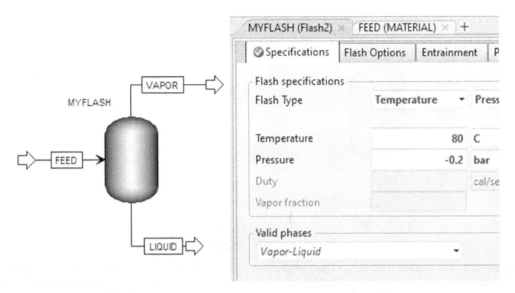

Figure 0.1 A simple model of a flash drum in Aspen Plus. The left side of the image shows the flowsheet model and the right side shows the model parameters that I specified for the flash drum model. The minus sign on pressure means it is a pressure drop.

known, the remaining unknowns in those equations are essentially the temperatures, pressures, compositions, and flow rates of the liquid and vapor streams. Although you personally might be able to solve the equations "by hand" or use a generic equation solver in another program, the built-in algorithm is custom-tailored to solve those particular equations in that particular format very quickly and reliably. This can be very beneficial for complicated models with built-in logic or models with discrete elements. For example, FLASH2 has some code that determines if the fluid in the drum will be a liquid, a vapor, or a mixture of both. This is not too difficult to implement using a computer program, but can be more difficult to handle in a general equation solver. Some modules even let you choose different algorithms or fine-tune the algorithm details in case the default algorithm does not solve it or is slow to solve it. Although speed may not be a huge factor for just one unit operation, when you have a complex flowsheet with hundreds of unit operations and a spaghetti of recycle streams, this speed and reliability can be incredibly important and sometimes the only practical way to model a flowsheet. In any case, I can now run the simulation, meaning Aspen Plus will execute the algorithm associated with the FLASH2 module. It dutifully computes the outputs, which as shown in Figure 0.1 are the temperature, pressures, flow rates, and compositions of the liquid and vapor streams, and the heat duty that needs to be provided to the flash drum to bring it up to 80°C.

The big tradeoff, though, is that because each module uses a pre-programmed algorithm, the simulation can only go in one direction: downstream. In Aspen Plus, each and every module *must* have all of the details of the input streams and the model parameters provided to it, and it can *only* compute the output streams and other performance information for the block because the algorithm is written that way. Suppose you wanted something else: you want to know which flash drum temperature will get you a certain composition of water in the liquid stream. Unlike a general equation-based model, you cannot specify the water composition in the liquid stream and have it solve for flash drum temperature, because the algorithm only goes in one direction. Instead, you must guess the temperature of the drum, run the algorithm, check to see if you ended up with the liquid composition that you wanted, and if not, guess a new temperature and repeat. This is the

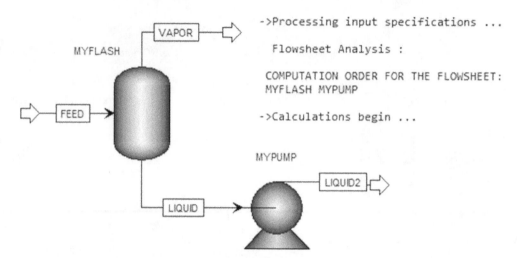

Figure 0.2 The Aspen Plus flowsheet now contains a pump model in addition to the flash drum. The text at the right is a portion of the program output that appears in the program's control panel when the simulation is run. It shows the computation order of the flowsheet, noting that the flash drum block runs first, and then the pump model runs next, but only after the flash drum simulation has completed.

big downside to having modules because it can make it harder to solve some kinds of problems, but the gains in terms of solver speed and reliability are usually worth it, especially for complex flowsheets. Fortunately, there is a tool built into Aspen Plus to automate this guess-and-check process for you in an intelligent way called a Design Spec (which you will learn about in Tutorial 3).

Now where does the "sequential" part come in? Suppose now I wanted to take the liquid product from the flash drum and pump it to 1.2 bar pressure. I can add a PUMP module (I called it MYPUMP) and add its outlet stream as shown in Figure 0.2. I can also set the outlet pressure of the pump by specifying that as one of the model parameters. Given this information, I want the PUMP module to compute the outlet conditions of the stream (the temperature should go up a little bit) and the electricity required. So now my simulation has two modules, MYFLASH and MYPUMP. Now, when I attempt to run the simulation, Aspen Plus analyzes the flowsheet and considers the order in which the blocks should be executed, since only one can be executed at a time. It is clear that you cannot run MYPUMP first before you run MYFLASH, because in order to run MYPUMP it needs to know everything about its input stream (LIQUID), but in order to compute LIQUID, it must first run MYFLASH. Therefore, it can only execute the modules in sequence: first MYFLASH and then MYPUMP. Hence the name Sequential Modular.

The sequential modular approach makes intuitive sense for simple systems such as in the above example. Typically, this means the program begins with the primary input stream (the first stream in the simulation) and works its way down the process, following the flow of material, energy, or information. However, things start to get trickier the moment you introduce recycle streams (be they material, energy, or information). For example, suppose I wanted to recycle a portion of the liquid stream back to the flash drum (e.g., this might happen in quench cooling applications). To do this, I can first add an FSPLIT block called MYSPLIT, which just divides a stream into parts (like a pipe tee), and specify that I want to send 80% of the liquid to the stream called RECYCLE, with the rest going to PURGE. Then I can connect the RECYCLE stream to MYFLASH as a second feed, as shown in Figure 0.3. Now, when I run the simulation, what happens? Which block is the first to run in the sequence? This is tricky since MYFLASH cannot run because we do not know what the contents

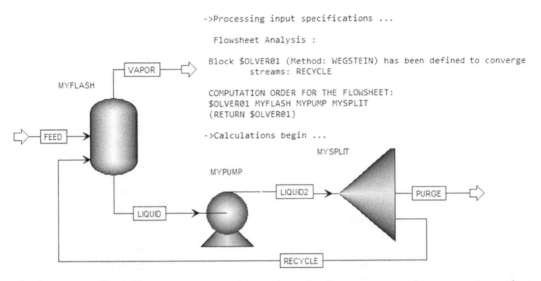

```
                        ->Processing input specifications ...

                        Flowsheet Analysis :

                        Block $SOLVER01 (Method: WEGSTEIN) has been defined to converge
                                    streams: RECYCLE

                        COMPUTATION ORDER FOR THE FLOWSHEET:
                        $SOLVER01 MYFLASH MYPUMP MYSPLIT
                        (RETURN $SOLVER01)

                        ->Calculations begin ...
```

Figure 0.3 The flowsheet now has a stream splitter and recycle. The computation order now contains a solver loop. In this case, the tear stream is the recycle stream, so it will first run the flash drum, then the pump, then the splitter, and then make new guesses for the recycle repeatedly until it converges.

of RECYCLE are *a priori*. So MYSPLIT must run before MYFLASH can run. But MYSPLIT needs to know what is in the pump output, so MYPUMP needs to run before MYSPLIT can run. But MYPUMP cannot run until MYFLASH is run, because it needs to know what is in LIQUID. And thus we are back to square one. This is called a *convergence loop*.

To get around this problem, one must *tear* a stream.[1] This means you select an appropriate stream in the convergence loop and literally just guess all missing information about the stream. For this example, suppose I choose to tear LIQUID. I would then guess all of the information about LIQUID, such as temperature, pressure, and the flow rates of each chemical, knowing that I might be totally wrong. I chose this tear in this case because I pretty much know the temperature and pressure exactly and can guess at some not unreasonable flow rates for the water and methanol. Then what Aspen Plus will do is use the guess as the input to the MYPUMP module. It dutifully executes the algorithm using this information and then computes the outputs accordingly. If the guess is terrible, then those output numbers will also be far off, but at least we have some numbers to work with for LIQUID2. Then, Aspen Plus will execute the MYSPLIT module using those LIQUID2 numbers as input, storing the output into the two streams. Then, MYFLASH can run since there are numbers for RECYCLE now available (FEED was already known). Now, since MYFLASH computes LIQUID as an output, we can check this output against my initial guess for LIQUID. If they match, then my guess was correct, and all of the equations in the flowsheet model have been solved. This means the loop has *converged*.

However, it is extremely unlikely that my original guess was exact, and in fact is probably off from the true solution by quite a bit. In that case, Aspen Plus will then compute a new guess for the conditions of LIQUID and repeat the cycle again, checking against the new guess. The algorithm that Aspen Plus uses to generate new guesses and check the result is called a *solver*. The solver will keep guessing new conditions for LIQUID each time until either the LIQUID guess matches the computed liquid output from MYFLASH within some

---

[1]That is, tear as in ripping in half, not as in crying, though I won't tell anyone if you do.

small error (called a *tolerance*), or it has decided that it has tried enough guesses and just gives up. If the loop converges, then we say that the flowsheet has been solved, and we know that the numbers computed by the different modules satisfy the model equations within tolerances. If it gives up, this results in an *unconverged loop*, meaning that none of the results calculated by any of the blocks in that loop have any meaning whatsoever, and should be discarded.

Since the act of choosing tear streams and guesses manually can be difficult to grasp, Aspen Plus contains a built-in algorithm that automatically chooses tear streams for you, as well as a menu of built-in solvers that use different strategies for generating new guesses at each iteration. In many cases, using the default settings (i.e., paying no attention to what Aspen Plus is doing) works sufficiently well such that you do not have to think much about tear streams while creating flowsheet models, even when the initial guesses Aspen Plus generates on its own for tear streams are quite terrible. Note, for example, in Figure 0.3, you can tell that Aspen Plus chose to tear RECYCLE because MYFLASH is the first model to run in the convergence loop sequence (which, by the way, converged quickly and without warnings or errors).

It is only when you get into trouble that you need to think about changing the convergence algorithm parameters, selecting your own tear streams, or giving it better initial guesses. In some cases, a flowsheet might not converge because you have created a flowsheet design that is mathematically impossible, which is squarely your fault (PEBKAC[2]). For example, suppose I did not have MYSPLIT and wanted to instead recycle 100% of the liquid stream back to the flash drum. This is a terrible design because there is no place for the liquid to leave the system. Were someone to actually build that, the liquid that was recycled would just accumulate in the drum, filling it up until something spills or breaks. As such, there is no way for that system to operate at steady state, and thus there is no solution for the convergence loop solver to find. When a solution technically does exist, but the solver just cannot find it, it can sometimes be corrected by providing better initial guesses based on your engineering intuition, choosing better tear streams (usually choosing where you can make the best guesses), or tweaking solver settings as a last resort. In this book, we discuss the basics of these for some common situations.

Although it is extremely effective and commonly used, the sequential modular approach is nothing new. The concept developed from the early days of computing in which there simply was not enough computing power or memory to consider large problems, and so having an iterative solution that allowed a large flowsheet to be broken into many tiny, customized pieces in sequence and solved with just one processor was essential to being able to tackle even moderate problems at all. As processes got faster, it became possible to converge larger and larger flowsheets with increasingly rigorous models. Today, however, computer processor speeds have essentially leveled off, and growth in modern computing power is now achieved instead by adding more processors that operate in parallel. Although commercial software is improving in order to take advantage of parallel computing (such as writing the algorithms for individual modules to take advantage of parallel processors), at its core, the modules themselves still require computation in sequence. As such, the speed at which we can solve flowsheets with the sequential modular approach has essentially peaked. True, we can now run several flowsheet simulations in parallel (and the new Aspen Multi-Case tool helps you do exactly this—see Bonus Tutorial 2), but the speed of each individual flowsheet solution has not changed much in the past decade. Although research into this is ongoing, there is another, more complex way of approaching the flowsheet model: the *equation-oriented* approach.

EQUATION-ORIENTED MODE

The equation-oriented approach uses a completely different way of solving the flowsheet model. The goal is essentially the same: given a set of model equations and certain known values and model parameters,

---

[2]Problem exists between keyboard and chair. Maybe get a standing desk?

find the values of all unknown variables. However, instead of using a sequence of individual modules, the equation-oriented approach takes all of the model equations from all of the flowsheet units and creates one gigantic system of equations. These equations describe everything in the entire process, such as the mass balances, energy balances, fugacity balances, physical property correlations, reaction kinetics, and so forth. Then, a generic equation solver made to solve arbitrary systems of nonlinear equations can be used (e.g., some might be familiar with the classic Newton-Raphson method for solving systems of nonlinear equations, which is a primitive form of one of the solver options available within Aspen Plus). Most equation solvers usually contain powerful algorithms that analyze the structure of the equations and look for patterns and symmetries that they can exploit in order to find the solution as quickly and reliably as possible.

This approach has some key advantages. First, the restriction that models can only be solved in one direction is gone. For example, if you wanted to solve the problem of finding the flash drum temperature that achieves a certain composition in the liquid product (even in the midst of the recycle connection), you simply include an equation that specifies the desired liquid composition and allow the flash drum temperature to remain an unknown variable that is solved along with all of the other unknown variables. In addition, because all equations are solved together, there is no need for tear streams, and so adding many recycle loops (even loops-inside-loops-inside-loops...) does not increase solution time or difficulty in the general case. This is because the entire equation-oriented solver is essentially one large guess-and-check convergence loop. There are other advantages, too, such as the ability to use powerful equation-based optimization algorithms that you cannot use with sequential modular formats (you will learn how to do basic optimization in sequential modular mode in Tutorial 6).

However, there is a huge negative to this approach. For large flowsheets with many chemicals, especially with rigorous models for unit operations like distillation and reaction, the number of equations (and number of unknowns that need to be solved simultaneously) can number in the millions. Even the best algorithms available today can have extreme difficulty solving the equations quickly (or even at all) when the system is that big without extremely good initial guesses. And what is the best way to generate initial guesses? You guessed it: sequential modular mode. So in practice, in order to use equation-oriented mode in Aspen Plus, you must first converge the flowsheet in sequential modular mode, thus essentially requiring you to solve the problem before you solve the problem. This is slightly less pointless than it sounds, because once the flowsheet has converged, you can start making changes from there and then find new solutions in equation-oriented mode using the original one as the initial guess, with a much higher rate of success. As a result, users need to learn sequential modular mode anyway. Therefore, in this book, we only use sequential modular mode, since it is generally adequate for almost all beginners and even advanced users.

GARBAGE-IN, GARBAGE-OUT

As a final note, it is important to recognize that Aspen Plus is not magic. It does not know what you are trying to do and does not provide much advice to you about sound chemical engineering practice. In many cases, it is easy to use the software without really understanding what it is doing or even the piece of equipment you are trying to model. As such, many beginners tend to overly trust the results of the program as the truth, without stopping to consider whether the results have meaning. Therefore, it is important to remember the principle of *garbage-in, garbage-out* (GIGO). Aspen Plus is primarily a set of equations (usually based on mass, energy, equilibria, and various chemical phenomena) and algorithms to solve them. If you give it garbage as input conditions (such as through bad input streams, bad model parameters, or bad physical property models), it will dutifully go through the motions and compute the outputs, but those outputs will be garbage, and you might not even know it. If you have a sequence of 10 modules in series, and you give garbage inputs to module 3, well, the garbage output of module 3 becomes the garbage input for module 4, thus giving garbage output for module 4 and so on for 5 through 10. If that garbage module is inside a convergence loop, everything in that entire loop will be garbage. If your system is poorly designed from the start, you will also get garbage. This knowledge can be extremely helpful in debugging your flowsheets and ensuring that your model results are meaningful and useful.

Other Competing Software

In addition to Aspen Plus, there are many other CAPE tools for general chemical process steady-state flow-sheeting available on the market. Other popular competitors include HYSYS (also owned by Aspen Technology, Inc., "AspenTech"), ProMax (by Bryan Research & Engineering), Pro/II (by SimSci/Schneider Electric), gPROMS (by Process Systems Enterprise), UniSim Design Suite (by Honeywell), IDAES (by the U.S. National Energy Technology Laboratory), and COCO (by AmsterCHEM). Each one has its own strengths, weaknesses, and special features. Some of the major differences are outlined next, which you may find useful if you are already familiar with process modeling using some other competing software or are trying to decide which software to select for your needs.

HYSYS was developed by Hyprotech, but now that AspenTech owns it, it is starting to merge with Aspen Plus in terms of function and capability. The visual style and user interfaces are quite different, and users will generally prefer one over the other based on personal preference or simple historical familiarity. Ultimately, although both use a sequential modular flowsheeting framework by default, Aspen HYSYS has some ability for information to move "upstream." For example, in HYSYS, if you want the outlet stream of a heat exchanger to be at a certain temperature, you can simply type the temperature you want into the form for the outlet stream itself, rather than into the form for the heat exchanger. In HYSYS, the heat exchanger model is smart enough to recognize this and consider this information in running the heat exchanger simulation, whereas in Aspen Plus, the user's information entered into the stream would be ignored, requiring the user to enter the desired output temperature directly into the heat exchanger model instead. In addition, HYSYS will by default automatically compute information for some blocks as soon as sufficient data are available without having to run the full simulation, which can be both good and bad depending on what the user wants.

The UniSim Design Suite is essentially a fork of HYSYS from an older Hyprotech version of the software and has developed independently from HYSYS for about a decade now. There are differences in terms of integration with other software and various internal details owing to their separate more recent developments, but on the whole UniSim and HYSYS are similar in appearance and function.

One other major difference, however, is that Aspen Plus will automatically select tear streams based on flowsheet structure and hide them from the user (well, you have to know where to look), whereas HYSYS and most other competing programs do not. For example, in HYSYS, the user must add a model block known as a "Recycler" which indicates the point in a convergence loop in which a stream is torn (a great many students have mistaken a recycler for an actual piece of chemical equipment instead of an abstract point in which the solver starts guessing). HYSYS does have a Recycle Advisor which can help you determine the best places for them, but ultimately they appear directly on the flowsheet and are normally a part of a user's model building thought process. ProMax has a similar Recycle block. Usually, Aspen Plus chooses tear streams which function reasonably well, and it contains a mechanism to override this with the user's own preference, should the need arise. There are strengths and weaknesses to this, but not always having to think about tear streams can be an attractive feature to some users.

Pro/II is similar to Aspen Plus in certain key areas such as sequential modular flowsheets, automatic tearing of streams, and the general form-based layout for model construction. There are differences in some of the libraries of physical property models and chemicals, but the core functions are similar. Both have a rich set of features and can be integrated with other software for the purposes of dynamic modeling, process control, and optimization.

Although ProMax is similar in functionality to HYSYS, its main advantage over other software is a niche market in gas sweeting and CO_2 removal applications, such as in natural gas processing, syngas cleaning, or power plant carbon dioxide capture. ProMax developed out of in-house software created by Bryan Research & Engineering specifically for these kinds of applications. It contains proprietary models for how different solvents (such as monoethanolamine, monodiethanolamine, diglycolamine, piperazine, and Selexol) interact with CO_2 and H_2S, although I cannot personally attest as to whether they are any more or less accurate than the models available in other software. ProMax also includes proprietary convergence algorithms for absorber and stripper models that are optimized for this separation. This is important because CO_2 and H_2S absorption models tend to be very numerically challenging to solve (see Tutorial 12 to learn how),[3] and my personal experience has shown that ProMax models tend to converge with far less fuss and require much less fiddling with algorithm parameters. ProMax also contains other flowsheet models and can function as a general process flowsheet simulator. One key practical difference is that ProMax exists as an add-on module to Microsoft Visio, which can be both a strength and a weakness depending on access to and familiarity with Visio.

gPROMS is both a chemical process flowsheeting tool and an advanced ordinary differential equation integrator in one. gPROMS operates in an equation-oriented environment. Modelers can use the graphical user interface to select models from a library, connect them together with streams, and set parameters just like any other flowsheeting software. However, the act of doing so creates a model made of one large system of equations that can be solved with a general equation solver built into the software. The equations can then be modified to make custom changes to models, or alternatively, models can be built from scratch using custom equations. The models can also be either steady state or dynamic, meaning the system changes over time. Some of the key difficulties encountered with the equation-oriented approach involve the high degree of difficulty involved in solving the model, owing to well-known problems associated with getting the simulation to initialize (meaning just to get it started). This is similar to how Aspen Plus works in equation-oriented mode, except that Aspen Plus has a significant advantage for steady-state simulations, in that the sequential modular mode can be used to initialize the flowsheet rather effectively. gPROMS also competes directly with Aspen Plus Dynamics (AspenTech's equation-oriented dynamic process simulator) and Aspen Custom Modeler (AspenTech's general equation-oriented ordinary differential equation integrator), although both are out of scope for this book.

COCO is interesting essentially because it is free (and still actively maintained). It is built in the context of the CAPE-OPEN framework, which is an information-exchange standard that different CAPE tools can use to interact with each other. COCO contains a graphical user interface for flowsheeting, a basic thermodynamics and physical properties package for a few hundred chemicals, some basic unit operation models, and basic numerical capabilities. Users can download plug-ins for extra features or create their own. Although it cannot compete with commercial packages in terms of features, model choice, and depth of embedded physical property packages, it can be useful for basic tasks with common chemicals. It can also be integrated with software that is able to communicate via the CAPE-OPEN interface, which includes most of the software mentioned above. For example, one can purchase commercial modules such as rigorous distillation models or thermodynamic property packages, and then use a CAPE-OPEN compliant interface to integrate them into the free COCO flowsheet environment at potentially a lower cost than a full commercial solution. For example, you can use the CAPE-OPEN interface to link COCO with Aspen Properties, such that COCO can make calls to the Aspen Properties thermodynamics engine. This is an advanced feature

---

[3]For more information about CO_2 and H_2S capture modeling in Aspen Plus, ProMax, Pro/II, and HYSYS, see Adams TA II, Khojestah Salkuyeh Y, Nease J. Processes and Simulations for Solvent-Based CO_2 Capture and Syngas Cleanup. In, Reactor and Process Design in Sustainable Energy Technology, 1st ed. (Fan Shi, ed.). Elsevier, 2014. But finish Tutorial 12 first!

which can be useful for constructing complex and unique models, and is beyond the scope of this book. In fact, the COBIA framework is now in development for the specific purpose of making it easier for chemical engineers to interact with chemical process software through CAPE-OPEN and develop CAPE-OPEN compliant components by reducing the computer programming skills requirement barrier that currently challenges chemical engineers in the CAPE-OPEN environment.

IDAES, the eponymous software by the Institute for the Design of Advanced Energy Systems at the U.S. National Energy Technology Laboratory, is an exciting new modeling framework currently in development. It is built on Pyomo, a model optimization framework written in Python and developed at Sandia National Laboratories. IDAES is released to public domain, which is even more free than open source, because a for-profit company can legally take the underlying code and develop and sell its own commercial products from it. Its premise is to re-think the underlying structure and framework of model development, with steady state, dynamics, control, and, most importantly, algebraic equation-based optimization (also called mathematical programming) in mind. The major commercial packages are mature and excellent in the modeling space, but what they do not do well is algebraic optimization, especially when considering things like probabilities and uncertainties. Optimization algorithms have developed separately from commercial process modelers and do optimization tasks with state-of-the-art methods, but are notoriously difficult to use in the context of rigorous chemical process modeling. IDAES is an ambitious, potentially paradigm shifting attempt to bridge this gap, and commercial vendors should take note. At the moment it is still in its early stages of development and lacks library richness, a robust graphical user interface, and user-friendly features that are required to be competitive with existing commercial products. However, it is growing quickly, led by a crack team of experts (including some of my former Ph.D. students), and I expect great things of it in the future.

In my own work, I have used nearly all of the above programs at one time or another, depending on the need. The bottom line is that you should choose the software that is right for you for each specific case. In my case, Aspen Plus has been by far my most popular choice. Although the legacy of its 1970s-era FORTRAN roots becomes apparent the further you get down the rabbit hole, it is an extremely powerful tool. It is loaded with features, and can connect to many other programs such that in the hands of a skilled practitioner it can be used to solve some extremely difficult problems. In this book, you will learn how to use the most important features so you can start solving many of those difficult chemical engineering problems quickly and effectively.

♫ Music break[4]

---

[4]Recommended listening: *Closer* by Deadmau5.

Learn Aspen Plus in 24 Hours

Getting Started

Objectives

- Get your feet wet with Aspen Plus V12, a chemical process simulator
- Convert a flowsheet drawing into a simulation to find the missing pieces of information
- Create a new flowsheet
- Add chemicals
- Choose physical properties
- Insert unit operations
- Connect the streams
- Enter block parameters
- Successfully execute the simulation
- Get results from simulations and use that to solve problems

Prerequisite Knowledge

This tutorial assumes that you have a very basic understanding of distillation and pumps. If you need a refresher on what distillation is, see this video on distillation[1] (and the rest of the distillation section there, if that helps) and this website on distillation.[2] If you do not know what a pump is, well, it moves a liquid from one place to another by increasing its pressure.

Why This Is Useful for Problem Solving

In this book, Aspen Plus will be your bread and butter. It is essential that you are able to do basic functional tasks such as creating new flowsheets, adding chemicals, choosing physical properties, adding unit operations and connecting them with streams, and running the simulation. In this tutorial, you will not be

---

[1] https://www.youtube.com/watch?v=GPDd5qXPKpo. If the link is broken after press time, see the learncheme.com screencasts on distillation. This is peer-reviewed material produced by the University of Colorado, Boulder.

[2] http://encyclopedia.che.engin.umich.edu/Pages/SeparationsChemical/DistillationColumns/DistillationColumns.html. This material was prepared by the University of Michigan.

thinking too much about the details yet, just learning how to use the software to enter in given simulation data and run the result. But you need these basics to be able to solve problems at all!

Tutorial

BACKGROUND

We will do a simple walkthrough of using Aspen Plus to simulate a process to separate *n*-hexane and *n*-decane, using distillation at high pressure, as shown in Figure 1.1.

In every simulation, you must define the **components** (a.k.a. *species*) that will make up the simulation. In other words, you must specify which chemicals Aspen Plus should take from its large database of chemicals and use in the simulation. In this case, we have only *n*-hexane and *n*-decane. When we specify the components in the program, this provides access to properties such as boiling point, thermal conductivity, phase equilibria, molecular weights, density formulas, and equations of state.

PART 1: GETTING STARTED WITH PROPERTY DEFINITIONS

Start the Aspen Plus program and create a new Blank Simulation (Figure 1.2).

Use Save As to save your simulation to a new (.apw) file. I suggest you save your work every few minutes because crashes can occur and you definitely don't want to start all over again.

Let's start by specifying some basic information about the process. On the left panel, there are several folders with names like "Setup, Components, Methods," etc. when the Properties tab on the bottom left is highlighted. This is a collection of forms that you have to fill out about the simulation.

First, in the Setup | Specifications form, type in a title for your project. In addition, in Units of Measurement, make sure that your Global Unit Set is set to MET (metric). These are the default units of measurement and are there for convenience. You can always change this later (Figure 1.3).

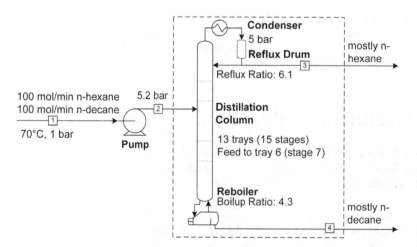

Figure 1.1 A process to separate hexane and decane using distillation at pressure. The dashed lines indicate that Aspen Plus models the distillation column, condenser, and reboiler together as one block.

Figure 1.2 Making a new, blank simulation.

Note that there are blue checkmarks and red half-circles. The blue checkmark means that there is enough data already entered into the forms such that the simulation can continue. The red half-circle means that you have more to enter before you can run a simulation.

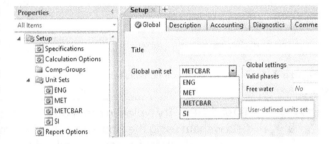

Next, let's choose the components. Click on the Components folder on the left side of the data browser (or click the Components button under Home | Navi-

Figure 1.3 Selecting a units set.

gate). Here you will see a list of the chemicals used in your simulation. It is currently empty. You can add chemicals into the simulation in a number of ways. The most general way is to use the Find button at the bottom of the form (the other buttons are sort of advanced, we won't go there for now). Anyway, click Find and then do the search for *n*-hexane (see Figure 1.4). This will search the Aspen Properties database. Notice that two chemicals come up because they both have *n*-hexane in the title. You can identify the correct one by the full name, chemical formula, molecular weight, or other factors (Figure 1.4).

Q1) Report the molecular weight of *n*-hexane contained in the Aspen Plus database.

Add *n*-hexane to your simulation by selecting it and clicking the Add selected compounds button (or double-clicking it). Now, repeat to add *n*-decane. Close the Find window to go back to your components list (see Figure 1.5). Notice that the Components folder and Specifications subitem now have blue checkmarks.

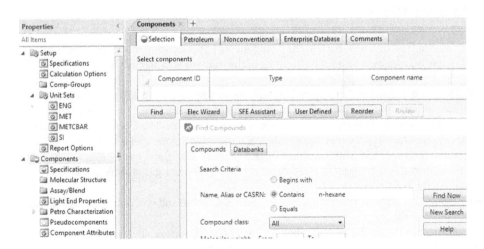

Figure 1.4 Using the Find feature to locate chemicals in the database.

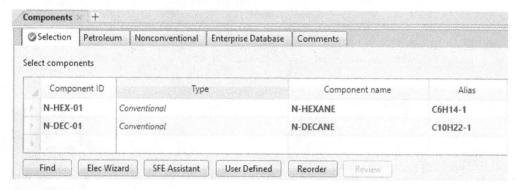

Figure 1.5   The completed components form.

TOM'S TIP: Try typing in common chemicals directly into the Component ID field without using the Find feature to save time. For example water, H2, CH4, and ethane will all yield expected results. Be sure to double-check the component name / alias to make sure it worked.

TOM'S TIP: Once you have imported a component, you can always rename its Component ID, as shown in Figure 1.6. I'll say this more than once

Figure 1.6   Usually it makes sense to rename your chemicals based on your own meaningful standards.

because people always seem to forget. It is unprofessional (and confusing) to have carbon dioxide called carbo-01 and carbon monoxide called carbo-02 just because you entered the carbon dioxide first!

TOM'S TIP: Empty circles or folders without checkmarks are optional extras that can be ignored for now.

Next, add *n*-decane in the same way. Back in the Components | Specification form, you should see the two components in the list, as shown below. The meaning of the columns is explained next.

Component ID: This is the name that Aspen Plus will use in your simulation. You can rename it by double-clicking it as you please, as long as it is unique, as shown above. (See, I said it twice.)

Type: This is the classification of the model used to represent this chemical. The *Conventional* classification is a typical pure chemical in either liquid, vapor, or supercritical states and is the most commonly chosen. Other options are *solid* (solids are hard—pun intended), *blend,* and *hypothetical.* Aspen Plus is able to model mixtures of things (like a messy collection of polymers, gasoline blends, air, etc.) and treat them like a chemical.

Component Name: This is the name in the Aspen Properties database and you cannot change it. More on Properties in Tutorial 2!

Alias (a.k.a. Formula): This is the second name for the chemical that usually includes the chemical formula. The alias for *n*-hexane is C6H14-1. The -1 at the end of the alias means that it is variant 1, since there are other chemicals which have the same formula.

Q2) Use the Find feature to determine how many components are in the APV120.PURE38 Databank (the default) with the name, alias, or CASRN (Chemical Abstracts Service Registry Number) of C_6H_{14} and report the number. Note that you can search by chemical formula. It will give more than you need, so click on the tabs at the top to sort them in a useful fashion.

Ok! We have added what we want. Now, choose which physical property package we want to use. Physical property packages are collections of data, equations, and models, which describe all sorts of information about the chemicals you have selected. For example, they have equations of state (which relate pressure, molar density, and temperature, like the ideal gas law); collections of physical properties like heat capacity, thermal conductivity, latent heats of vaporization; and vapor-liquid equilibria to predict how stuff mixes and separates.

The correct choice of physical property model is absolutely critical for a valid result. However, selecting the correct physical property model is a lesson for another day. For now, select the PRSK (Predictive Redlich-Kwong-Soave) method as your base method, as shown in Figure 1.7. To do this, go to Methods | Specifications and choose it from the drop-down box in the Base Method section. The stuff on the right side will fill in automatically. This tells you which equation of state, data set, enthalpy, and volume models this base method will use. In advanced situations, you can change these, but let's not do that now. Leave the rest at the default.

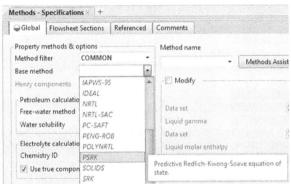

Figure 1.7 Choosing a physical property method.

🎵 Music break[3]

PART 2: SETTING UP THE FLOWSHEET

Now, let's go make our flowsheet. Switch to the flowsheet, the white area, by clicking the Simulation tab on the left bottom, below the Properties tab.

First, let's add the pump. In the bottom, you should see the model library. These are all of the chemical process models contained inside Aspen Plus. Go to the Pressure Changers tab, and click the Pump model. Then click somewhere in the flowsheet to add a pump, as shown in Figure 1.8A.

In this case, it has a default name of B1. Change it and call it something else. Right-click on the pump and select Rename Block. Give it a new name (eight characters maximum). Once you've renamed it, you can grab the name text with the mouse and move it somewhere else if you like. You can also resize the icon if you please. This is just for the sake of appearance.

Ok, that gives us the pump. Let's add some streams going to and from the pump. Click on the STREAMS model in the bottom left of the model library. It should say *Material*. This means a material stream model. If you click the little down-arrow, you can change which kind of stream you add, namely work or heat. Add a material stream by making two clicks on the main area, once in the whitespace to specify where the stream should start and then once to specify where it should end.

---

[3]Recommended listening: *Da Funk* by Daft Punk.

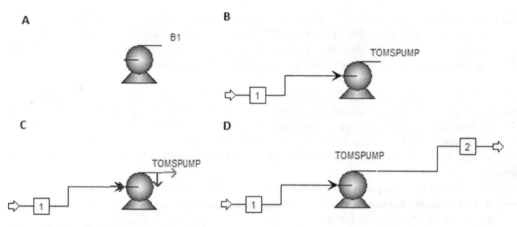

Figure 1.8   (A) Adding the first pump. (B) A pump with an inlet stream attached properly. (C) The red (horizontal) outlet stream appears when adding another material stream, showing where you have to click if you want your material stream is actually connecting to the model. (D) The completed model with inlet and outlet streams.

You will notice that the first time you click, a red (horizontal) arrow will suddenly appear on the pump. What you want to do is make your second click on this arrow. Hover your mouse over the red (horizontal) arrow until it highlights. Then click. This will connect the outlet of the stream to the pump inlet, as shown in Figure 1.8B.

 TOM'S TIP: If you screw up, and miss, your stream will be going from nowhere to nowhere, and what's worse is that it may *look* like it is connected, when it is not. This is a common error for beginners (well, for everyone actually). If this happens, right-click on the stream and select Reconnect | Reconnect Destination, then try clicking on the red (horizontal) arrow again. Alternatively, you can double-click on the ending arrowhead of the stream to do the same thing.

Note that the stream is given a name of either S1 or 1 by default (depending on your version or settings). You may rename it as you please, like you did the pump (either right-click and choose Rename Stream or left-click and hit CTRL-M). I prefer 1.

Make another material stream at the exit of the pump. After you click on the *Material* model icon in the model library, make sure your next click is on the red (horizontal) pump exit arrow and the second into whitespace. Notice that there are several arrows to choose from when you do this (see Figure 1.8C). The red (horizontal) arrow means that in order for the pump model to run, you must have a stream connected to the red (horizontal) exit port. The blue (vertical) arrow means that this is an optional connection. Hover your mouse over it to see what it means. Anyway, connect to the red (horizontal) arrow (Figure 1.8D).

 TOM'S TIP: When adding outlet stream connections to the pump, it is easy to make a mistake and click on the blue outlet arrow instead of the red one, and not even know it. If this happens, you might get a message from Aspen Plus asking you if you want to add water to your simulation (you don't want to in this case). This is because the blue arrow is an optional "free water decant" port that is used for certain kinds of models that are not covered in this book. If you say yes to this question or if you already have water in your simulation, then you will discover later that you cannot run the simulation because the flowsheet section is incomplete. It can be hard to diagnose the problem because on the flowsheet it will visually look like you have attached the stream correctly, with the only clue being a small stray line on the right that only an experienced user would recognize. To make matters worse, in your simulation browser, it may show that your pump input is complete, making it hard for you to diagnose exactly where you made a bad connection.

Expand the streams and block folders on the left. The red half-circles mean it does not have enough information to run the model, as shown in Figure 1.9. Note that we have to add the input (stream 1) and the pump parameters but not the output (stream 2). Aspen Plus always computes output streams from a block. Information flows along with the material in the process.

Let's do it! Click on the folder for your input stream. This is where you specify the state variables for that stream. Enter in the appropriate information based on the flowsheet in Figure 1.1. Note that you may have to change the units to meet your needs. In the Composition section, you may either enter the individual flow rates of both components or use mole fractions. It's up to you. `Mole-Flow` means molar flow rates; `Mole-Frac` means mole fractions. The other options are not particularly helpful at the moment. When you are done, you should get the blue checkmark in this form. Specify everything completely. Don't assume the computer knows what you mean.

Figure 1.9 The blue checkmarks indicate that Aspen Plus has enough information in that section in order to run the model. The red half-circles indicate that more information is needed before the model can be run.

TOM'S TIP: See how we can only choose two out of the three of temperature, pressure, and vapor fraction? And that we must also completely specify flows of each component in some fashion? This gives the minimum amount of information needed to do a flash calculation. A flash calculation is when you know enough state information about a mixture such that you can calculate all the rest of the state information. For example, if you know enthalpy, pressure, and composition of a mixture, you can then compute everything else such as temperature, vapor fraction, and the amount of each chemical in each phase. Flash calculations are an integral part of almost every unit operation model in Aspen Plus.

Ok. Let's set the data for the pump. Select the *Pump* block from your list of blocks. You have a number of options. In your case, you want to set the discharge pressure to the value shown in Figure 1.1, since you know what you want it to be. You can also set the pump efficiencies. If you leave it blank, it uses a default value, which you often have to find by digging in the documentation. In this case, the pump efficiency is sometimes 0.9, sometimes 0.65, and sometimes something else, depending on the outlet pressure and amount of solids in the stream. Use `0.85` for now for the pump (85% pump efficiency) and `0.96` for the driver (the default driver efficiency is 1).

Great, now every relevant form should have blue checkboxes instead of red half-circles. When you have no red half-circles, the status on the bottom right of the screen changes from Required Input Incomplete (with the red background) to Required Input Complete.

Now, let's run! You can hit F5, select Run from the Run menu, or click Play (the triangle in the title bar or in the Home ribbon). If it worked without errors, you will get the beautiful Result Available message in the bottom-left status area. This is truly a beautiful message, as you soon shall discover.

Ok, so what happened? The simulation has executed successfully, meaning that all of the blocks have been executed and all stream data have been calculated. Right-click on any stream or block and choose Results to see what it calculates. You can also single-left-click to select it and hit CTRL-R which is faster.

When you do this, you'll see the results of the stream. For example, my results for stream 1 are shown in Figure 1.10. You'll see how you entered the temperature, pressure, and flows, and it calculated the rest. For example, it calculated mass flows, enthalpy, entropy, density, molecular weight, and standard liquid volume. It is reporting them in the default units rather than the units I entered. We can change this later.

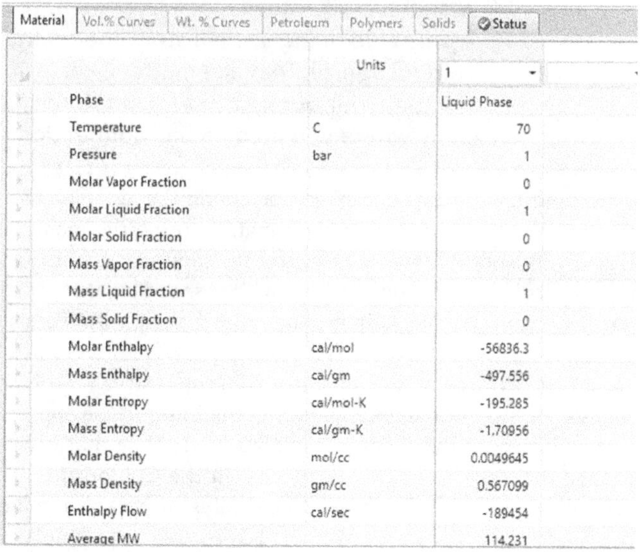

Material	Vol.% Curves	Wt. % Curves	Petroleum	Polymers	Solids	Status
		Units			1	
Phase					Liquid Phase	
Temperature		C			70	
Pressure		bar			1	
Molar Vapor Fraction					0	
Molar Liquid Fraction					1	
Molar Solid Fraction					0	
Mass Vapor Fraction					0	
Mass Liquid Fraction					1	
Mass Solid Fraction					0	
Molar Enthalpy		cal/mol			-56836.3	
Mass Enthalpy		cal/gm			-497.556	
Molar Entropy		cal/mol-K			-195.285	
Mass Entropy		cal/gm-K			-1.70956	
Molar Density		mol/cc			0.0049645	
Mass Density		gm/cc			0.567099	
Enthalpy Flow		cal/sec			-189454	
Average MW					114.231	

Figure 1.10 Stream results for stream 1 after successful execution of the flowsheet.

Q3) Report the temperature of stream 2, the pump outlet, in kelvin.

Q4) Report the total electricity required to operate this pump, in kW.

Great, you have now done something useful! Now, go back to your main flowsheet and let's add the distillation column section. To do this, first go to the Columns tab on the model library. Note that there are many kinds of models. These are all different ways of modeling a distillation column which use different approaches and assumptions. We'll do more with some of these in later tutorials so you know how to choose and use them correctly.

The Columns tab has from left to right:

DSTWU: Winn-Underwood-Gilliland model. It is a shortcut model that uses lots of assumptions, akin to what a student might do when they are first learning distillation. This is covered in Tutorial 5.
Distl: The Edmister method (another shortcut model) for distillation.
RadFrac: A rigorous model for distillation with a wide range of options and features. See Tutorials 5 and 8.
Extract: This is for liquid-liquid extraction.
Multifrac: This models complex distillation columns with side units, used in the petrochemical industry. It is essentially a collection of several RadFrac models in one block.
SCFrac: A shortcut version of Multifrac.
PetroFrac: A complex distillation/separation system used in the petrochemicals industry for petroleum refining. It is similar to a RadFrac model with some auxiliary units integrated.
ConSep: Perform feasibility and conceptual design calculations for distillation columns.
BatchSep: This is a batch still model, for making booze (and other stuff I suppose). This is covered in Tutorial B3.

For this one, we'll use the RadFrac model as our distillation column model. It doesn't matter which icon you use (you can hit the little down-arrow next to the RadFrac button in the model palette to see the different options); the model is exactly the same regardless of the icon chosen. If you pick the wrong icon, you can change it later by right-clicking on it and selecting Exchange Icon, or left-clicking it and hitting CTRL+K repeatedly until you get the one you want.

 TOM'S TIP: Serious modelers should make sure that the icon matches the kind of unit operation you intend to model, in terms of the kind and position of the condenser or reboiler (if either even exists). This will help others avoid confusion by making it explicitly clear what unit operation exists in the real plant when they look at your flowsheet. Choosing the wrong icon is a great way to tell others that you do not really know what you are doing and that your work should not be trusted!

Add a RadFrac model to your flowsheet and give it a new name. Then connect your pump outlet to the inlet of the column (perhaps by using Reconnect Destination). Note that the RadFrac model by default already includes the condenser, reflux drum, and reboiler. You have the ability to remove these if you choose, but we want them. Therefore, the dashed lines in Figure 1.1 represent all of the equipment modeled in Aspen Plus using one RadFrac model.

Add the distillate (stream 3) and bottoms (stream 4) streams to your flowsheet. The final sheet should look like Figure 1.11.

Finally, it's time to enter parameters into the RadFrac block according to Figure 1.1. Set the number of stages to 15 (Aspen Plus counts condensers and reboilers as stages, if they exist; so there are 13 trays in our case), the condenser type to Total (i.e., what is shown in Figure 1.1, but we can choose other things), and the reboiler to Kettle. In the operating specifications, set the reflux molar ratio to 6.1, and the boilup molar ratio to 4.3. In the Streams tab, set the feed stage to 7 (i.e., stream 2 is fed to just above stage 7, which is tray 6).

In the Pressure tab, set the condenser pressure to 5 bar and leave everything else at their default values. At this point, all of our data browser folders should have no more red half-circles and our status indicator should be the elegant yellow message saying Input Changed. Run the simulation again.

Q5) Report the total flow rate of n-hexane in the distillate in kmol/hr.

Q6) Report the total heat duty in the reboiler in GJ/hr.

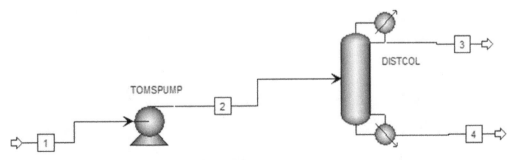

Figure 1.11 The completed flowsheet for Part 2.

PART 3: A LITTLE MORE ON YOUR OWN

Add a heat exchanger to the outlet of the bottoms product (the *n*-decane product) which cools the temperature of the bottoms product down to 25°C. Use the HEATER model which lets you issue a temperature change without worrying about how exactly that will happen (more on that in Tutorial 4). However, you will need to specify the pressure drop; so, assume that there is no pressure drop. You can do this either by specifying the pressure of the heat exchanger to the same as the inlet pressure, or you can specify a pressure drop directly entering 0 (zero) into the pressure field of the HEATER block parameters input. However you do it, verify that the outlet pressure is the same as the inlet pressure after you have run the sim. Note that it is usually a much better practice to specify pressure drops instead of absolute pressures when working with large flowsheets where the components or pressures upstream and downstream can change as you work with the flowsheet or run different simulations with different conditions.

Q7) Report the heat duty required to do this in GJ/hr. A negative number represents cooling.

Q8) Report the mole fraction of hexane in the distillate if you feed the mixture to the column above stage 3 instead of stage 7.

TOM'S TIP: In older versions of the software, and for some templates, basic properties like mole/mass/volume fractions or flows do not show up by default when looking at the stream results. If this is true for your case, go to Setup | Report Options | Stream, check the `Mole` checkbox in the Fraction Basis section, and rerun to add them to the stream results.

♫ Music break[4]

---

[4]Recommended listening: The Contortionist's cover of *1979* by Smashing Pumpkins.

Physical Property Modeling

Objectives

- Utilize the physical property estimation procedures in Aspen Plus for components
- Use the stream analysis tools to estimate stream properties
- Use theoretical model blocks such as duplicators
- Use a pump to manage pressure differential
- Tweak convergence criteria for blocks to overcome convergence problems
- Use distillation models to simulate a pressure swing distillation system
- Construct flowsheets with recycle in an intelligent way by working piece by piece rather than creating one big flowsheet and hitting Run

Prerequisite Knowledge

This tutorial assumes that you have a basic understanding of distillation, pumps, and phase equilibria. If you need a refresher on what distillation is, see this video on distillation[1] (and the rest of the distillation section there, if that helps) and this website on distillation.[2] If you don't know what a pump is, well, it moves a liquid from one place to another by increasing its pressure. If you do not remember what vapor-liquid equilibria (VLE) phase diagrams are, such as T-xy and P-xy diagrams, try this video on phase equilibria.[3] If you do not know what azeotropes are, try this video on azeotropes.[4] Of course, you should complete the previous tutorial first before attempting this one.

---

[1]https://www.youtube.com/watch?v=GPDd5qXPKpo. If the link is broken after press time, see the learncheme.com screencasts on distillation. This is peer-reviewed material produced by the University of Colorado, Boulder.

[2]http://encyclopedia.che.engin.umich.edu/Pages/SeparationsChemical/DistillationColumns/DistillationColumns.html. This material was prepared by the University of Michigan.

[3]https://www.youtube.com/watch?v=-XcTEknC9Aw. From University of Colorado, Boulder.

[4]https://www.youtube.com/watch?v=28WWKdf3h1o. From University of Colorado, Boulder.

Why This Is Useful for Problem Solving

One of the fundamental problems a conceptual process designer might have to face is how to determine the best way to separate a mixture of chemicals. VLE diagrams are one of the best places to start. For example, you would need to recognize whether an azeotrope exists or not and then determine the strategy for separation, such as ordinary distillation, pressure swing distillation, and absorption. Or, you might need to estimate the distillate and bottoms temperatures, or choose what operating pressure to use for distillation based on a number of factors. The stream analysis feature can help predict bubble and dew point temperatures and other key metrics. Using this information, you can sketch out a process in which a given feed is separated into a certain set of products at various desired purities. You can also predict quantities or ranges for temperatures, pressures, compositions, and/or flows.

In addition, to use the software effectively, you will find it necessary to understand convergence, how to tweak the convergence parameters and why, and how to interpret control panel output in order to actually construct and complete simulations at all.

Tutorial

PART 1: PHYSICAL PROPERTY BASICS

In this section we will experiment with different physical property models and use some of the special property tools which are included with Aspen Plus. We will use the tools to help us synthesize certain separation flowsheets.

Let's synthesize a process which will separate a stream containing 70 mol% methanol and 30 mol% chloroform into high-purity methanol and high-purity chloroform. At this point, we know nothing else, so let's use Aspen Properties to get some useful information about these two chemicals and their VLE.

Make a new, blank simulation. For convenience, use the METCBAR units set (either check in the Properties | Setup | Specification form, click the Unit Sets icon in the Home ribbon to go there, or just select from the drop-down box right above the Unit Sets icon). Under the Properties | Components form, add methanol and CHCL3 (chloroform) to the components list.[5] Now, we are going to use the Non-Random-Two-Liquid model with the Redlich-Kwong equation of state (NRTL-RK). This is one of the most popular models and is often a great choice when dealing with phase equilibria for mixtures, especially with azeotropes.[6] In Methods | Specifications, if you look in the base method, only NRTL exists (which is different from NRTL-RK; NRTL assumes ideal gas). This is because, by default, Aspen Plus filters out the property models to be only the common ones (see the process type drop-down). Thus, change the process type to All to be able to see all possible selections, and then go back to the base method box and select NRTL-RK.

Now what happens next is a little irritating, but incredibly important to understand. If you look in the Methods | Parameters | Binary Interaction | NRTL-1 folder of the data browser, you'll see that the red half-circle has appeared in the Binary Interactions folder, as shown in Figure 2.1.

---

[5]If you get a message that adding chemicals will cause the parameters on the form to update, click either Yes or No. Because the template has the NRTL method already chosen for you by default, it is trying to tell you (in not so clear words) that it is going to add the methanol-chloroform binary parameters to the simulation for you, and is asking for your permission to do so. Since I'm asking you to change the physical property model, it doesn't matter what you pick here.

[6]Of course, in the real world it's up to you to select the appropriate model; sometimes NRTL is horrible for certain systems.

What has happened is that we now need to add the binary interaction parameters between methanol and chloroform. The NRTL part of the NRTL-RK model is an activity coefficient–based model, which is used to predict liquid-phase activity coefficients γ_i as a function of temperature and composition. I am sure you remember activity coefficients? They form the basis for writing *fugacity balances*. Yeah, I just went there.[7]

At VLE, the fugacity of each component i in the liquid phase equals the fugacity of each component i in the vapor phase.

$$f_i^L = f_i^V$$

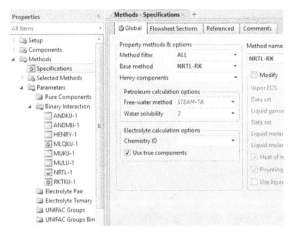

Figure 2.1 Choosing a physical properties model.

So, let's say I have a mixture of water and ethanol at VLE. The fugacity of water in both the liquid and vapor phase might be 6.5 bar, and the fugacity of ethanol in both the liquid and vapor phase might be 2.5 bar. I made up those numbers, but you get the idea.

An activity coefficient model like NRTL lets you compute *liquid*-phase fugacities like this:

$$f_i^L = x_i P_i^{\text{sat}} \gamma_i$$

where x_i is the liquid mole fraction, P_i^{sat} is the saturation pressure (a.k.a. vapor pressure), and γ_i is the activity coefficient of i. The vapor pressure is a known function of temperature (e.g., you could use Antoine's equation).

The activity coefficients are also a function of temperature and composition. The model that NRTL uses in particular to compute this is as follows:

$$\ln \gamma_i = \frac{\sum_j x_j \tau_{ji} G_{ji}}{\sum_k x_k G_{ki}} + \sum_j \frac{x_j G_{ij}}{\sum_k x_k G_{kj}} \left(\tau_{ij} - \frac{\sum_m x_m \tau_{mj} G_{mj}}{\sum_k x_k G_{kj}} \right)$$

where

$$\tau_{ij} = A_{ij} + B_{ij} T^{-1} + E_{ij} \ln(T) + F_{ij} T \quad \forall i \neq j$$

$$\tau_{ij} = 0 \quad \forall i = j$$

$$G_{ij} = \exp(-\alpha_{ij} \tau_{ij})$$

$$\alpha_{ij} = C_{ij} + D_{ij}(T - 273.15K)$$

Ok, that's a lot to handle. For now, just worry about this: the terms A_{ij} through F_{ij} are *constants* that are determined by regression of experimental data. They are the same for each pair of chemicals at any temperature, pressure, or composition. They are just fixed numbers and Aspen Properties has a nice database containing thousands of these constants for many different pairings of chemicals. To load them, click on the red half-circle in the Binary Interaction Folder in the subheading NRTL-1. If the data exist, then they will

---

[7]Oh Dr. Adams, that's the evilest thing I can imagine!

automatically load from the databank. On the right-hand side, you should see the numbers fill in. Note, however, that we have no idea how good these actually are, but it's a start.

So in the end, it's a lot of number crunching, but as long as I know the liquid mole fractions and the temperature, I can compute the activity coefficients (well, we let Aspen Properties do it).

Q1) Report the value of C_{ij} contained in the Aspen Plus NRTL-RK databank, where i is methanol and j is chloroform.

Ok, so what does the RK part of NRTL-RK mean? That is the equation of state (Redlich-Kwong) used to describe the properties of the *vapor* phase. It is also used to compute the vapor-phase fugacity, like this:

$$f_i^V = y_i P \varphi_i^V$$

where y_i is the mole fraction in the vapor phase, P is the pressure, and φ_i^V is the fugacity coefficient of the vapor phase. I won't get into the equations for it now, but the RK method is used to predict φ_i^V as a function of composition and temperature.

Note that if you had just chosen `NRTL` instead of `NRTL-RK`, Aspen would use the ideal gas law instead of the Redlich-Kwong equation. In that case, $\varphi_i^V = 1$. However, since the computer is doing all the work for us, it's just as easy to use a more rigorous model instead of the ideal gas law, so you might as well (assuming it is accurate).

♫ Music break[8]

PART 2: RETRIEVING PHYSICAL PROPERTY DATA

Aspen has a lot of physical property data that aren't shown on this form. You can get to it by pressing Retrieve Parameters (see Figure 2.2) and then hitting OK on the next form (you do not need to copy or overwrite any data right now, so leave those boxes unchecked). Then head down to the Methods | Parameters | Results tab to see all of the different physical properties or property parameters that are in the database. They are all stored in a sort of coded form. For example, in the Pure Components section, you can find PC, which is critical pressure.

In order to figure out what they mean, try searching for them in the help file (F1). These are usually legacy variables from very early versions of the program, which is why they are usually all-caps and have six characters or less.

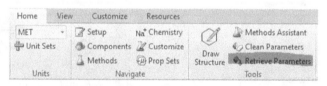

Figure 2.2 Find additional properties about your chemicals with the Retrieve Parameters feature.

Look in the T-dependent tab of this form to find out what the equation used for the vapor pressure is (PLXANT-1). Elements 1 through 7 are the coefficients c_1 through c_7 in Antoine's equation:

$$\ln P_i^{\text{sat}} = c_{1i} + \frac{c_{2i}}{T + c_{3i}} + c_{4i}T + c_{5i}\ln T + c_{6i}T^{c_{7i}}$$

where T is in K.

Q2) What is c_{2i} for i methanol in Antoine's equation as used by NRLT-RK?

---

[8]Recommended listening: *Nobody Told Me* by Andrew Bayer.

PART 3: CREATING A VLE DIAGRAM

One of the first stages of answering "how do we separate these chemicals?" is to look at the VLE. This will tell you a lot about how difficult it will be to use distillation. Let's first use Aspen Properties to do this.

Aspen Properties has a collection of tools that lets you use the physical property models to make physical property estimations without having to create a flowsheet. Go to the Analysis section of the menu and then select Binary. This will bring up a little dialog where you can perform equilibria calculations.

To understand the VLE, it is usually most convenient to consider a T-xy diagram. Set the analysis type to Txy. You then choose the two components you want to compare (your only two chemicals should already be selected by default). You can also select the range of mole/mass fractions of the primary component that you want to look at. Usually, you can just leave it at 0 (0% species A, and 100% species B) and 1 (100% species A, and 0% species B). However, sometimes you need to focus on just a particular range, so you can change that here. For example, you may need a higher resolution in a particular range, or the temperatures/pressures at certain compositions are extremely high or low such as for normal gases, or for the separation of chemicals in which some are not very volatile.[9]

You can choose the level of resolution for the analysis within the range you specify. The default is 50 intervals, which since you chose a range between 0% and 100% is every $(100\% - 0\%)/(50\,\text{intervals}) = 2\%$. What the binary analysis will do is run a loop, using liquid-phase mole fractions starting from the low end of your range (0% methanol in the liquid phase, by default) to the high end (100% methanol in the liquid phase, by default) in steps of 2 mol%. For each iteration of that loop, it will run a flash calculation at that concentration and at the given pressure (1.01325 bar by default). The flash calculation will determine what temperature and vapor mole fractions will result in a VLE such that the liquid mole fractions are equal to the mole fractions of that iteration. For example, suppose the current iteration is to run the flash calculation at a liquid mole fraction of 5% methanol and 1.01325 bar. The analysis would determine that the equilibrium temperature under these conditions is about 330 K (about 57°C) and that the vapor phase is about 16 mol% methanol. This is important to understand, because if you set your range such that it will perform flash calculations for liquid mole fractions that are not realistic or do not exist in practice, you may get errors or garbage results.

Change the interval of your analysis to 100 so that we get one data point at every 1 mol%.

In the upper right of the form, we can pick the phases we are looking at. Leave it at vapor-liquid-liquid for now, but note that you can also choose other variants if you know for sure that you want to consider them.[10] The screen capture in Figure 2.3 demonstrates this.

In the pressure box we perform the T-xy analysis at a certain pressure. We can also choose multiple pressures. Let's use 1.01325 bar, 5 bar, and 10 bar. This goes in the little list of values. (Just type them into empty cells to the right of 1.01325 bar. It is ugly, but it is what it is.)

Finally, we could change our property method here, but let's leave it at the default of NRTL-RK. Changing it is useful if you have a collection of candidate property models that you are investigating and want to see how each of them performs (see Bonus Tutorial 4 for more about this), without changing your default property model for the flowsheet. Hit the Run Analysis button. What you should get are three different T-xy diagrams on the same plot, one for each of the pressure ranges, as shown in Figure 2.4.

---

[9]See Bonus Tutorial 4 for more information about how to deal with these complex situations and how to choose the correct physical property models in the first place.

[10]Even if your system does not have two liquid phases (such as oil and water), it is ok to leave it at vapor-liquid-liquid. With this setting, Aspen Properties will automatically check to see if there are two liquid phases before checking for a single liquid phase. There is no harm in leaving it as is, unless you really need to shave off the miniscule additional run time and you know for sure that there will definitely not be a second liquid phase.

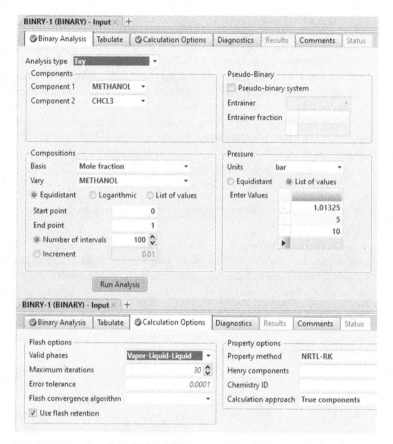

Figure 2.3 Setting up a binary phase-diagram analysis.

This tells us a lot! For each of the T-xy curves, the top line is the dew point and the bottom line is the bubble point. This also means that at 0% methanol (100% chloroform) on the left side, they come together at the boiling point of chloroform. On the right side, the two lines come together at the boiling point of methanol. Note how it changes with pressure.

When the saturated vapor and liquid curves are close together, you need more stages for distillation. When they meet at a place other than the left or right side of the diagram, that's called an azeotrope, or commonly a "pinch point." But we're serious professionals here, so we'll use the term "azeotrope." This forms two separate phase envelopes on either side of the azeotrope. A distillation column can only operate within one phase envelope at a time. That means we cannot make a single distillation column that produces both high-purity methanol and high-purity chloroform. We can have only one high-purity product with the other near the azeotrope (so we operate on either one side of the azeotrope or the other). Notice also how everything changes with pressure (see Figure 2.4). As pressure increases, the overall temperature gets higher and the azeotrope shifts toward higher methanol concentrations.

In addition, all of this data appears in table form. Just expand the BINRY-1 folder, go to the item called Input, and go to the Results tab. It's the same data; you could copy-paste it into a spreadsheet program like Microsoft Excel and make your own plot if you wanted and it would look the same. The table has extra data though such as the K-values and activity coefficients (called gamma) which are calculated from the NRTL-RK method. Use the table to answer these questions.

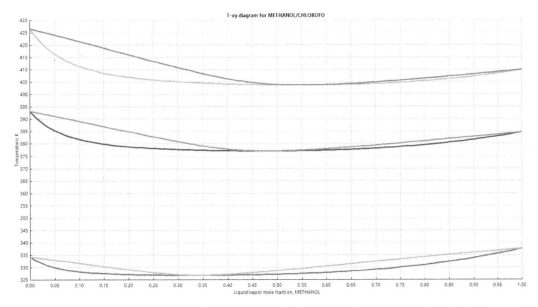

Figure 2.4 T-xy diagrams at three different pressures for the system of interest as predicted by the NRTL-RK model.

Q3) What is the boiling point of chloroform at 10 bar?

Q4) What is the liquid-phase activity coefficient of methanol for a mixture of 70% methanol and 30% chloroform at 5 bar?

Q5) At what mole fraction of methanol does the azeotrope occur at 5 bar? There are several ways to find this:

1. Where the vapor and liquid mole fractions are about equal
2. Where the temperature is the lowest (it's a low-boiling azeotrope)
3. Where the K-values are closest to 1.000 since:

$$K_i = y_i / x_i$$

So there are some key separation points we can take away from this analysis.

- If we do distillation at any pressure, we will have the azeotrope to deal with and can operate on only one side of the azeotrope or another in a single column.
- The azeotrope moves "to the right" with increasing pressure and moves significantly.
- For the 10-bar case, the "left side" of the azeotrope has the fattest phase envelope between the vapor and liquid lines.
- For the 1-bar case, the "right side" of the azeotrope has the fattest phase envelope compared to the other pressures.
- The highest temperatures, even at 10 bar, are still low enough to use steam in the reboiler if we use distillation. That's good because we don't want to build a furnace if we don't have to.

PART 4: PRESSURE SWING DISTILLATION

You can use the VLE diagrams directly to help you design the system. Figure 2.5 takes the original VLE diagram and shows an example of how the streams could flow between the different process sections. Here, I have chosen to collect high-purity chloroform (D) at about 153–155°C in the bottoms of the 10-bar column because that part of the VLE envelope is the fattest, and therefore would require the least number of stages and/or lowest heat and cooling duties. The distillate (E), which will be near azeotropic conditions, is fed to the 1-bar column; the rightmost envelope is the fattest at that pressure (we could even think about vacuum pressures, but that brings its own challenges, so let's keep it atmospheric). There, methanol (B) is recovered in the bottoms near its normal boiling point, and the distillate on the 1-bar column (C), which will be close to the 1-bar azeotrope in composition, is fed back to the first process. The feed stream (A) is sent to the 1-bar column because its composition (70% methanol) falls in the working range of that column (35–100% methanol). It could still potentially work if it were fed to the 10-bar column instead, and similarly, the two columns could theoretically be swapped (methanol is recovered from the 10-bar column instead of the 1-bar column). However, this is unlikely to be better.

Figure 2.6 shows how you can overlay the process operations on top of that. You can tell which part of the column that a product will leave from (the distillate or the bottoms) by looking at the temperatures. For example, the azeotropes will always leave through the distillate because the azeotrope temperature is lower

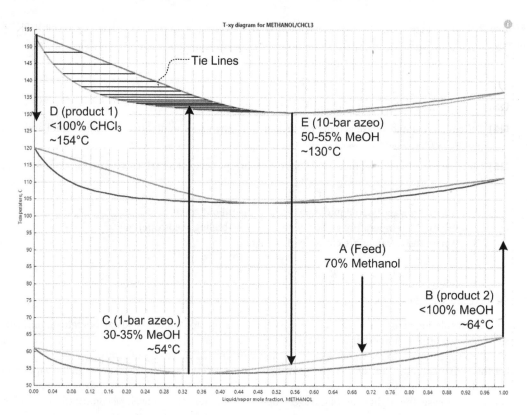

Figure 2.5   VLE diagram for methanol/chloroform process. High-purity chloroform will be recovered in the bottoms of the 10-bar column, and high-purity methanol will be recovered in the bottoms of the 1-bar column. The horizontal tie lines inside the phase envelopes show how vapor and liquid mole fractions are connected.

than the boiling point of the pure components. For example, the temperature of the azeotrope is about 54°C at 1 bar, but the boiling point of the methanol is about 65°C. This is why it is called a "low-boiling azeotrope."

Figure 2.7 then gives the final process in a nice form. Note that pumps and valves do the job of changing pressures. In addition, there are two feeds to the first column because the azeotrope stream E can be recycled. It just increases conversion because otherwise there will be considerable waste.

Well, look at that! The process pretty much writes itself from the VLE diagram. Note that we have a choice to feed E to the same tray as A, or feed to some higher tray. And, by looking at the graph, it shows us a great guess as to where to feed the streams (i.e., about halfway between the top and the bottom).

Go to the flowsheet window and simulate the first column using a feed of stream A with the 70 mol% methanol (see beginning of tutorial) and 200 kmol/hr at 30°C and 1.01325 bar. (The actual flow rate doesn't really matter though right now.) Use `RadFrac` for the column with 35 stages, a total condenser, and reflux and boilup ratios of 3.2 each. Feed stream A to stage 15. Set the pressure of the top stage in the condenser to 1.01325 bar (also, the feed stream should be at the same pressure).

Run the simulation. You'll notice you get error messages, including the general red "Results Available with Errors" message, a "blocks were completed with errors" message in the Run Summary, and possibly a red X

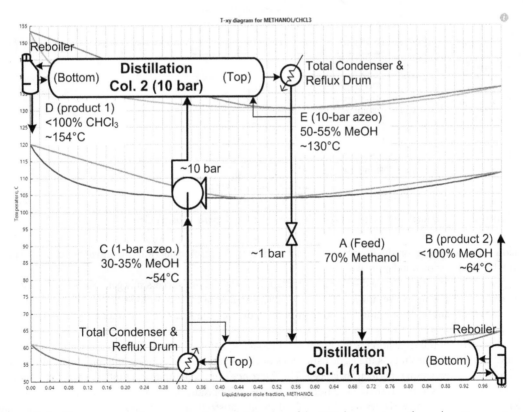

Figure 2.6 The process can be conceptually overlaid on top of the VLE diagram to easily see how it connects.

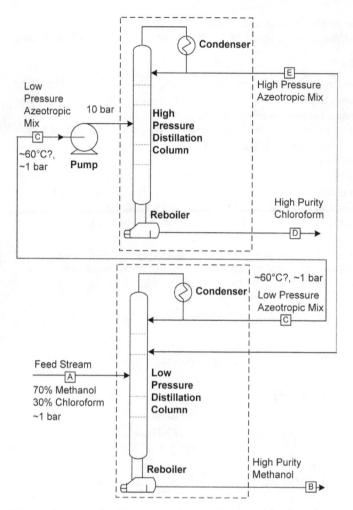

Figure 2.7 The final process as designed by using the T-xy diagrams as a guide.

icon near the column on the flowsheet. A quick look at the control panel (Run | Control Panel or F7) gives you the details shown in Figure 2.8.

Basically, what has happened is that the simulation model did not successfully get a result. Why not? The Fortran routine which solves the RadFrac block uses an iterative guess-and-check procedure, and after 25 guesses it did not find a solution (did not converge). However, the Err/Tol number tells us how close it is to converging.

This *error* of the simulation is the norm of all model equation residuals. The residuals are the left-hand sides of the equations minus the right-hand sides of the equations. If the residual of an equation is exactly zero, then the equation is perfectly balanced. The *tolerance* is the maximum amount of error that is allowed. So if Err/Tol is above 1, then we're not done converging because some of the equations have too much error, so not all of the variables have been solved to our satisfaction. If Err/Tol is below 1, then we have solved the problem within tolerances. The point is that if it is heading toward 1, we are on the right track. It can be a bit of an art form to look at a sequence of Err/Tol numbers and decide whether the solver is approaching a solution, or, it is going nowhere. A good rule of thumb is that if the `Err/Tol` is staying below 1000, is generally decreasing (perhaps going up and down but generally is tending toward 0), and has not recently risen above 100,000, then it is probably on the right track.

So, let's tell the program to keep trying. To do this, go to Blocks | Column Name | Convergence | Convergence (or double-click on the column to get to Blocks | Column Name). Under Basic, you will see that "maximum iterations" is at 25 by default. Change it to 200 (the maximum), as shown in Figure 2.9. Rerun the simulation. It should converge now.

Verify that the bottoms stream should be high-purity methanol, and that the distillate should be close to the azeotrope composition. If your stream results have empty spaces for mole fractions, go to Setup | Report Options | Stream, check the appropriate boxes, and then rerun the simulation.

Q6) What is the purity (mole fraction) of methanol produced? It should be close to what we predicted from the diagram.

Q7) What is the temperature of the azeotrope (distillate) in °C? It should be close to what we predicted.

```
Block: B1        Model: RADFRAC

Convergence iterations:
  OL   ML   IL    Err/Tol
   1    1    3     89.394
   2    1    4     65.722
   3    1    3     66.593
   4    1    3     101.88
   5    1    4     296.25
   6    1    6     971.16
   7    1    4     357.86
   8    1    4     592.65
   9    1    5     922.15
  10    1    4     577.58
  11    1    4     659.87
  12    1    5     473.02
  13    1   10     1015.1
  14    1    8     935.68
  15    1    5     659.55
  16    1    3     158.94
  17    1    3     72.626
  18    1    4     61.937
  19    1    4     87.187
  20    1    4     236.01
  21    1    6     1025.4
  22    1    4     484.35
  23    1    5     831.46
  24    1    6     1006.1
  25    1    4     786.39
** ERROR
   RADFRAC NOT CONVERGED IN  25 OUTSIDE LOOP ITERATIONS.
```

Figure 2.8 Control panel output for a RadFrac simulation that did not converge, and so the model outputs are meaningless, and you should not use them.

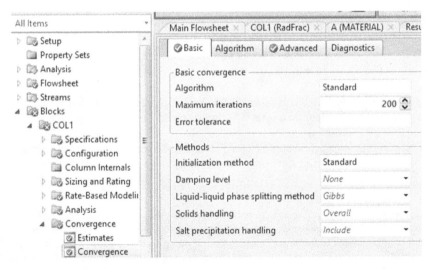

Figure 2.9 The Convergence form for the RadFrac block. Use this to change the maximum iterations and other parameters of the underlying numerical methods used to solve the model equations.

Ok, great! Let's add the second column. Let's also use 35 stages, with a reflux ratio of 3.2 and a boilup ratio of 7. The condenser pressure should be 10 bar (again, assume no pressure drop for the rest of the column). Let's guess 10 for the feed stage. You'll also need to add the pump appropriately, because if not you will get warnings that you are trying to put a low-pressure stream into a high-pressure column. Don't recycle the distillation stream to the 1-bar column yet. *Never add the recycle stream until everything else is working to expectation! Remember this.* If you have problems with convergence, try getting the column to converge with 10 stages, then 20, and then 30. The program uses the previous results as guesses for the next run, so this is why you want to (a) start with something small that works and then work your way up, and (b) not add recycle until everything else is working. Alternatively, you can try changing convergence algorithm (i.e., the way in which it guesses and checks its way to the solution of the model equations for the column). The default setting is `Standard`, which you can see on the Configuration page for the second column. I switched it to `Azeotropic` and found better performance (go figure).

Q8) What is the purity (mole fraction) of chloroform produced from the second column? It should be close to what we predicted from the diagram.

Q9) What is the temperature of the azeotrope (distillate of the second column) in °C? It should be close to what we predicted in Figure 2.5.

Verify that the mole fraction of the azeotrope is what you expected, and then connect the recycle stream to the first column (try stage 8[11]); do not mix it with the main feed! Using a valve to reduce the pressure is optional. As long as the pressure of the stream going into the column is higher than on the stage to which it is fed, the model assumes it will automatically flash to the lower pressure anyway through the inlet nozzle. Rerun the simulation.

Q10) What is the new purity of the chloroform?

Q11) What is the new purity of the methanol produced? It should be close to what we predicted but not quite as good. This could be fixed with more stages, higher reflux ratios, or better positioning of the feed trays. We'll worry about this at another time.

Ok, so even though both purities are not what we wanted, we have a wonderful starting point with which to improve the system. We will not explore this, but we will learn techniques for how to do this in later tutorials.

Ok, now let's learn a few more things about this system. The physical property system lets us find out considerably more information about a stream than what is in the stream results (by the way, clicking on a stream and hitting `Crtl+R` is a fast shortcut).

To get more properties, click on a stream, and under Home | Stream Analysis you can see many options. Do this with your chloroform product stream, and choose `Bubble and Dew Point curve` (Figure 2.10).

---

[11] If you look at Figure 2.7, it appears that the recycle should appear at a stage higher than the feed stage because it has a lower methanol content, so generally it would be more efficient to feed it higher up. However, since it is not so different in composition from the feed, it could be cheaper to just have one feed stage with both streams going to it, so you don't have to have two separate feed distributor systems. The final decision could be made using a detailed technical and economic analysis to compare the two options.

Leave the options at the default, but you can see they are similar to your previous analysis. Click go and you'll see the plot of the bubble and dew points for temperature and pressure. For example, pick a pressure, and then where the vapor line intersects that pressure is the dew point, and where the liquid line intersects that pressure is the bubble point. Similarly, at a constant temperature, you can get the dew/bubble point pressures.

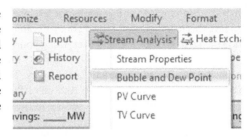

Q12) What is the bubble point temperature in °C of the chloroform product stream at 5.5 bar?

Figure 2.10 Stream analysis tools provide a convenient way to predict useful properties about a stream.

You can get additional information not in the results section, such as viscosity of the mixture (MUMX, MU is μ), thermal conductivity (KMX), or surface tension (SIGMAMX). To do this, close the previous analysis windows, select the appropriate stream, choose Home | Stream Analysis | Point, and select both thermodynamic and transport properties options.

Q13) What is the thermal conductivity in J/sec-m-K of the chloroform product stream?

Additionally, you can add other properties to your stream summaries (i.e., the stream results form) by customizing your stream template. This is nice because it is persistent across runs and shows up for all streams in the future. Click on <add properties> at the bottom of the stream results form for any stream. You are then presented with a search dialog where you can find properties that you want to add to your stream template. As shown in Figure 2.11, I have searched for "fugacity" and added the pure component fugacity and fugacity coefficients to my template. You can see that I have already added the higher heating value as well. There are a lot of properties to search from, so hopefully you will find what you need. Note that when you

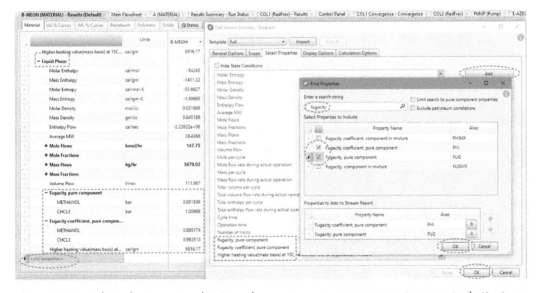

Figure 2.11 Editing the stream template is another way to access stream properties conveniently. Key items to note are boxed and circled.

modify this template, and you attempt to close the software, you may be prompted as to whether you want to save this template or not. If you do, you can reuse this for other simulations you make in the future, so you don't need to go back and add everything yourself again.

You can find all sorts of physical property information on just specific individual properties such as heat capacity, density, enthalpy, and latent heats. This works at any time and does not require you to select a stream. Under Properties, go to Analysis | Pure. The Property drop-down box contains lots of options. Hover your mouse over them to see the full-text description in the status bar at the bottom. The rest of the form should be familiar.

Q14) What is the molar density in mol/L of liquid chloroform at 50°C and 12 bar?

♫ Music break[12]

---

[12]Recommended listening: *My Friend* by Groove Armada.

Problem Solving Tools

Objectives

- Learn to use Sensitivity in Aspen Plus
- Learn to use Design Specs in Aspen Plus
- Understand pressure level heuristics for compressors and turbines
- Understand the difference between heat, material, and work streams

Prerequisite Knowledge

It is advisable to complete both prior tutorials before you begin this one. At this point, you'll need to understand the basics of compressors, turbines, etc., so you should be familiar with those before trying this tutorial. You'll also need to have an understanding of the concept of the degrees of freedom (DOF). Check out this video on DOF,[1] if you need to refresh your knowledge.

Why This Is Useful for Problem Solving

The Design Specs and Sensitivity features are key tools in using flowsheet simulations. In the prior tutorials so far, flowsheet simulations have been used to answer the question "What are the stream conditions or unit operation results when I have these inputs and parameters?" However, this tutorial will help you ask questions such as "What inputs or parameters do I need in order to get a certain stream condition or unit operation result?"

For example, without any special tools, you can set up a simulation of a heat exchanger using a HeatX model (see Tutorial 4) in which you enter the inlet conditions of the coolant (the cold stream) and inlet conditions of the hot stream and run the simulation to determine the outlet temperature of the hot stream. But what if you have a different problem, in which you know what the outlet temperature for the hot stream should be, and you want to figure out how much coolant it will take to achieve it? One way to solve this problem "by hand" is by guessing different flow rates of the coolant, running the simulation for each guess, and checking the outlet temperature. You would

---

[1]https://www.youtube.com/watch?v=tW1ft4y5fQY. This is peer-reviewed material produced by the University of Colorado, Boulder.

change the guesses each time in some intelligent way, getting hot stream outlet temperatures that are closer and closer to your goal until you finally find the right coolant flow rate.

The Aspen Plus Design Specs feature automates this guessing-and-checking of flowsheet parameters to achieve a certain flowsheet objective. For example, the flow rate of coolant into a heat exchanger can be adjusted by a Design Spec to achieve the desired hot stream outlet temperature (we'll get into this more in Tutorial 4). Similarly, the Sensitivity feature can be used to determine how flowsheet variables change with respect to changing flowsheet parameters. This automated process avoids the lengthy time that running numerous simulations manually might entail.

The Design Specs and Sensitivity tools can be used among other things to make certain flowsheet design decisions about selection of operating conditions of unit operations (i.e., pump pressures, heat exchanger, heat duties, etc.), and flow rates and properties of chemicals (i.e., composition, temperature, pressure, etc.). They can also be used to suggest better flowsheet designs given different circumstances, predict what the flowsheet inputs and outputs might be, identify errors such as violations of the laws of thermodynamics on a given flowsheet, identify certain limitations, and other such concepts.

Tutorial

PART 1: DESIGN SPECS

In this section, we will be working on a process involving an exothermic reactor whose products exit at 420°C. These need to be cooled before downstream treatment and separation. Rather than just rejecting this heat to the environment, this high-temperature heat can be used for something more useful: electricity generation. To do this, let us consider the addition of a steam power plant to the process, as shown in Figure 3.1. In this process, boiler feedwater (BFW) just below the boiling point at 95°C and 1 bar (stream 1) is pumped to high pressure at 20.5 bar (stream 2). The high-pressure BFW enters a heat exchanger where it is boiled to high-pressure steam (HPS) at 360°C (stream 3) using heat from the reactor effluent. The reactor effluent is subsequently cooled to 150°C. The HPS is then sent through a series of two turbines, which produce electricity in each. The steam exits the second turbine at low pressure again (1 bar) and at a temperature just above its boiling point (still a vapor, stream 5). Then, cooling towers are used to condense the steam into a liquid at 95°C and provide a little additional subcooling.

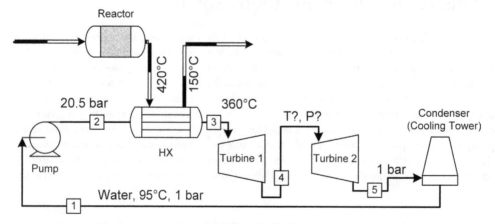

Figure 3.1 A process to generate electricity by using heat from a reactor.

The problem is that we don't know what intermediate pressure to select for the two stage turbines; that is, we don't know what the discharge pressure of turbine 1 should be. Clearly, anything between 1 and 20.5 bar will theoretically work. The question is, which is best? In addition, we do not know what the flow rate of steam should be. So how can we find these answers?

The strategy is as follows:

(1) Create a model in Aspen Plus for the steam plant using what we know.
(2) Use a model with a Design Spec to figure out how much steam we need to achieve a steam temperature of 360°C and a hot outlet temperature of 150°C.
(3) Complete the model to determine how much power is produced for one specific guess of the intermediate pressure.
(4) Use a sensitivity analysis (Part 2) to vary the pressure and determine how the power produced changes with the intermediate pressure.

Let's do it! Two more things we need to know before we start:

(1) Assume for now that we know that 200 MW of cooling is needed to take the reactor effluent from 420°C to 150°C.[2] This means that we can use a Heat stream to model the heat transfer without needing to model the reactor effluent or the reaction.
(2) Aspen Plus does not handle processes that are completely a single loop like the steam cycle well (or at least not without some trickery). In many cases when you have a loop like this, you can find a place where you can "break" it, so you actually only model a single, one-directional flow, but you know that, in practice, the loop is really closed. The best place to break a loop is at a point where you know or specify everything about a certain stream (i.e., temperature, pressure, flow) because then you can be sure that the end of the loop is identical to the beginning.

Therefore, with these two pieces of information, our model, for the purposes of Aspen Plus, should look like Figure 3.2. Note that stream 1 has been broken into 1 and 1B. It is desirable to do this because Aspen Plus is a sequential-modular flowsheeting program, which means that the simulation is solved one block at a time in some logical sequence (see the Introduction for more on this concept). In practice, Aspen Plus usually starts with an input stream and then solves each block in order, tracing the pathway of the stream. So, by splitting stream 1 into 1 and 1B, you are telling Aspen Plus that it should start at stream 1, then compute the model for Pump, then HX, then Turb1, then Turb2, and then Condenser. We as the designer know that

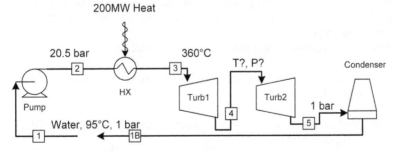

Figure 3.2 Process flow diagram adapted for Aspen Plus.

[2]We would know this from other simulations or calculations, such as using a Heater block, which is based on the enthalpy difference between the hot inlet and hot outlet streams.

streams 1 and 1B should be exactly the same, as long as we have set up the simulation correctly, so we can check this after the run has completed. If we did not break the loop into two pieces, then Aspen Plus can sometimes have trouble because it creates a cycle of dependency in which each block depends on all others, which can require lengthy iterative solutions and lead to convergence problems. Of course, this can be done correctly with experience and know-how, but it is complicated. In many cases, breaking the loop at a point in which you know everything about a stream is often the best way to go, for both beginners and experts.

Let's address the simple question of how much water should be in our steam loop. To rephrase the problem more specifically, we need to find the amount of water that provides 200 MW of cooling while exiting the heat exchanger at 360°C and 20.5 bar.

First, start a new model in Aspen Plus (I suggest making sure your units are set to METCBAR). Since we're going to use only water, use STEAMNBS as the physical property choice (it's the best choice when water is your only chemical). Create a model of the process from streams 1–3. Assume no pressure drop across the heat exchanger. In order to build this model, you'll have to add a Heat stream of 200 MW, which feeds into the heat exchanger. To do this, you go to the Material icon on the left-most side of the Model Palette toolbar and hit the drop-down arrow to select it. Note that a Heat stream is just a model. It is the magical Q that shows up so famously in chemical engineering equations. This is how we add the 200 MW of heat from the heat exchanger. We could model the other half of the heat exchanger if we wanted, but we only need the heat portion so we just use the heat stream here with a half exchanger (Heater) block.

When you add the Heat stream, you have to enter the heat duty as shown in Figure 3.3. However, you may also fill in the starting and ending temperature of that heat to correspond to what the real heat exchanger is doing. However, the start and end temperatures will have no bearing on our model. You can type in whatever you want and the results of the Heater wouldn't change. It's there so that you as the engineer can check for temperature crossover or use it for other purposes. You can even leave it blank for our case, as shown in Figure 3.3.

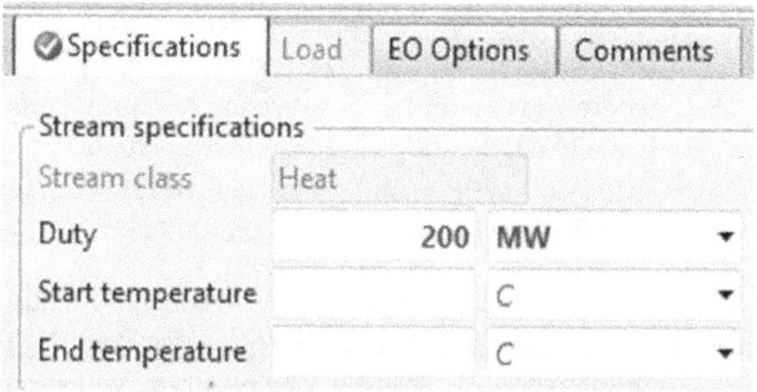

Figure 3.3 Setting up the heat duty stream.

Notice also that when you connect the heat stream to the heat exchanger, that uses one of your DOF. Therefore, when you go into the heat exchanger and specify the pressure drop, the other drop-down box is disabled (the text "Inlet heat stream" appears instead, telling you that this has taken up one of the DOF).

Note that because we are specifying the heat duty and pressure (or alternatively, zero pressure drop), we cannot specify the temperature. This makes sense. Given a known pressure and heat input duty, the temperature will be calculated, instead of specified.

Let's just see if the model works for now. Guess a water flow rate of 14,000 kmol/hr.

Q1) For a guessed water flow rate of 14,000 kmol/hr, what is the temperature in °C of stream 3? If done correctly, it should be in the 390–410°C range.

Q2) For a guessed water flow rate of 15,000 kmol/hr, what is the temperature in °C of stream 3? If done correctly, it should be in the 315–325°C range.

Ok, so we know that the flow rate of water that will achieve a stream 3 temperature of 360°C will be between 14,000 and 15,000 kmol/hr, but you can see how tedious this is going to be if we keep changing and checking by hand until we get 360°C to exact precision. So let's automate the process by using the Aspen Plus Design Specs tool.

Under the Simulation tab, go to Flowsheeting Options | Design Specs. Here, you will see an Object Manager that lists the set of design specifications you have created. Click New to make a new one, and give it a name (or leave it at the default of DS-1). The Design Specs tool works like this:

- You tell it what output you want to achieve. For example, you want to achieve 360°C in stream 3. You do this with a combination of the Define and Spec tabs.
- You tell it what input specification or block model parameter you want Aspen Plus to change until your specification is met. For example, you want to change the water flow rate of stream 1. You do this in the Vary tab.
- The other tabs are advanced. For example, in the Fortran tab, you can write a program to make complicated decisions. We won't do that here.

Let's start with item 1. Go to the Define tab. This is where you define variables to be used later in the Spec tab. This is like defining a variable in a programming language, except instead of making a blank variable we will be getting the value from Aspen Plus.

Click New to make a new variable. Give it a name. You are going to make a variable that is the temperature of steam in stream 3, so perhaps T3 might be a good name. Then when you click OK, you get another dialogue that shows you more details. This is where you search for all the variables in your model that you can get. Here, you can access anything that can be seen on the Results tab of a stream or block (Crtl+R), or typed into an input box for a block.

In the Reference section at the right, choose Stream-Var from the Type drop-down. This filters out the variables to be only stream variables. Then in the Stream drop-down just below it, choose your stream 3.

Whatever names you used on your flowsheet will appear here. Leave the substream as MIXED (this book does not cover substreams). Then, in the Variable drop-down, select TEMP, and then select the appropriate units. Now, you have selected the temperature of stream 3. (See Figure 3.4.)

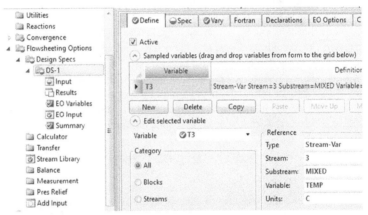

Figure 3.4 Setting the Define tab of a Design Spec.

Go to the Spec tab. This is where you tell Aspen Plus the exact specifications you want. We want the temperature of stream 3 to be 360°C. To do this, type T3 into the Spec box and 360 into the Target box. The meaning should be obvious. Note that you cannot change the units on the Spec tab, so your units will be whatever you defined on the Define tab.

But what is not obvious is tolerance. Since this is a guess-and-check algorithm, and floating points[3] are imprecise, Aspen Plus will never get exactly 360°C, or at least take a very long time to get there. However, it

---

[3]Most modern computers store numbers as a pattern of bits (ones and zeros). The particular way in which noninteger numbers are usually stored is called the Floating Point format. This format can represent both very large and very tiny numbers, but it cannot store all numbers precisely, because it has only 64 bits (sometimes 32 or 128). Usually, this means that decimal numbers can only be stored up to the first 10–15 decimal places or so, and after that, the number is rounded off. With many calculations, this error can compound and become significant.

could get 359.938382°C rather quickly, for example, and you have to decide if you are ok with that. You need to define your tolerance, that is, you need to tell Aspen Plus how close to 360 is acceptable. Type 0.1 into the tolerance box. This means that anything within 0.1°C of 360°C is acceptable. So once Aspen Plus has reached a value between 359.9°C and 360.1°C, it will stop.

Last, we go to the Vary tab. This is where you tell Aspen Plus what to change, which is the molar flow rate of stream 1.[4] Use the Stream-Var type, select your stream 1, and choose MOLE-FLOW as the variable. Then you have to change your manipulated variable limits. You have to tell Aspen Plus what is the lowest guess it can make (the Lower field on the right) and the highest guess it can make (the Upper field on the right). From Q1 and Q2, we know that the range will be from 14,000 to 15,000 since one was too high and one was too low when we were exploring "by hand." So type those in here now, as shown in Figure 3.5.

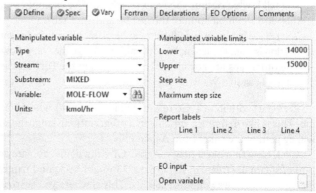

Figure 3.5 The Vary tab of a Design Spec.

The other tabs can be left at the default. You could make changes such as limiting the step size of the guess to a certain amount (i.e., how big of a jump Aspen Plus is allowed to make between the previous guess and the next guess), but it is almost always better to use the default settings except in very special cases.

That's it. If you've done it correctly, rerun the simulation. You can see the stream results (Crtl+R) of the input or any of the other streams to find out the final water flow rate. Or, you can go to the results tab of the Design Spec that you made and see where the variables ended up. Make sure you get the Results Available message! Also, verify that your Design Spec (temperature) was met within tolerances.

Q3) What is the flow rate in kmol/hr of water that exits the heat exchanger at 360°C (within 0.1°C) and 20.5 bar?

It can be useful and interesting to look at the actual guesses that the solver took when arriving at the final value. Go to Convergence | Convergence and look for the solver that goes with your Design Spec (for this example, it should be the only one there). Go to the Spec History tab on the form for this solver, and you will see what guesses it made in each iteration. In my case, it only required four guesses. Readers who are savvy about numerical methods will notice that the solver is using a classic Secant method, where the second guess is exactly 1% of the range away from the initial point. If you do not know what this means, it suffices to know that if the system of equations of interest is linear, this means it will always converge on the third guess as long as a feasible solution exists and the specification is well posed. The more the system deviates from linearity (which happens most of the time), the more guesses will be required.

Sometimes, the Design Spec never converges. This can be either because the solver did not have enough iterations to find the solution, the system is so nonlinear or stiff that it just simply cannot find it without very good guesses, the specification is attempting to vary something that does not impact what is being

---

[4]Why not the molar flow rates of stream 2 or 3? This is because in a sequential modular program information carries forward only (see Introduction). So if you changed stream 3 then you would still have to type a guess for stream 1. This would lead to a mass balance error. If you change stream 1, then streams 2 and 3 would have their flow rates calculated automatically.

measured, or the solution simply may not exist at all within the bounds that you specify and so there is nothing to find. If your Design Specs do not converge, you should focus on identifying which of the three this could be before continuing on with the rest of the flowsheet. For example, it is quite common for beginners to create Design Specs that are thermodynamically impossible without realizing it.

🎵 Music break[5]

PART 2: SENSITIVITY ANALYSIS

Ok, now that we know the steam flow rate, we can design the rest of the system. First, add the remaining streams and blocks into the model according to Figure 3.2. For the turbine models, you'll find them in the Model Palette under the Pressure Changers | Compr model section drop-down. Aspen Plus uses the same model for both compressors and turbines, so it actually does not matter which icon you select, but try to get into the habit of choosing the correct icon anyway. Make sure the models use an isentropic turbine and leave the efficiencies empty (meaning that Aspen will use default efficiency correlations which are somewhat complex). Let's make a guess at the outlet pressure of the first turbine of 5 bar. You can specify the outlet pressure of the second turbine according to the process diagram in Figure 3.2.

You should be able to handle the condenser already. (You know the requirements for the hot stream and don't know anything about the cold stream, so which block do we use?) Assume no pressure drop in the condenser.

 TOM'S TIP: In the convergence tab of turbine setup, change valid phases from Vapor to Vapor-Liquid. This will allow the model to function properly in case a liquid phase is formed in the output since the stream will get colder after expansion. In most real cases we don't want liquid formation in our turbines, but we can always go back and see if this is a problem and avoid it. Do this for both turbines, as shown in Figure 3.6.

Q4) What is the total cooling duty in MW required in the condenser when the first turbine has an exit pressure of 5 bar?

Finally, we are interested in the total work produced by the system. To find this conveniently, use Work streams (like Heat streams, but different). Add Work streams to the outlets of the two turbines and the pump (which consumes some of the power). This represents what in reality might be a shaft for a compressor/turbine system to transmit mechanical work, or an electrical connection for electrical work. Now we can model the magic that appears in calculations.

To get the total work, we can either add them together by hand, or have Aspen Plus add them for us by using a Work Mixer. This is not a physical thing in itself (don't go around asking people for a work mixer), it just lets us add the work together to get a sum easily. The Work Mixer icon is in the regular Mixer section, but you have to get it from the drop-down arrow, as shown in Figure 3.7. Unlike other cases, the icon for the work mixer represents a completely different model than the others in this case. Add a third Work stream to the outlet of the Work Mixer to make a stream that has both turbine works combined. So it's just like a mass mixer, but for work.

---

[5]Recommended listening: *Satellite* by Oceanlab.

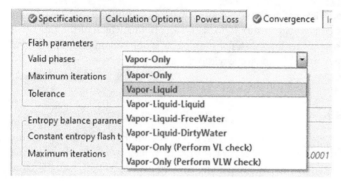

Figure 3.6 It can be handy to change compressors from vapor-only mode to vapor-liquid mode. Normally, you do not want any liquid droplets in your compressor because it may damage the equipment. However, when running many of the simulations programmatically (such as with a Design Spec or Sensitivity), you may need to allow two phases in the compressor just to keep the algorithms from stopping prematurely due to bad guesses.

Run the simulation, using the correct water flow rate that resulted from the Design Spec and the assumed 5 bar outlet pressure in the first turbine.

Q5) What is the total electricity produced in MW by the system when the first turbine has an exit pressure of 5 bar? The negative is the software's way of saying that work is produced, rather than consumed.

As a check, the results to Q4 and Q5 should add up to 200 MW by an energy balance. Double-check to make sure that stream 3 is still at the proper temperature (360°C).

Next we want to find the turbine 1 discharge pressure that maximizes our work produced by the turbines. We can't use a Design Spec because we don't know what the exact power output we want to produce actually is, we just know we want the highest possible.

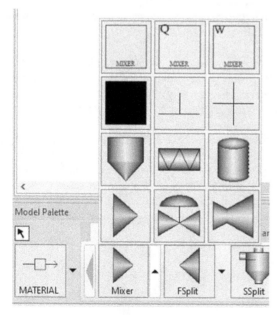

Figure 3.7 Material, heat, and work mixers are found in the mixers model, but are actually different models, not just different icons for the same model.

So, we'll use another tool called Sensitivity. This is basically just the "guess" part of the guess-and-check. It just reruns your simulation a bunch of times and tells you the results. We will use a Sensitivity to run many different simulations and different turbine outlet pressures and record the net work produced in each case. Then, we can look at the result and choose the one that has the highest net work. In other words:

Design Spec: You tell it exactly what you want and it changes something in your simulation until you get it (or it can't find it and it gives up). The thing you change is almost always something you normally type into a box by hand. It does the work of figuring out the right value to type in the box for you and actually uses that value in the simulation. Only the final result is reported.

Sensitivity: This changes a value in a box, just like the Design Spec, but it shows the results in a separate place because it doesn't pick any one of them for you. You get a nice table of results instead and you can decide what to pick later.

Let's do it! Make a new Sensitivity in Model Analysis Tools | Sensitivity. This is going to look a lot like the Design Spec, but now we have the Define, Vary, and Tabulate tabs. The Define and Vary tabs are just like in a Design Spec. The Tabulate tab is where you tell Aspen Plus what you want it to report.

Start with the Vary tab. In this tab, we can have Aspen Plus vary one or more variables. We'll just do one for now: the specified outlet pressure for turbine 1. Select this variable just like you did for the Design Spec | Vary case. Select <New> from the Variable drop-down button, choose Block-Var for the type, and then select your turbine 1 unit. For the variable, choose PRES. If you hover your mouse over the long list of options, you'll see that PRES is "Specified outlet pressure." You can see that there is a lot here you can mess with. Once selected, you should see the units pop up in bar. If not, change it here, and/or make sure your simulation units are set to METCBAR. For the range, vary from 2 to 20 bar in increments of 0.1 bar. You should be able to specify this on the right side since it is similar to the vapor-liquid equilibria (VLE) stuff we did in Tutorial 2. Leave the Report labels blank. See Figure 3.8 for final form settings.

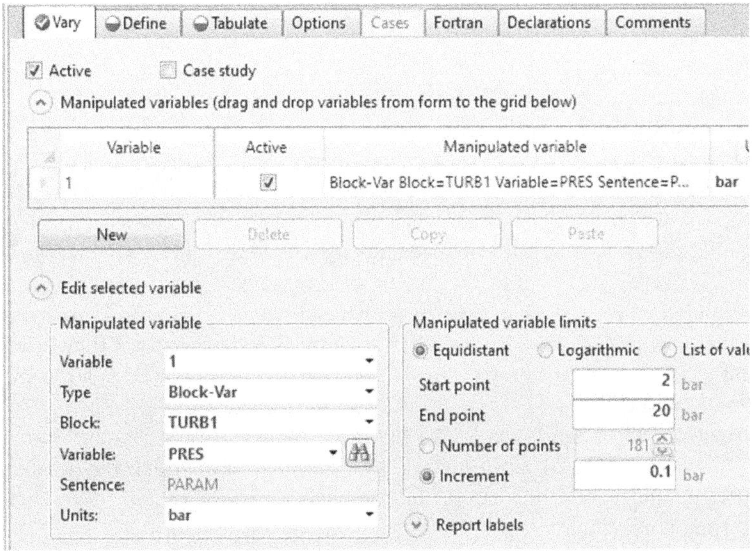

Figure 3.8 The Vary tab of the Sensitivity run.

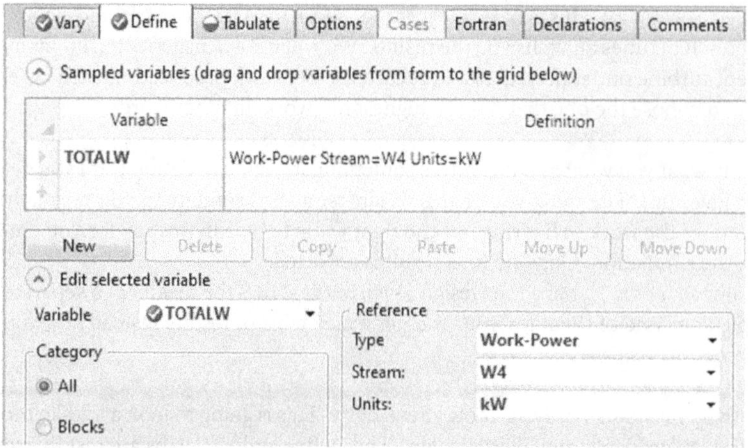

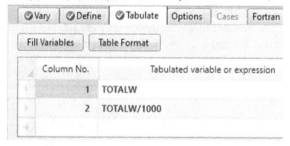

Figure 3.9 The Define tab of the Sensitivity run.

Ok, we told Aspen Plus what to vary. Now we have to tell it what to report to us, that is, what do we care about? We care about the total work produced by the turbines. So to do this, first go to the Define tab, and make a new flowsheet variable and give it a name (I called it TOTALW for total work, as shown in Figure 3.9). You want this variable to be the total work produced by the turbines, so select Work-Power from the drop-down for type and select the Work stream that is leaving your WORK Mixer. (See why we did this now?)

Once you are done, click Close and go to the Tabulate tab, as shown in Figure 3.10. This is where you tell Aspen Plus which values it should report for each iteration of your Vary variables. To do this, you pick the variable name or expression on the right side and select which column you want it to go into on the left. The column number doesn't really matter much; it's just the order in which you want to see the results.

For the tabulated variable or expression, you can start by just typing your variable name. For my case, I would type TOTALW because that's what I called it in the design tab.

Figure 3.10 The Tabulate tab of the Sensitivity run.

You can also write whole mathematical expressions. For example, I know that TOTALW is in kW but I want to see the results in MW. I could type TOTALW/1000 to do this. It uses Fortran syntax, but it's just like Microsoft Excel equations without the = sign. So it's not scary.

Ok, when you're done, you should see the yellow Input Changed message, and then run the simulation. It may take a little while. If everything worked, you'll get the Results Available message. Now, if you were to go look at the stream results in the simulation, you would still see the same results as your Q4 and Q5 answers. This is because by default, after the sensitivity analysis is finished, it runs the flowsheet one last time using your original settings, so nothing will look different on your flowsheet. What you want to do is go to the special place where sensitivity results are held.

So, on the left go to Model Analysis Tools | Sensitivity | S-1 | Results. Now, you should get a little table showing each of the Vary values (going from 2 to 20 bar in steps of 0.1), the values of anything you put into the Tabulate tab, and a status message under the Status tab saying completed normally (if it is not ok then there was an error or warning in your simulation).

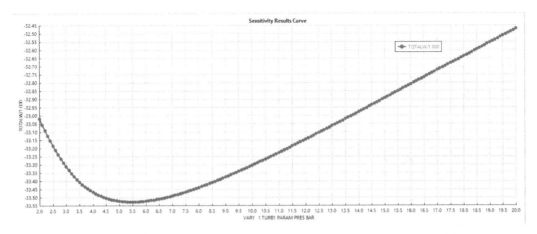

Figure 3.11   A plot of the sensitivity results, showing the total net work as a function of turbine outlet pressure. The pressure with the lowest net work is the best choice because the work numbers here are negative, and the largest negative means the most work (electricity) produced!

If you want to see the results visually, you can copy-paste the table into Microsoft Excel or some other software and make a plot there if you like. If you click the little grey area on the upper left hand of the results table (just to the left of the "Row/Case" column header), it will highlight the whole table. You can right-click and choose Copy or hit CTRL+C, then paste into your desired software.

Q6)   What is the net power produced in MW when the discharge pressure is 4.8 bar? Ignore the negative like before.

Q7)   Which discharge pressure (to the nearest 0.1 bar) provides the most power?

Now, if you're being observant, you'll notice that the very sneaky Aspen Plus has added another menu option in the menu bar that only appears when you are looking at sensitivity results! The Plot menu has just appeared in the Home ribbon to help you plot the data in a table. Although we can copy-paste into Excel and make plots there, it is often convenient to use the Aspen Plus plotting tool to plot the results quickly.

Ultimately, we would like a plot similar to the one in Figure 3.11. There are a few ways you can do this easily:

(1)   Click the header for VARY 1 (turbine 1 outlet pressure) so the whole column is highlighted. Select Results Curve from the Plot menu and then the following window pops up showing you have selected the column as the X-Axis (see Figure 3.12). Then you can make selections of the curves to show (in our case, the power in MW only). Then click OK.

(2)   Click the header for VARY 1 to select the whole column and use Ctrl+Alt+X to define that variable as the X-Axis variable for the plot. There is no feedback! You just have to hope it works. Then highlight the column for your total power produced and use Ctrl+Alt+Y to define this as the Y variable. There is still no feedback. Use Ctrl+Alt+P to show the plot. Now you can see it.

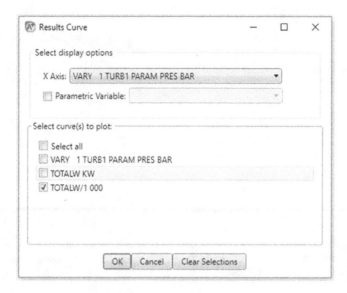

Figure 3.12 Use the Result Curve option of the Plot menu to make a plot of the sensitivity results.

Lastly, let's update our final simulation using the results from the Design Specs and the sensitivity block. (Remember, what's on the current flow-sheet does not reflect the sensitivity results, only your initial guess.) Type your final value for the water flow rate into the parameter specifications box for stream 1 (i.e., where you normally type flow rates and tempera-ture). Type the final value for the pressure into the input box for discharge pressure in turbine 1.

Now, go back to the Design Specs and Sensitivity Tabs and disable them. It is not obvious how. Go to Flowsheeting Options | Design Specs. Look on the left-hand side where it lists the different Design Specs, right-click your Design Spec (whatever name you gave to it in Part 1 or DS-1 by default), and choose DEACTIVATE. It will then have a grey symbol and all related folders will be grey (see Figure 3.13). This means that Aspen Plus will ignore the Design Spec completely. You can always reactivate it again later. It's a nice way of saving you from the work of deleting and remaking it when you are playing around. Do the same for the sensitivity analysis. Rerun your final design and answer the following questions:

Q8) What is the pump electricity usage in kW of the final design?

Q9) What is the electricity produced in MW by the second turbine?

Q10) What is the total electrical efficiency (power produced/total energy input) of the final power plant?

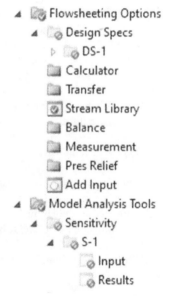

Figure 3.13 The Sensitivity block and Design Specs have been deactivated.

♫ Music break[6]

---

[6]Recommended listening: *Trahison sur ma peau* by Marie-Mai.

Heat Exchangers

Chinedu O. Okoli and Thomas A. Adams II

Objectives

- Develop a basic understanding of heat exchangers
- Learn to use the `Heater` model in Aspen Plus
- Learn to use the `HeatX` model in Aspen Plus

Prerequisite Knowledge

It is advisable that you have completed a basic course in thermodynamics or heat transfer and have an understanding of the first and second law of thermodynamics, as well as a basic understanding of heat exchangers. You should be able to differentiate between streams that require heating (cold streams), and streams that require cooling (hot streams), and understand when the second law of thermodynamics is violated in heat exchangers (temperature crossover).

There are also links to some useful video material on heat exchangers at the LearnChemE website covering the basics of heat exchangers such as how to calculate heat duties, heat transfer coefficients, and sizing parameters.[1]

Although it would be useful to have completed Tutorials 2 and 3 before attempting this one, it is not necessary.

Why This Is Useful for Problem Solving

Heat exchangers are chemical engineering unit operations used to transfer heat from one fluid to another by taking advantage of a temperature gradient between the fluids. They are very common in chemical engineering processes as they are used to increase or decrease the temperature of fluids in a process, or cause phase

---

[1]http://www.learncheme.com/screencasts/heat-transfer. This is peer-reviewed material produced by the University of Colorado, Boulder. On the page, navigate to the heat exchangers section for the video links.

change such as boiling or condensation. For example, in a steam power plant a heat exchanger is used to heat up cooling water before it is sent to a boiler. A boiler is another kind of heat exchanger in which a liquid (usually water) is converted into vapor, usually with heat provided by combustion of some kind of fuel.

Heat exchangers are also important for temperature regulation in process plants, and contribute to the efficiency and safety of many processes. Furthermore, the effective design and use of heat exchangers can lead to a significant reduction of utility costs in process plants. As a chemical process design engineer, you should know where to use a heat exchanger in your process design and what the objective of the heat exchanger is. For example, you might need a heat exchanger to heat a fluid to a certain temperature or to condense another fluid from a gas phase to a liquid phase.

It is also important to know what fluid streams should be used in the heat exchanger. For example, it might be preferable to use a process stream to heat or cool another process stream instead of using a utility to perform that function. At other times, using a utility might be the only available option.

Besides the fluids in the heat exchanger, the choice of heat exchanger and its design is important. Aspen Plus provides different options for modeling heat exchangers, ranging from a simple model such as the Heater model, to the HeatX model which can be adapted to model detailed heat exchangers and performs rigorous calculations. For example, the HeatX model allows for the effects of heat transfer coefficients, fluid phase, heat exchanger geometry, and temperature crossover to be considered in the heat exchanger design.

Types and Classifications of Heat Exchangers

There are many kinds of heat exchangers, but two of the most common types are as follows:

SHELL AND TUBE HEAT EXCHANGERS

This heat exchanger consists of a cylindrical shell which houses a large number of tubes, as shown in Figure 4.1. The tubes contain one of the fluids which has to be heated or cooled while the second fluid which is on the shell side flows over the tubes, thus providing the heating or cooling needed by the tube side fluid.

Advantages
- Can handle high-operating temperatures and pressures
- Easy to control and operate

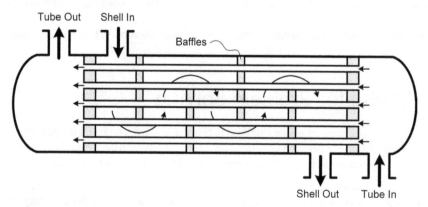

Figure 4.1   Shell and tube heat exchanger.

Disadvantages
- Lower heat transfer efficiencies than plate heat exchangers
- High maintenance costs

PLATE HEAT EXCHANGERS

This type of heat exchanger is made up of lots of thin plates which are stacked in series with small separations between them, as shown in Figure 4.2. The plates have small fluid flow passages and very-large surface areas for heat transfer. The spaces between the plates alternate between hot and cold fluid zones.

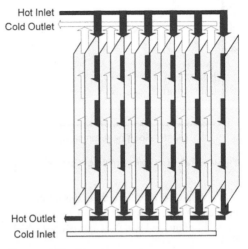

Advantages
- Simple and compact size
- Good heat transfer efficiency

Disadvantages
- Not suitable for high-operating temperatures and pressures
- High capital costs

Figure 4.2 Plate heat exchanger.

Another way to classify heat exchangers is by the direction of flow of the two fluids in the heat exchanger relative to each other. There are two main flow arrangements.

In cocurrent flow (or parallel flow), the two fluids enter the heat exchanger at the same end and move parallel to each other to the exit (note that baffles may cause some twists and turns on the way). In this flow arrangement, the highest temperature difference between the two fluids is at the heat exchanger inlet, while the lowest temperature difference is at the exit, as shown in Figure 4.3.

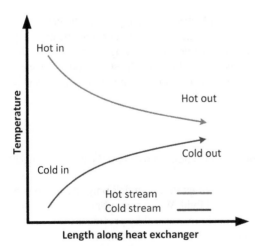

Figure 4.3 Cocurrent flow arrangement.

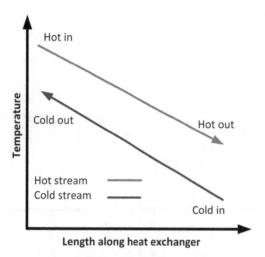

Figure 4.4 Countercurrent flow arrangement.

In countercurrent flow, the fluids enter and exit at opposite ends of the heat exchanger. They are the most efficient of heat exchanger flow arrangements because the cooler fluid exits the heat exchanger at the inlet of the hot fluid and will thus approach the higher inlet temperature of the hot fluid (Figure 4.4).

Tutorial

CASE STUDY

Pure *n*-butanol produced at a flow rate of 1000 kg/hr and 2 bar from an upstream distillation process needs to be cooled down from 117.7°C to 40°C before it can be stored in a tank. Cooling water at 25°C and 2 bar is available to provide cooling. In this example, we would like to model a heat exchanger in Aspen Plus to find out how much cooling duty is needed to get the butanol to the desired outlet temperature, and how much cooling water would be required to provide it.

Create a new simulation in Aspen Plus using the Chemicals with Metric Units template. In the Properties window add N-BUTANOL and WATER to your flowsheet and choose the WILSON model as the property method. You will probably have the red Required Input Incomplete notification showing. This is because the binary interaction parameters for *n*-butanol-water have not been loaded in the WILSON model. Click on the Methods | Parameters | Binary Interaction | WILSON-1 form to load it. Once that is done, go to Simulation mode to design the heat exchanger.

PART 1: USING A HEATER MODEL

In this first part of the tutorial, you will learn to model the heating and cooling process of the heat exchanger by using the Aspen Plus Heater model.

The Heater model is what we call a "half-heat exchanger," meaning that it models only one of the two fluids in the heat exchange process: that is, either the hot or the cold fluid. When computing the outlet conditions, it does not take into consideration the utility or other medium that will be providing the required heating or cooling. Typically, you provide two of the following degrees of freedom (DOF) as model parameters: outlet temperature (defined explicitly or relative to the inlet temperature or the bubble or dew points), outlet pressure (or pressure drop), vapor fraction (ranging between 0 which equals saturated liquid and 1 which equals saturated vapor), and heat duty (the negative of which is cooling duty). When executed, Heater computes the other variables and outlet stream conditions. The most common use is to specify the desired temperature, a small pressure drop, and then use the model to find the heat duty required and the corresponding outlet conditions.

It is common in the early stages of conceptual process design to build a flowsheet simulation with only Heater models for heat exchange. Usually, this happens when you want to get a process stream from one thermal condition to another, and you are not particularly worried exactly how this will be achieved yet. At later stages of the model development, Heater blocks might be replaced with more rigorous HeatX models (discussed in the next section), or used as part of a complex heat exchanger network (which you will learn about in Tutorial 11).

In the model library at the bottom of your screen, click on the Exchangers tab. For now, select the Heater model (see Figure 4.5) and add it to your flowsheet. We will use it to figure out how much cooling is required by the butanol stream.

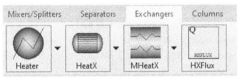

Figure 4.5 Available heat exchanger models.

Add input and output material streams to your Heater (mine is called COOL with BOH-IN and BOH-OUT as the inlet and outlet butanol streams), and specify the given information for butanol in the input stream as shown in Figure 4.6.

Now go to your Heater and specify its conditions as shown in Figure 4.7. Choose the pressure as 0 bar (which is how you tell Aspen Plus that you want there to be no pressure drop, it does *not* mean 0 bar absolute or 0 bar gauge pres-

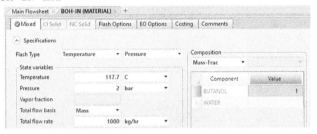

Figure 4.6 Butanol feed conditions.

sure) and the temperature as the final temperature we want for the butanol stream (40°C). Once that is done, run the simulation, check the results for the block to see the cooling duty required, and verify that the stream does indeed reach 40°C. Note that this block does not care at all how the cooling duty is actually provided; it only computes how much needs to be delivered.

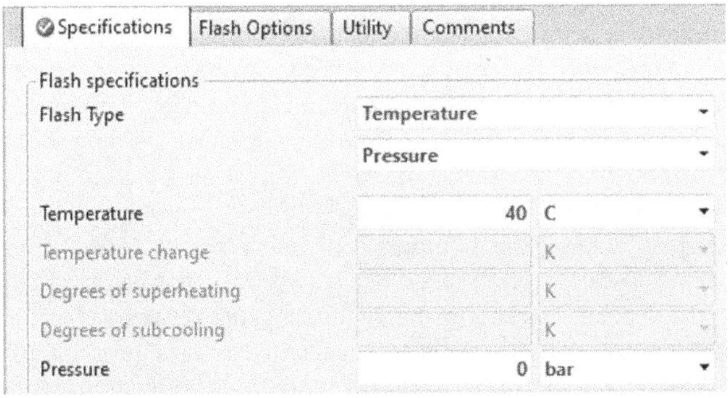

Figure 4.7 Cooler specifications.

Q1) How much cooling is required by the process in kW?

Now, that we know how much cooling is required, we can figure out how much water we should use to provide it, and what the outlet conditions of the water stream would be. The problem is not so simple though because we actually have many choices. For example, you could imagine using a bare minimum amount of water to do the job, creating steam, or using an incredibly large amount of water such that the water outlet temperature is just slightly higher than the water inlet. So we will have to make some design decisions.

Add another `Heater` model to your simulation (mine is called HOT) as well as input and output streams. We know the temperature and pressure of the inlet water stream, but we don't know the flow rate. Let us choose something random for now, say 600 kg/hr. Now, enter the temperature, pressure, mass fraction, and flow rate into the inlet stream of your HOT `Heater`. We use up one DOF when we assume there is no pressure drop in the `Heater` (type 0 as Pressure in your HOT `Heater`). The second DOF we can specify is its heat duty, which we know from the results of the COOL `Heater` model. Instead of manually entering the duty, we can specify it in the HOT heater by connecting the two half-exchangers using a `HEAT` stream. In your model library, click on the drop-down for Material and select the `HEAT` icon. Now connect the `HEAT` stream from the COOL `Heater` to the HOT `Heater` and run the simulation. My final Aspen Plus diagram looks like Figure 4.8.

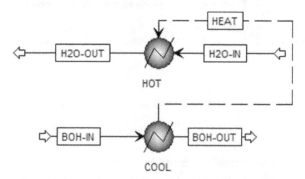

Figure 4.8 Two Heater blocks that make one complete heat exchanger model.

Q2) What is the outlet temperature of the water stream in °C?

If you have modeled it according to the above instructions, the outlet temperature of the water stream will be higher than the inlet temperature of the butanol stream. It is a violation of the second law of thermodynamics because heat will never spontaneously flow from a cold source to a sink of a higher temperature. This impossible situation is illustrated in Figure 4.9, where you can see that the temperature profiles cross over in the middle of the heat exchanger (this is called a temperature crossover, unsurprisingly). This means that our design makes no physical sense, although Aspen Plus does not issue any particular warning about it. To make our design feasible, we need more cooling water to ensure that the outlet temperature of the water is lower such that there is no temperature crossover.

When designing a heat exchanger, it can be helpful to use the minimum approach temperature (ΔT_{min}) as a guide. The *approach temperature* is the smallest temperature difference between the hot process fluid and the cold process fluid at any point inside the heat exchanger. The *minimum* temperature approach is the smallest approach temperature that you as the designer will allow to occur. In general, as the approach temperature decreases, the heat transfer rate slows down, and thus you need more and more surface area (requiring more steel and thus more capital cost) to transfer the same amount of heat, resulting in

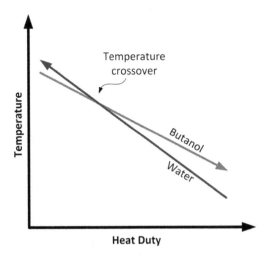

Figure 4.9 Temperature crossover of streams.

diminishing returns. For example, you need an infinite amount of surface area in order for the approach temperature to approach zero.

ΔT_{min} is therefore the smallest approach temperature that you will allow to occur in your heat exchanger because you have decided that anything smaller would probably cost too much capital for very little extra heat exchange duty. In practice, most people choose to use a ΔT_{min} between 5°C and 10°C, which is based purely on heuristic advice provided by experience, as opposed to a rigorous study of the optimal number for your particular circumstances.

For our design, let's use a ΔT_{min} of 10°C. Then using this heuristic, we can use a design spec to determine the amount of flow rate that gives us this approach temperature. Anything below this flow rate and the outlet temperature would either be too high for our liking (meaning it would require too much steel) or even so high as to be thermodynamically impossible. Anything above this flow rate would give lower water outlet temperatures, but that is certainly feasible and does not violate any laws of thermodynamics. However, we would rather not do that because then we have to buy a greater amount of cooling water than we really need.

Setup the design spec in Aspen Plus such that the inlet temperature of the butanol stream is higher than the outlet temperature of the water stream by ΔT_{min}. In other words, you are using countercurrent flow, and you know that this will occur at this point in the real heat exchanger. My design spec setup looks like the one shown in Figure 4.10.

The design spec is set up to compute the temperature approach, and then vary the water flow rate until the temperature approach reaches 10°C. This is a nice way to do it in the general sense because if you have a situation where you are running many different simulations in which the hot inlet stream temperature is different in each simulation, and you want to find a new water flow rate, you can just keep the design spec the way it is. However, you probably realized that for this specific problem, since you know the inlet temperature is exactly 117.7°C, you could have just specified the outlet water temperature to be 107.7°C and just skipped the math. That works too, except for more complex flowsheets, if anything changes upstream, then typing the exact temperature may be rather inconvenient.

Once you are done setting up the design spec, run the simulation.

Q3) What is the flow rate of water in kg/hr that ensures ΔT_{min} is not violated?

🎵 Music break[2]

---

[2]Recommended listening: *Riptide* by Vance Joy.

Figure 4.10   Setting up a design specification that varies the water flow rate until the temperature approach is 10°C.

PART 2: USING A HEATX MODEL

The whole process of using two HEATERs to model one heat exchanger may seem a little ridiculous, but there is a reason for it that we will discuss later. For now, let us address the same problem by using a HeatX model.

The Aspen Plus HeatX model is a complete heat exchanger model because it models both the hot side and cold side of the heat exchanger. In the HeatX model, you may also specify more detailed information about your heat exchanger such as the heat transfer coefficients and heat transfer area.

Save your flowsheet, then save again as a new copy with a new name. Delete the two heaters, the heat stream connecting them, and add a HeatX model to your flowsheet. Connect inlet and outlet streams to your model, while taking note of which inlet and outlet streams correspond to the hot and cold streams (hover your mouse over the blue arrows for the block when connecting the stream sources and destinations to see a tooltip pop up that tells you which blue arrow is which). The hot stream is the butanol stream which requires cooling, and the cold stream is the cooling water stream that will be heated up. See Figure 4.11 for the completed flowsheet.

Specify the hot and cold stream inlet conditions like we did in Part 1. Even though the design spec from the previous part helped us find a good cooling water flowrate, let's start with 600 kg/hr as the flow rate of cooling water. We know that will not be enough to achieve our specifications, but let's just try it to see what happens.

From the Specifications sheet you can see that the HeatX model allows for modeling simple heat exchangers by using the Shortcut Model fidelity option or more rigorous heat exchangers such as the Shell & Tube, Plate, and Air Cooled heat exchangers. We will focus our learning only on the Shortcut Model fidelity option as the other rigorous options are out of the scope of this book.

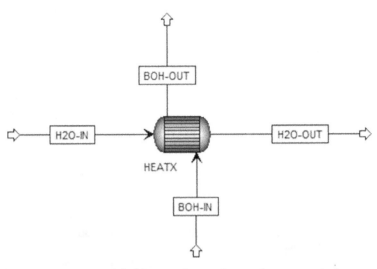

Figure 4.11 A model of the same heat exchanger, but using a single HeatX block.

After selecting Shortcut as the Model fidelity, choose Counter-current as the Shortcut flow direction and Design as the Calculation mode. Under the Specification drop-down we have different options from which to choose. Since we know the outlet temperature of butanol we want, select Hot stream outlet temperature, and enter 40°C as its value. Furthermore, enter 10°C as the minimum temperature approach (ΔT_{min}), as shown in Figure 4.12. Assume that there is no pressure drop in the heat exchanger and leave the heat transfer coefficient (in the U Methods tab) at the default.

After the simulation run has completed, Aspen Plus shows the Results available with Errors notification. If you got to the Thermal Results | Status tab of your model you should see the following message:

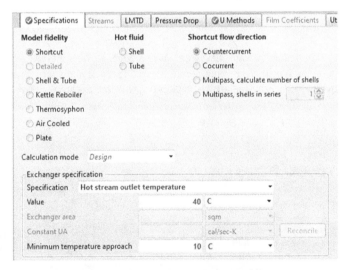

Figure 4.12 Directly specifying the temperature approach in a HeatX model.

```
** ERROR
   TEMPERATURE CROSSOVER DETECTED
   RE-CALCULATING WITH MINIMUM APPROACH TEMP. SPEC
```

Unlike the Heater model, the HeatX model is able to detect that there is a violation of the second law of thermodynamics leading to a temperature crossover. It then tries to avoid this problem by re-simulating the model but instead uses the specified ΔT_{min} (10°C) as the heat exchanger objective instead of the hot stream outlet temperature that we specified.

If you check the results under the Thermal Results | Summary tab, you will see that the Hot stream inlet minus Cold stream outlet temperature difference is 10°C which corresponds to ΔT_{min}. Furthermore, the butanol outlet temperature is 51.73°C. This avoids the temperature crossover, but we are unable to reach our desired outlet temperature of 40°C. The heat duty of the heat exchanger is also less than what we obtained in Part 1 of the tutorial (because it is not doing as much of the job as we want).

Q4) What is the heat duty of the heat exchanger in kW?

The `HeatX` model also contains a great feature that allows us to see where the temperature crossover occurs in our original heat exchanger design. To do this, we override the minimum temperature approach constraint by running the simulation with the Allow temperature crossovers option in the Options | Convergence tab of our `HeatX` model activated (see Figure 4.13). Also, go to the TQ Curves | TQ Curves Setup tab and check that the Calculate TQ curves option is activated (see Figure 4.14). This will allow us to plot a temperature versus heat duty diagram and see where the crossover occurs.

When you rerun the simulation it will complete with warnings, but now we get interesting results about the temperature crossover. Under the Home menu ribbon click on the TQ Curves icon under Plot to generate a temperature versus heat duty plot (you may have to left-click to select the `HeatX` block first or go to the `HeatX` form). On the generated plot you should be able to see where the temperature crossover occurs.

Q5) At what temperature in °C does the temperature crossover occur?

The result indicates that we require a larger flow rate of water to meet our desired butanol outlet temperature. Normally, it is not recommended that you allow temperature crossovers because then you have to manually check to make sure that it is not violating the laws of physics. However, with the Allow temperature crossovers option active, we can use a design spec similar to the one used in Part 1 to determine the

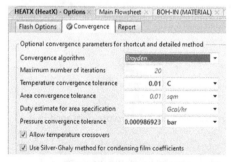

Figure 4.13 Allowing temperature crossovers in a `HeatX` model. If there is a crossover though, it means that your results are meaningless because it violates the second law of thermodynamics. However, it can be useful to allow crossovers temporarily while you are building a flowsheet or using advanced tools in order to aid in convergence and usability.

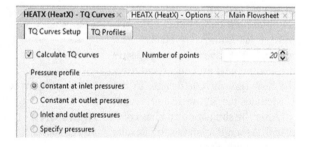

Figure 4.14 Setting up TQ curves for your model.

minimum water flow rate which ensures that the ΔT_{min} is not violated. Allowing temperature crossover is mostly a matter of convenience because it allows the design spec to guess flow rates that might be infeasible without triggering errors on the way to the true solution. If we did not allow temperature crossovers (the default), you would have to ensure that the bounds of the design spec are such that they ensure that all guesses are feasible. But usually you do not know that *a priori*. Now, go do it!

Q6) What is the flow rate of water in kg/hr that ensures ΔT_{min} is not violated? If your answer is correct you should have the same answer as in Q3.

Q7) What is the heat duty of the heat exchanger in kW? You should now be able to meet the desired outlet temperature of butanol and be at the ΔT_{min}.

In the Thermal Results | Exchanger Details tab, we can see more information about the heat exchanger such as the log mean temperature difference (LMTD). Mine is 12.3°C. You can also see the default heat transfer coefficient (U) that Aspen Plus uses for calculation and the required heat exchanger area.

Q8) What is the required heat exchanger area?

From what we have done so far, it is already easy to see that using the `HeatX` model to design a heat exchanger offers more benefits than using two `Heater` models connected by a heat stream.

So far we have learned how to model a shortcut heat exchanger in Aspen Plus and have explored the Design Calculation mode, where we specify our design objective such as the outlet hot stream condition and allow Aspen Plus to determine the heat exchanger size to meet this objective.

Now we will quickly look at the `Rating` and `Simulation` options in Calculation mode. Deactivate your design spec, and use a water inlet flow rate of 700 kg/hr.

The `Rating` option allows us to determine if a specified heat exchanger for a given design objective is over/under sized, that is, is it too big or too small to meet our design objective. Let's try an example. Go to your `HeatX` Specification tab and select Rating in Calculation mode. Leave the `Hot stream outlet temperature` at 40°C. Now, you have blank options for Exchanger area and Constant UA. Enter an Exchanger area of 8 sqm. Notice that the Constant UA options become grayed out. This is because the exchanger area and UA value are correlated. Thus if we know one value, the other one can be calculated. See Figure 4.15. Run the simulation and answer the next question.

Q9) Is the heat exchanger over designed or under designed, and by what percentage? You can see this result in your Thermal Results | Exchanger details tab.

Next let's take a look at the `Simulation` option. The `Simulation` option allows us to determine what the heat duty and outlet conditions of the hot and cold streams will be for a given Exchanger area. In the Exchanger specification form change the Exchanger area to 5 sqm, as shown in Figure 4.16.

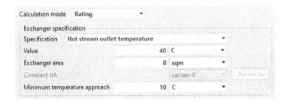

Figure 4.15 The setup for Rating mode.

Figure 4.16 Example setup for Simulation mode.

Q10) Run the simulation. What is the outlet temperature of butanol in °C?

In this tutorial, we have learned how to use the Aspen Plus `Heater` and `HeatX` models to design simple heat exchangers. In particular, the `HeatX` model is very versatile and can be used to design more rigorous heat exchangers such as Shell & Tube, Plate, and Air Cooled heat exchangers. Furthermore, it is also possible to include more details in the `HeatX` model design like considering pressure drops on the hot and cold side of the heat exchanger, and using more accurate values or methods to calculate the U value of the heat exchanger.

Why Choose Two HEATERs versus HeatX?

So far, we've seen that you can emulate a `HeatX` by using two `Heater` blocks connected by a Q stream with special checks put in place to ensure that there is no temperature crossover. But clearly, when you have two streams that you know are going to exchange heat, the `HeatX` block is much easier to use for modeling this than two HEATERs. So why use two HEATERs?

The answer is that based on your use of tear streams (see the Introduction if you need a refresher on tear streams), flowsheet convergence may be much more successful if you use two HEATERs rather than one `HeatX`. In most cases, `HeatX` will work just fine. But in some circumstances, using two HEATERs will make things much easier. For example, consider the common circumstance of using an economizer before a distillation column. An economizer is a common name for a heat exchanger that uses some waste heat from some downstream source to provide heat immediately upstream. In the case shown, the hot bottoms product is used to preheat the feed to the distillation column, which both reduces the reboiler load (saving money) and cools the bottoms product down (saving more money when cooling is desired, which often is). There are two ways of modeling this, as shown in Figure 4.17.

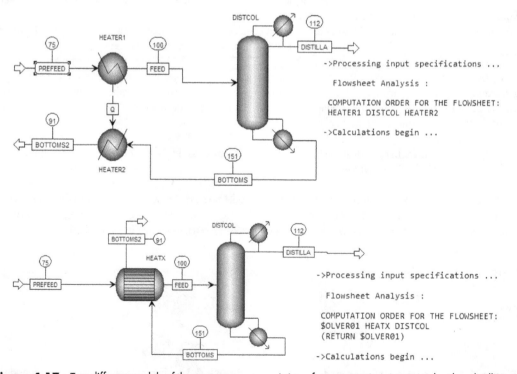

Figure 4.17 Two different models of the same process consisting of an economizer integrated with a distillation column. The results of both models are exactly the same.

Suppose that both HEATER1 and HeatX are set such that the pressure drops are small and the outlet temperature of the cold stream is at 100°C. With the two-Heater configuration, Aspen Plus can run HEATER1 immediately without knowing anything about what is going on in HEATER2. Once HEATER1 is run, DISTCOL is run next, and then finally HEATER2 is executed, such that all results are completed in three steps.

But with the single HeatX configuration, it is not possible to run the HeatX immediately away because in order to execute a block in Aspen Plus, all of the input streams to that block must be known first. However, in order to find the BOTTOMS stream conditions, the DISTCOL model needs to be run, but that cannot be run until the FEED conditions are known, meaning that HeatX must be run first (but it can't!). And thus, a convergence loop is created. In some cases, this convergence loop is not a major problem, but sometimes getting distillation column models to work can be very difficult or time consuming, and so knowing how to use the two Heater approach may be critical to get certain stubborn flowsheets to converge quickly and reliably. However, if a HeatX works to your satisfaction, just use it.

♫ Music break[3]

---

[3]Recommended listening: *Wake Me Up* by Avicii.

Equilibrium-Based Distillation Models

Objectives

- Use shortcut distillation models to get good estimates as a first step in a conceptual design or simulation process
- Use rigorous distillation models to get more detailed and accurate results
- Use the duplicator block to help with "what-if" scenarios
- Create Property Packages which you can import/export between spreadsheets or share with colleagues
- Use the UNIFAC method to predict activity-coefficient model parameters

Prerequisite Knowledge

If you are still not familiar with distillation, there are a variety of resources you can try. *Product & Process Design Principles* has chapters on distillation, distillation sequencing, and distillation modeling.[1] *Separation Process Principles* is a popular textbook that covers distillation in great detail.[2] If you are already familiar with the basics of distillation, I recommend *Perry's Chemical Engineers' Handbook*, which has a section specifically on how distillation processes are simulated that can be very useful. It includes several videos which are available via AccessEngineering.[3] There are other videos[4] on distillation in general which I also recommend. I suggest you start with the binary flash distillation example and then look at some of the others like the one

---

[1]Seider WD, Seader JD, Lewin DR. *Product & Process Design Principles: Synthesis, Analysis and Evaluation.* John Wiley & Sons; 2009. (See especially Chapter 19 in the third edition, and Chapter 13 in the fourth edition.)

[2]Seader JD, Henley EJ, Roper DK. *Separation Process Principles.* John Wiley & Sons, 3rd ed.; 2010.

[3]Green DW, Perry RH. *Perry's Chemical Engineers' Handbook.* McGraw Hill Professional; 2008. (See Chapter 13 in the eighth edition, especially 13.6. Available for free to AccessEngineering members. Many universities and professional organizations pay for unlimited access for their members.)

[4]http://www.learncheme.com/screencasts/separations-mass-transfer. This is peer-reviewed material produced by the University of Colorado, Boulder.

with multiple feeds or the ones with partial condensers, just to give you a flavor of all the different variants that exist.

Why This Is Useful for Problem Solving

Separation by distillation is one of the most common chemical engineering unit operations. It also accounts for a significant portion of the energy used in a process plant; thus, the ability to model it correctly can lead to significant energy savings for real processes.

In this tutorial, you will learn how to design a distillation column using Aspen Plus shortcut distillation models, such as DSTWU, and more rigorous models, such as RadFrac. If you have done some of the earlier tutorials, you have already used RadFrac many times now. Now we're going to take a closer look.

As a process engineer, there is a decent chance that you will be required to use some sort of distillation modeling (or extraction, absorption, stripping, rectification, or other similar operations) at some point in your career. That means you will need to be able to select the appropriate model, implement it, run it to successful conversion, and interpret the results. For example, you may need to use a shortcut column like DSTWU to help give good predictions about what parameters to use for RadFrac. This is especially true when you don't have good column design parameters up front (like reflux and boilup ratios), which have typically been provided in earlier tutorials. So you might need to know to try out a simple shortcut model first just to give you an idea of what to try before you wade into the deep waters of RadFrac.

On the physical property side of it, you'll need to understand a little bit about UNIFAC and how it is used to predict the binary interaction parameters of activity-coefficient models. Furthermore, it will be helpful to learn how to use Aspen Properties Backup Files to ensure that you and other design collaborators are all using the same physical property packages. That way, when you put the pieces together, it works.

Tutorial

PART 1: ASPEN PROPERTY PACKAGES

In this tutorial, we will learn how to design a separation column using the various models and tools we have available in Aspen Plus. The basic strategy is:

- Use a shortcut column to get an approximation for the optimal value of key parameters, such as reflux ratio and number of stages.
- Use a rigorous column to get more accurate results using the suggested values.
- Use optimization to narrow in on the best choice.

For this example, we will separate an equimolar mixture of acetone, isobutyraldehyde, ethyl acetate, and *n*-heptane, with normal boiling points shown below (note that it may differ from Aspen Plus's Databanks).

Acetone	56.1°C
Isobutyraldehyde	64.1–64.3°C (uncertainty)
Ethyl acetate	77.06–77.6°C (uncertainty)
n-Heptane	98.4°C

The mixture is at 30°C and 1 bar, at a rate of 200 kmol/hr. (I am not sure you'd ever find this mixture in the industry, but it makes for a good, contrived example with the precise properties I was looking for.)

In this example, we will create a properties backup file. This means that all of your physical property settings are loaded in one file. You can then import that file over and over again into new simulations and flowsheets so that you can keep reusing the same physical properties package on all of them.

Create a blank simulation and add the four above components. For the Methods type, choose UNIQ-RK. Verify that it worked by going to the Binary Interaction folder and looking at the UNIQ-1 parameters. Verify that the source is VLE-RK. If it is VLE-IG, it means you did something weird earlier, such as setting the wrong physical property package, and then when you switched to UNIQ-RK, these parameters did not change. If this is the case, change the source to VLE-RK and verify that the numbers changed. There should also be five pairs.

Notice that one of the pairs, Isobutyraldehyde–Ethyl Acetate, does not exist in the databank. When this happens, you can either find your own in the literature or you can try to use the UNIFAC method to estimate the vapor-liquid equilibria (VLE) parameters for you. The UNIFAC method is a group-contribution method that looks at the shape and structure of a molecule and then uses certain heuristics to try to predict how it will interact with other molecules. This is built into Aspen Plus, and so we're going to use it here to predict the missing UNIQUAC coefficients.

So, let's get the Properties feature to finish the job for us. Go to the Properties | Estimation | Binary tab. The Binary tab is disabled unless, on the Setup tab, you select Estimate all missing parameters (see Figure 5.1).

Click New to create a new parameter to estimate and choose UNIQ. For the method, choose UNIF-DMD (the UNIFAC method with the Dortmund modification—the most modern and accurate in the general case).

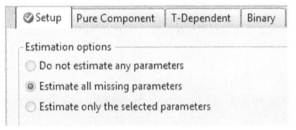

Figure 5.1 Estimating missing property parameters.

Then below that, select the two missing binary pairs. To the right, you select the temperature range of validity. The UNIFAC method may generate different parameters that are better at different temperatures. So, you can choose to have it be optimized for one specific temperature, or sort of averaged out over a range of temperatures. Typically, it makes sense to choose the two normal boiling points (64.3°C and 77.6°C) as shown in Figure 5.2, because two-phase behavior will usually occur inside that range. However, you may want to try different temperatures if your application calls for higher pressures, if there is a low-boiling azeotrope, or if it will be a part of a VLE mixture with other chemicals at different temperatures.

Setup	Pure Component	T-Dependent	Binary	UNIFAC Group	Comments

| Parameter | < | UNIQ | ▼ | > | | New | | Delete |

| Method | UNIF-DMD ▼ |

Components and estimation methods

	Component i	Component j	Temperature	Temperature
	ISOBUTYL	ETHYLACE	64.3	77.6

Figure 5.2 Specifying the temperatures at which UNIQUAC parameters should be estimated.

Ok! Next, in order to use the UNIFAC method, we have to tell Aspen Plus what the molecular structure of the components is, since they are not kept in the databank. To do this, go to Properties | Components | Molecular Structure. Click on Isobutyraldehyde for now.

The idea is that you list the atoms and how they are bonded to their neighbors. You can make up any numbering system you want. Ignore the hydrogens. For example, isobutyraldehyde is shown in Figure 5.3. I've made up my own numbering system, and you are free to change it. I would then enter this into Aspen Properties, as shown in Figure 5.3.

Molecular Structure - ISOBUTYL ✕ **+**

☑ General | Structure and Functional Group | Formula | Comments

Define molecule by its connectivity

Atom 1 Number	Atom 1 Type	Atom 2 Number	Atom 2 Type	Bond type
1	C	2	C	Single bond
2	C	3	C	Single bond
3	C	4	O	Double bond
2	C	5	C	Single bond

Figure 5.3   An example numbering system for a molecule and the corresponding way of describing the molecule accordingly.

Note, there is a shortcut you can apply. Often the structure is already stored in the databank for many molecules, even though the table is empty. So you can do a trick to get the structure to be automatically loaded into the table for you, so you don't have to type it. We're going to do this for ethyl acetate. Go to the Structure tab of ethyl acetate and ensure that the picture of the molecule is there. If it looks right, click on Calculate Bonds to convert it to the same kind of table as before (see Figure 5.4).

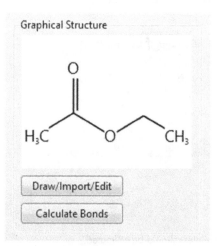

Graphical Structure

Draw/Import/Edit

Calculate Bonds

Figure 5.4   The graphical structure of a molecule contained within the Aspen Properties database.

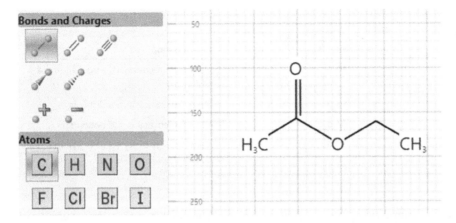

Figure 5.5   The graphical molecular structure editor allows you to modify the structure of a molecule (or just make really nice drawings).

You can also draw a molecule graphically with the Draw/Import/Edit button (see Figure 5.5). This is really useful later in life for drawing molecules if you just want nice images to use, or if you are making a new molecule that isn't in the database. Although the Aspen Properties database is quite extensive, you may need to do this someday for specialty chemicals, pharmaceuticals, or just rare chemicals. When you are done looking at the molecule, make sure ethyl acetate has data in the General tab too.

Almost done! Next, we just need to check a box that tells Aspen Plus to put the results in a convenient place rather than burying them in a text file somewhere. Go to Setup | Report Options | General, and make sure the box "Generate a report file" is checked.

Ok, now click Run (or hit F5). This will compute your estimate. You should see the result in the Methods | Parameters | Binary Interaction | UNIQ-1 form, with a new column added for Isobutyraldehyde–Ethyl Acetate. Double-check and make sure the TLOWER and TUPPER are in the temperature units you expected. Otherwise, you need to revise your Estimation | Input entry.

Q1)   What is the value of A_{ij} with i = isobutyraldehyde and j = ethyl acetate?

Q2)   What is the value of B_{ij} with i = isobutyraldehyde and j = ethyl acetate with the temperature units set to Celsius?

Now, save your result. Then go to File | Export | File, and export to an Aspen Properties Backup File (.aprbkp). This file can be used for all future Aspen Plus files (and even likely future versions of the program) so you don't have to do this again every time you want to make a new file.

🎵 Music break[5]

PART 2: DSTWU AND SHORTCUT COLUMN MODELS

Ok, let's start the distillation design process with a shortcut distillation column model. Although you could just go to the Simulation tab to get started, let's do it a different way so we can practice how to use the backup

---

[5]Recommended listening: *Good Luck* by Basement Jaxx.

Table 5.1   Desired Separation for a Mixture of Four Chemicals for This Example

Chemical	B.P. (°C)	Specification
Acetone	56.1	Very high recovery (>>96.5%) in the distillate
Isobutyraldehyde	64.3	96.5% recovery in the distillate
Ethyl acetate	77.6	96.5% recovery in the bottoms
n-Heptane	98.4	Very high recovery (>>96.5%) in the bottoms

file. So, for example, in a team design project, someone might probably want to make a master properties backup file that everyone else uses for their own flowsheets. That way, everyone on your team has the exact same physical property models with the same chemicals named exactly the same way and in the same order.

Create a new blank simulation. Once it opens, go to File | Import | File and import the .aprbkp file that you just made. To verify that it worked, check out the binary interaction pairs and the property methods.

DSTWU is a shortcut distillation model in which an estimate of the reflux ratio or the number of stages can be made given the desired separation result. In this case, you tell Aspen Plus what the recovery factors should be and it computes the rest. However, this assumes ideal behavior, which never happens in reality. Therefore, it serves mostly as a great starting guess for something more rigorous down the road.

Simulate the separation of a 200 kmol/hr of 30°C, 1 bar equimolar mixture of the four components shown in the table. Use the DSTWU model. Assume the column is also at 1 bar throughout. Let's assume the goal is to obtain 96.5% recovery of isobutyraldehyde in the distillate. In addition, we want 96.5% recovery of ethyl acetate in the bottoms. This is summarized in Table 5.1.

DSTWU requires the desired output conditions to be specified in terms of the molar recoveries of the heavy and light keys. This can be a little confusing at times. Looking at Table 5.1, you can see that the chemicals are arranged from top to bottom in terms of increasing boiling points, meaning that acetone is the lightest and n-heptane is the heaviest. For this example, we desire that in the ideal case, we want all of the acetone and isobutyraldehyde to leave via the distillate and the ethyl acetate and n-heptane to leave through the bottoms. The *light key* characterizes what is leaving in the distillate: it is the *heaviest* (or least volatile) of all of the chemicals that we want to leave through the distillate, so in this case, it is isobutyraldehyde. Similarly, the *heavy key* characterizes the bottoms product: it is the *lightest* (or most volatile) of the chemicals leaving through the bottoms, so in this case, it is ethyl acetate.

So, we enter into DSTWU that isobutyraldehyde is the light key, and we want 96.5% of it to leave through the distillate (we are being realistic that we can't get all of it). Note that molar recovery is not mole purity! We also must specify ethyl acetate to be the heavy key, but instead of saying that we want 96.5% of it to leave through the bottoms, we have to actually specify the opposite, that is, we want 3.5% recovery of ethyl acetate in the distillate. Tricky, but that's how it is. Then, the model already knows that acetone is more volatile than isobutyraldehyde, so it will also leave mostly through the distillate (with much higher recovery than isobutyraldehyde, in fact), and that n-heptane will leave through the bottoms since it is less volatile than ethyl acetate.

The DSTWU model uses shortcut calculations developed over seven decades ago, and is limited in accuracy because it uses certain assumptions to greatly simplify the calculations to make it possible to design a distillation column "by hand." Despite this, it is still useful in the modern computer age because it can be used to make predictions about the trade-offs between reflux ratio and the number of stages required to achieve a certain purity in the products. In general, the higher the number of stages, the lower the reflux ratio required, and vice versa.

Often, at the beginning stage of designing a distillation column, the designer has little information about what the number of stages and the reflux ratio should be or even a feasible range. DSTWU is useful to estimate these numbers. In general, you specify either the reflux ratio or the number of stages, and then it will estimate what the other value should be in order to achieve the separation objectives that you require.

Be careful though: it is possible to guess a number that is too low. There is a certain minimum number of stages and minimum reflux ratio that are necessary to achieve your desired separation. So if you guess too low, you'll get an error message. Since it's hard to know what that minimum is when you are first getting started, it makes sense to guess something extremely high, just so you avoid the error. Sure, you'd probably never want to design a column with that large a reflux ratio or number of stages, but we need something to work for our first run-through. So let's guess an extremely high reflux ratio of 45.

Run the simulation. Check the stream results to make sure they make sense, and then answer the following by looking at the Results tab of the DSTWU block.

Q3) What is the minimum number of stages required to achieve the desired separation (at infinite reflux)? Answer as a whole number. If it is a decimal, you have to round it up. This is because stages are discrete values; thus, fractional stages don't exist.

Q4) What is the actual number of stages required to achieve the desired separation using a reflux ratio of 45? Answer as a whole number.

Ok, so since 45 is high, let's look at what happens if we change the reflux ratio. Fortunately, DSTWU will give us a nice plot. In the Block | Input, go to the Calculation Options tab, then check "Generate table of reflux ratio versus number of theoretical stages." Sounds good! But we need a range. The lowest number of stages is going to be your answer from Q3, as this signified operation at total reflux. We don't really know the highest number of stages yet. Let's try something large, like 50. We can see if this is a good guess later. Change the "increment size" to be 1. Basically, Aspen Plus will rerun the simulation for stage numbers from Q3 to 50 in steps of 1 and then compute a reflux ratio required for each.

Rerun the simulation. Go back to the DSTWU Results form and then go to the Reflux Ratio Profile tab. Remember, you can use the Custom button in the Plot section of the Home ribbon, or the Ctrl+Alt+X, Ctrl+Alt+Y, Ctrl+Alt+P shortcuts to plot, hopefully getting something like Figure 5.6. The idea of the plot is to choose something that has a low reflux ratio, but before it gets "flat" (adding extra stages doesn't really help much). That's a good heuristic to use in the absence of cost data to have a reasonable trade-off between capital and operating costs. In general, increasing the number of distillation stages increases its capital cost and reduces its operating cost, while reducing the number of stages reduces its capital cost and increases its operating cost.

For example, if I have a column with 14 stages, I would require a reflux ratio of around 8, but if I pay for just one more stage, I can bring the reflux ratio down to about 5.5, for roughly a 32% reduction in energy costs. So it's worth it to go from 14 to 15. And I would argue it's worth it to go further to 16 stages. On the other hand, if I have a column that is 24 stages, I can only get a very small reduction in the reflux ratio if I add a 25th stage, so it might not be worth paying for the extra tray at that point.

Q5) What is the fewest number of stages in which the reflux ratio is below 2?

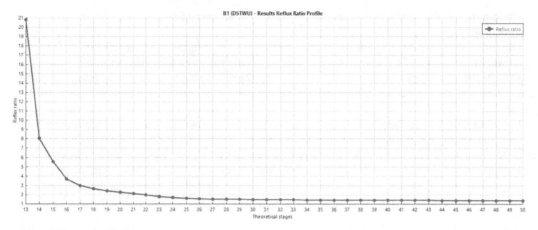

Figure 5.6   The reflux ratio versus the number of stages plot as predicted by DSTWU for a desired separation.

Choose your answer in Q5 as the final design condition. Now, let's predict the best possible location for the feed tray. Simply rerun the simulation again with your new choice. Verify that the actual reflux ratio calculated is the same as the plot from Q5.

Q6) What is the optimal feed location? Express this as a whole number since you only have integer amounts. Think, which way do you round? The feed stage is "above stage," meaning that the feed will be sprayed above its entering stage. Stage 1 is the condenser, while stage N is the reboiler.

Q7) What is the expected mole fraction of isobutyraldehyde in the distillate using these conditions as predicted by DSTWU? Also, write down the corresponding optimal distillate-to-feed ratio for later.

PART 3: RIGOROUS DISTILLATION MODELS

Duplicate your feed by using Dupl block from the Manipulators section of the Model Palette. Dupl basically takes an input and makes copies of it (all the same flow rate and everything). It's there for convenience in modeling and is usually used for "what-if" scenarios, where each branch downstream of the Dupl is a different model or a different option to consider. It is not an actual piece of equipment. Have one stream leaving the Dupl route to the DSTWU block (so it should be the exact same thing as before). Have another stream leave the Dupl block and route into a new RadFrac block. Use your answers from above as the new settings in the RadFrac block.

Q8) What is the more accurate prediction for the mole fraction of isobutyraldehyde in the distillate as determined by RadFrac?

Let's use Murphree vapor efficiencies to make our simulation somewhat less ideal. Go to Efficiencies of the RadFrac block. In the Vapor-Liquid tab, specify that the Murphree stage efficiencies from stages 1 to 10 will be 82% (0.82), and from 11 until the bottom of the column will be 73% (0.73). You need not type each stage in the box, and Aspen Plus will linearly interpolate between them. So, for example, if you enter just stage 1 as 0.82 and 10 as 0.82, then stages 2 through 9 will also be 82%. Note these are stage efficiencies, so even though stage 1 is the condenser/flash drum and the highest stage is the reboiler, they too have efficiencies.

Q9) What is the prediction for the mole fraction of isobutyraldehyde in the distillate as determined by RadFrac using your custom stage efficiencies? Techniques for predicting stage efficiencies exist. We'll get into that in Tutorial 8.

Now, the proper thing to do at this point would be as follows. Set up an optimization to adjust the key continuous design parameters, such as reflux ratio and distillate-to-boilup ratio, until you've achieved a certain separation factor, product purity, etc. Then, you'd run a sensitivity analysis on the feed stage to see which feed stage is best (factoring in both the capital and the operating cost of the column as it goes). Then you'll end up with the optimal operating conditions. But we are not ready for this yet. There is a lot more we can do with RadFrac; it's just too much for this tutorial.

 TOM'S TIP: There is a lot you can do with the RadFrac model. Some of this we will cover in Tutorial 8, such as using a special Design Spec/Vary feature for the RadFrac only, the use of many kinds of trays or packing, or the evaluation of flooding, weeping, or entrainment that could occur during operation. In addition to ordinary distillation, you can model rectifiers (i.e., the top half of a distillation column) by setting the reboiler type to None, strippers (i.e., the bottom half of a distillation column) by setting the condenser type to None, and absorbers (i.e., no condensers or reboilers) similarly. You can also have partial condensers such that you get a vapor distillation product (with or without a liquid distillate product). You can even have a reaction inside of the column. Tutorial 12 will have some examples of these advanced configurations.

Finally, we can do things with side duties (or heaters) and pumparounds. These are used in practice sometimes, especially in petroleum refining, but they are not standard to most columns. Go to the Configuration | Heaters and Coolers of the RadFrac block and add a side duty to stage 7 in the amount of 0.1 Gcal/hr. Rerun the simulation.

Then add a pumparound (Configurations | Pumparounds) from stage 8 to stage 14. Set the flow rate equal to 25% of the total liquid flow leaving stage 8, and assume the pump is adiabatic. Look at the block profile from your previous run (where the side duty has been added) to help determine what the flow rate leaving stage 8 is.

Q10) Using the custom stage efficiencies, the heater, and the pumparound, what is the predicted value of the isobutyraldehyde mole fraction in the distillate?

So yeah, we don't normally do that since it doesn't help here. But it is useful for complex columns, such as dividing wall columns.

🎵 Music break[6]

---

[6]Recommended listening: *Rockefeller Skank* by Fatboy Slim.

Advanced Problem Solving Tools

Objectives

- Learn to use the Utilities feature in Aspen Plus
- Use simulations to compute utility costs
- Learn the Optimization feature in Aspen Plus
- Combine the two features to design a process to have the lowest energy costs

Prerequisite Knowledge

It is advisable to complete Tutorials 1, 2, and 3 before you begin this one. At this point, you'll need to understand distillation, valves, pumps, and vapor-liquid equilibria (VLE) diagrams. If you still don't understand distillation, I suggest you try watching the four videos in the distillation section of the LearnChemE website.[1] You will also need to have a basic understanding of the first and second laws of thermodynamics as they relate to heat transfer (specifically, the basic concepts of energy balances and that heat cannot transfer from cold to hot spontaneously) as these concepts are very important for the selection of utilities. You may also be interested in a short video introducing the concept of optimization.[2]

Why This Is Useful for Problem Solving

The Utility feature in Aspen Plus is incredibly useful for design projects, because it can be applied so many times, and it makes it much easier to determine and optimize process costs when used in the correct fashion. To reduce the utility costs of a process, you will need to know what types of utilities (such as steam, fired heat, cooling water, and refrigeration) are available, their operating conditions (i.e., temperature and pressure), and which one

---

[1] http://www.learncheme.com/screencasts/separations-mass-transfer. This is peer-reviewed material produced by the University of Colorado, Boulder.

[2] https://www.youtube.com/watch?v=YSewtaL3tYY. This is a video from the AIChE Academy.

to choose for the different heat exchangers in your process. This is important as poor utility selections can lead to much higher process costs.

Optimization is critically important to any conceptual designer. Very often, a designer is faced with important choices such as "Which pressure is the best?" "What's the best recycle ratio?" and "How much energy should I use?" Optimization is the act of answering the question of which possibility is *best*. Most designers use optimization in some fashion to help create their designs, and thus it is very important to know how to set up and execute an optimization in Aspen Plus. This would require you to know what buttons to push, how optimization works, how to define the objective function and constraints, how to define the variables that use them, how to define variable bounds, how to set up a working initial guess for the optimizer to use, and how to determine if the results are good or not. To successfully complete an optimization, you would be required to complete all of those steps. Furthermore, it is good to know if optimization is the appropriate tool to use versus something like a design spec or sensitivity analysis.

There are some other skills you will need to know to use the software effectively. For example, best practices involving the use of reinitialize, estimates, or stream reconciliation will be very helpful, as will an understanding of strategies to maximize the probability of convergence when adding recycles to flowsheets.

Tutorial

PART 1: UTILITIES

In this tutorial, you will use the Utilities feature. This is basically a convenient way of tracking the costs and amounts of utilities that your plant requires. These include electricity, different types of steam, different types of cooling, fuels, refrigerants, and others. Basically, you create your own list of utilities available to your plant, along with appropriate temperature ranges and costs of use. Then, for units such as heat exchangers, pumps, compressors, etc., you simply select in each block which utility you are going to use. Then, Aspen Plus will figure out how much of that utility you need and how much it costs.

What Are Utilities?

Most everyday people understand what "utilities" are since they routinely pay utility bills as a part of living expenses. The term has a very similar meaning in a chemical engineering context. When a chemical process consumes a utility, it is essentially "purchasing" heating, cooling, refrigeration, or electricity from some other provider, even though that provider may just be another business unit of the same company. In practice, other services may be considered utilities, such as waste removal, water treatment, pressurized air, high-purity oxygen, process water, or solvents of various kinds. However, in Aspen Plus, "utilities" refer only to energy services: heating, cooling, and electricity. Conceptually, utilities are provided "on demand" and are not directly integrated into your chemical process. A chemical process designer is not often concerned with exactly how those utilities are provided; the designer instead determines what utility is needed and how much. It is important to understand that when you are purchasing utilities like steam, cooling water, or refrigeration, you are not actually purchasing the actual steam, water, or refrigerant—you instead purchase the heating or cooling service they provide. For example, when you purchase high-pressure steam as a utility, you are really only purchasing the latent heat it carries. In fact, your chemical plant would normally have a condensate return line where the used, condensed steam (still hot and at high pressure) returns to the steam generation plant. If you buy refrigeration utility, you are only buying the service of removing low-temperature thermal energy from your plant (making it colder). This is why utility accounting uses a cost-per-unit-energy framework, because you are only purchasing the energy.

Figure 6.1 Adding a new utility to your model.

Defining Utilities

To use utilities in Aspen Plus, you must first define what utilities are available for you to purchase. Start with a blank simulation in Aspen Plus, preferably with the Chemicals with Metric Units template. You can define a utility in the Utilities folder under the Simulation tab (or click Process Utilities in the Economics ribbon). Click New to make a new utility, and call it something meaningful (see Figure 6.1). Let's make LPS (low-pressure steam). In the Copy Form, choose LP Steam. Aspen Plus comes with default prices and settings for LPS. These are heuristics which may or may not be correct for your plant, and you'll see the actual numbers show up in the Specifications and Inlet/Outlet tab once you choose LP Steam and select OK. Note also that if you have not added water to your components list, you will be prompted to do so (do it).

These numbers are very convenient, but are they right? The energy prices given here are 1.9×10^{-6} per kJ, which is $1.9 per GJ. AspenTech sometimes updates these numbers, but of course the prices will be different for each specific circumstance and point in time.[3] The bottom line is that it will vary with the price of energy. At the same time, there are differences in the temperature/pressure ranges of the steams between Aspen Plus's default and heuristics given in other sources. There is no standard! Essentially, each plant will have access to steam levels depending on the other plants that sit nearby. For example, a company with lots of different processes on-site may have standard steam lines for that location. Or, individual plants may have custom steam pressure lines which are optimized to some particular values. My point is, don't just blindly go with the default numbers.

Now I've said that, let's blindly go with the default numbers since we're just learning how to use the tool and are less concerned about what the numbers actually are.

In the Inlet/Outlet tab, you'll notice that the vapor fraction is defined as 1 for the inlet and 0 for the outlet. This means that heat is provided by condensing steam (this is normal). As mentioned in earlier tutorials, Aspen Plus uses a convention that if you *specify* vapor fraction = 1 or vapor fraction = 0 in an input form, that means you are defining that the inlet is exactly saturated vapor at the dew point or the outlet is exactly saturated liquid at the bubble point. The temperatures are given, so the pressures will be computed from this information. The outlet is defined to have a slightly lower temperature than the inlet so that the outlet will

---

[3]These numbers have been the same since at least version 9 (2018), so they have not recently updated them over time.

have a slightly lower pressure than the inlet. In the event that you wanted superheated steam or subcooled liquid, you would enter the temperature and pressure combination instead of a temperature and vapor pressure combination. Usually, real systems superheat and subcool by a few degrees as a safety margin.

Notice that there is a Carbon Tracking tab. This is a new feature that can help with sustainability analyses. On this tab, you'll see that by default it chose a U.S. Environmental Protection Agency rule for computing the amount of global warming potential (measured in CO_2 equivalents). It also selected natural gas combustion to produce the heat to make the steam, and an efficiency of 85%. Aspen Plus has a default number for the average composition of natural gas and its heat of combustion. It takes the amount of steam duty you need and then divides that by 0.85 to determine the total natural gas combustion duty that is required to produce it (so only 85% of the chemical energy from the natural gas is used to make steam via combustion, the rest is assumed lost as waste heat). Then it assumes that all the carbon atoms in the natural gas end up as CO_2, and then outputs that number. So Aspen Plus computes the *total direct emissions* for this utility, but does not account for the CO_2 that was emitted in order to produce the natural gas and pipeline it across the country and into your plant.

Now, go ahead and add a utility for electricity, using the default. We are going to add our own electric utility that assumes the electric power is produced by a coal power plant with a solvent-based CO_2 capture system on it. The default electricity setting in Aspen Plus assumes classic power plants with no CO_2 capture, at a rather typical price of about 8 cents per kWh. However, there is a power plant by SaskPower in Saskatchewan, Canada which is the first of its kind to implement CO_2 capture at a meaningful scale from coal, which has about 80% lower emissions per kWh than traditional pulverized coal. Of course, it is more expensive than classic power plants without CO_2 capture. According to the U.S. Department of Energy (document DOE/NETL-2007/1281), we can predict that the prices for this type are higher, at 13.2 cents per kWh (in the United States), after adjusting for inflation. So make the change in the Specifications tab. Similarly, the CO_2 emissions should also be changed since they are much lower. A recent life cycle analysis study[4] determined that the total cradle-to-grave life cycle emissions for a power plant of this new type should be about 78.65 kg of CO_2 equivalent per GJ of electricity. That includes the construction of the power plant, the mining and transport of the coal, the construction and use of the CO_2 capture, pipeline and storage system, and even the electricity transmission losses from connecting the power plant to your plant. So it's a much better number to use than what Aspen Plus has because it includes more of the life cycle. Anyway, enter this number by setting the CO_2 emission data factor source to USER, typing the number into the CO_2 emissions factor box, and setting the efficiency factor to 1.0 since the efficiency factor is already included in the 78.65 kg/GJ number.

Finally, add utilities for medium- and high-pressure steam, and for boiler feedwater (BFW) streams at low, medium, and high pressure, using the defaults in Aspen Plus. Note that the BFW defaults are "Steam Generation" items. They are basically backward versions of the corresponding steam utility and actually make money and "consume" CO_2 instead of costing money because the energy price and CO_2 efficiency is negative. It is an accounting trick. The idea is that it prevents the cost of steam and the emission of CO_2 somewhere else by making it here.

Also, note the temperature ranges of these utilities for later. For example, cooling water comes in at 20°C and comes out at 25°C, which is rather generous but we'll go with it. It also has a temperature approach maximum of 5°C by default. That means that for countercurrent flow, the coldest anything can get is 25°C (cold in = 20°C + 5°C = 25°C). If you violate this by assigning the utility where it shouldn't be used (your heat exchanger comes too close to temperature crossover or actually does crossover, as discussed in Tutorial 4), Aspen Plus will throw you an error.[5]

---

[4]Nease J, Adams TA II. Comparative life cycle analyses of bulk-scale coal-fueled solid oxide fuel cell power plants. Appl Energy, 2015, 150 (15):161–175.

[5]"Throwing" is the correct computer science term here. I guess the idea is that computer programs are spoiled brats who throw errors at you when they don't get their way.

Using Utilities

Ok, let's see how utilities are used in the software. In this example, we are going to separate a 100 kmol/hr mixture of acetone (27 mol%, T_{bp} = 56°C), methanol (38 mol%, T_{bp} = 65°C), and n-butanol (35 mol%, T_{bp} = 117°C) at 25°C and 1 bar that we get after a biofuel production process. Acetone and methanol form a low boiling azeotrope at 55°C, so we're going to use pressure swing distillation again (see Tutorial 2).

The process shown in Figure 6.2 can be used, with all of the remaining design parameters given. Simulate this process in Aspen Plus. You should be able to do this, given what you already know. Remember, it is better to get one block working at a time before adding another one, and get all three columns working without recycle first before connecting the final recycle stream. When it is finished, you should be able to get at least 98 mol% purity in each of the three product streams. Assume the distillation columns and their supporting reboilers and condensers are at constant pressure throughout. Use RADFRAC for the distillation columns of course and the NRTL-RK property package. Remember the best practices we learned from earlier tutorials: *change the property package first before adding the chemicals!* (butanol is n-butanol). Go back to check and make sure that the binary parameters come from the VLE-RK database and not the VLE-IG one! Do this by looking at the Source drop-downs of the different chemical pairs under the Input tab of the Properties | Parameters | Binary Interaction | NRTL-1 form.

Q1) What is the purity (molar fraction) of n-butanol in stream 3?

Q2) What is the purity of acetone in stream 8?

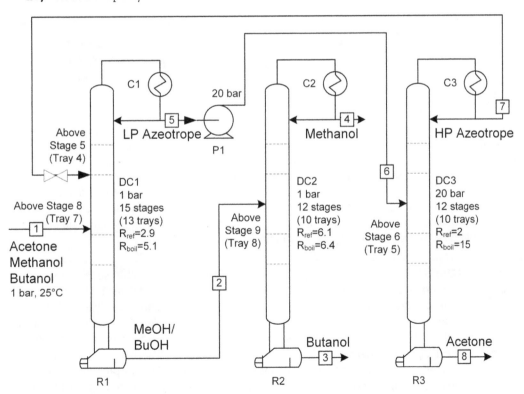

Figure 6.2 A process to separate a mixture of acetone, methanol, and butanol using pressure swing distillation.

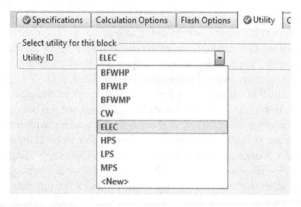

Figure 6.3 Where to assign utilities in a pump model.

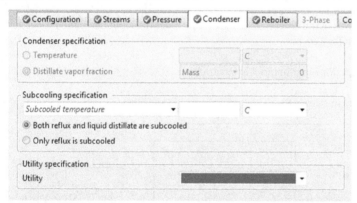

Figure 6.4 Where to assign utilities for the condenser in a RadFrac model.

Ok, let's go back and assign utilities to the process units. Each block has a different way of doing this. For example, in the Pump, there is a Utility tab, where you select which of the utilities that you created to be used with it (see Figure 6.3). In the RADFRAC blocks, you can find the utility specification in the Condenser (see Figure 6.4) and Reboiler tabs (Figure 6.5), respectively, at the bottom.

Go through and choose the correct utilities for each. You can use your simulation results to help you select. Remember not to violate the second law of thermodynamics! Remember, use BFW for cooling whenever you can because you'll generate steam instead of paying for cooling! Rerun the simulation. You can check the utility results in the block's results form, in the Utility tab or similarly named.

TOM'S TIP: When selecting utilities, you generally want to choose the cheapest utility that does the job. Suppose for example, you need to deliver 1.2 GJ/hr of heat to a stream that you want to heat up to 150°C. Your choices may include LPS (available at 125°C), MPS (175°C), HPS (250°C), and Fired Heat (e.g., inside a furnace with a minimum of 400°C). In this case, you cannot use LPS because it is not hot enough, so rule that out. MPS will work because it is 25°C hotter than your maximum needed heating temperature, which is well above the 5–10°C approach temperature we often assume (see Tutorial 4 for more about approach temperatures). HPS and Fired Heat are even hotter and can also be used, but they are more expensive per GJ. (The default in Aspen Plus for HPS is $2.5/GJ but MPS is only $2.2/GJ.) So we select MPS because it is cheaper, with a total cost of $2.64/hr.

Figure 6.5   Where to assign utilities for the reboiler in a RadFrac model.

   TOM'S TIP: Although Aspen Plus lets you select Electricity as the utility for almost any heat exchanger, this is very rarely the actual utility used in practice. Electric heaters (e.g. using resistors or actually shocking the target directly) are usually special cases only.

Q3)   What is the cost of operating the pump, in $/hr?

Q4)   What is the cost of operating the reboiler in DC3, in $/hr?

Q5)   What is the amount of money that you are making by using BFW?

Q6)   What is the total direct CO_2 emissions of DC2? Note that cooling water has no direct CO_2 emissions by Aspen Plus's default, which isn't actually true.

🎵 Music break[6]

PART 2: OPTIMIZATION

The Utilities feature provides a very convenient tool for calculating the total energy costs of running this plant. Right away, we can tell that if we make some changes, our energy costs will be different. For example, if I mess with the reflux or boilup ratios, I will require different utilities for each.

One way to figure out what is better is to guess new values for reflux and boilup ratios, ensure that we meet our purity specifications, and then recalculate the new energy costs. Then keep guessing over and over.

---

[6]Recommended listening: *Rango II* by Vulfpeck.

In the last tutorial, we learned how to do this iteratively with a Sensitivity block. The problem with that is that the simulation visits every single point you tell it to visit and nothing more or less. This means that if the optimal solution is not on one of the points you picked, you won't know the true optimum (e.g., it could be in-between points). And, if you have more than one variable you want to change at a time, you might require an impossibly large number of points if you use a Sensitivity block.

That's why Aspen Plus has the Optimization feature. Optimization is basically a sophisticated guess-and-check algorithm that helps you find the minimum value of a function (kind of like the Solver in Excel). It is generally way faster than trying a massive sensitivity analysis and hunting through it for the best result.

For this problem, let's allow ourselves to change these variables: the `boilup ratio` of DC1, the `reflux ratio` of DC3, and the `boilup ratio` of DC2. Let's find the set of variables which has the lowest total energy cost with the following constraint: all the three products (acetone, methanol, and butanol) must have purities of 98 mol% or greater.

You can start a new optimization by going to Simulation | Model Analysis Tools | Optimization. Make a new one with any name you want. The rest of the forms will look very much like the design spec tab. The key parts of an optimization problem are listed next.

Objective Function

This is what you want to minimize or maximize. For example, you might want to minimize operating costs, minimize total costs, maximize revenue, maximize profits, maximize yield, maximize efficiency, or minimize environmental emissions. You should be able to compute *one single number* for any of these by looking at the flowsheet results. For example, if I wanted to maximize butanol purity, I would write on a sheet of paper:

$$\max x_{\text{but},3}$$

where I would know that $x_{\text{but},3}$ is the mole fraction of butanol in steam 3. Or, if I wanted to maximize the revenue from sales, I might write something like this:

$$\max 2F_3 + 1.5F_4 + 3F_8$$

which might give me the total amount of money made per hour if I sell butanol at $2/kg (and F_3 is the mass flow rate of stream 3 in kg/hr), methanol at $1.5/kg, and acetone at $3/kg (these are made-up prices but you get my point).

Decision Variables

These are the variables you want to change in order to find combination that yields the best objective function. Typically, these are design variables such as reflux/boilup ratios in the columns, recycle ratios, and temperature/pressure settings in various units. In all cases, these should be degrees of freedom which you have the ability to change. Normally you can recognize these as something you would type into a form.

Variable Bounds

These are limits on the range a decision variable is allowed to change. For example, suppose I want to vary the reflux ratio. That number cannot be zero or negative, and very high reflux ratios (>100) are usually absurd unless you have only very small distillate flow rates. This prevents the optimizer from trying guesses outside the range which we already know won't work or won't be the best.

Constraints

These are limitations which must be satisfied that are not variable bounds. In Aspen Plus, they can be =, ≥, or ≤ relationships. For example, we might want to maximize revenue from sales while ensuring that all our products are at least 95% purity by mole. We could then write:

$$\max 2F_3 + 1.5F_4 + 3F_8$$

$$x_{\text{but},3} \geq 0.95$$

$$x_{\text{MeOH},4} \geq 0.95$$

$$x_{\text{acetone},8} \geq 0.95$$

You Try It

Let's start with the objective. In our case, it is to minimize the total cost of utilities per hour (or per year or whatever). Take a second and write down what the objective function should be here.

$$\min \textit{your stuff here}$$

Then let's define the variables that you need to compute the objective function in Aspen Plus. For example, you can get to the cost of operating the Pump, as shown in Figure 6.6. The cost of the reboiler utility in a RADFRAC model is the variable REB-UTL-COST. Similarly see COND-UTL-COST for the condenser. REB-UTL-USA is the rate of utility usage in the reboiler (kg/hr, for example) and COND-UTL-USA is the rate of utility usage in the condenser, if you need them.

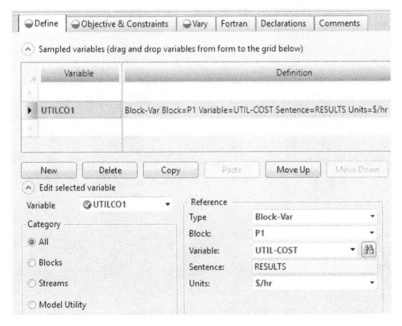

Figure 6.6 Selecting the utility cost of the pump as a variable to use in optimization.

Variable	Definition
CONDCO1	Block-Var Block=DC1 Variable=COND-UTL-COS Sentence=UTILITIES Units=$/hr
REBCOST1	Block-Var Block=DC1 Variable=REB-UTL-COST Sentence=UTILITIES Units=$/hr
CONDCO2	Block-Var Block=DC2 Variable=COND-UTL-COS Sentence=UTILITIES Units=$/hr
REBCOST2	Block-Var Block=DC2 Variable=REB-UTL-COST Sentence=UTILITIES Units=$/hr
CONDCO3	Block-Var Block=DC3 Variable=COND-UTL-COS Sentence=UTILITIES Units=$/hr
REBCOST3	Block-Var Block=DC3 Variable=REB-UTL-COST Sentence=UTILITIES Units=$/hr
UTILCO1	Block-Var Block=P1 Variable=UTIL-COST Sentence=RESULTS Units=$/hr

Figure 6.7 All of the variables used in the optimization example.

You can copy-paste from the define section and edit accordingly to speed this up. Or you can drag and drop from a block results form (it's tricky though to get the mouse clicks just right). Figure 6.7 shows the variable summary when finished.

In the Objective & Constraints tab, you use the variable names from the Define tab to mathematically express the objective function in the box. For example, I would start typing CONDCO1+REBCOST1 but then add more stuff. Make sure you choose Minimize since we want to minimize the objective function (find the minimum cost). Note that if you misspell the names of any variables and the variable you typed does not exist, Aspen Plus will use zero for that value, and you might not see the warning. So, be careful for typos, especially confusing 0 with O.

Let's go to the Vary tab. Here, you tell it which variables you want to change and their allowable range. Let's start simple: change only the boilup ratio of DC2, and set the variable bounds to change between 0.5 and 10 (the lower and upper limit boxes). (MOLE-BR is the variable name for boilup ratio in a RadFrac block.) You can drag and drop here as well.

Rerun the simulation. You can see the final boilup ratio by looking at the block's Results tab. Be careful: *if you look at the input form, you will still see your original entry, not the final value.*

Q7) What is the resulting boilup ratio determined by the optimization?

Q8) What is the purity of butanol in stream 3?

Ok wow, so that isn't good because we are definitely below the required 98 mol% purity. This is because we did not add a *constraint* to the optimization. We have to tell it to keep the purities high.

Go to Model Analysis Tools | Constraint; add the three constraints that require each of the product streams to have a purity of at least 98 mol%. Again, you define the flowsheet variables and then write the expression in the Spec section. Use a tolerance of 0.0001 (meaning that we are ok with anything in the range of 97.99–98.01 mol%). You will have to make three separate constraints. See Figure 6.8 for an example.

Here is a catch which is not obvious. When you make the constraints in the constraints section, nothing actually happens. You have to add them into your optimization. Go back into Model Analysis Tools | Optimization | Objectives and Constraints and notice now that there is stuff in the Available section of Selected constraints. Move that over to the Selected section to turn them on. Now, rerun the simulation. It is very helpful to reinitialize first (Shift+F5).

Figure 6.8 Defining a constraint for an optimization.

TOM'S TIP: If you get a message that the optimization didn't converge within 30 iterations, it means it tried 30 different values for the boilup ratio and decided that it could do better if it were to keep going. You can increase the maximum number of iterations of the SQP optimizer in the Convergence | Options | Methods | SQP Tab. Raise it to 400.

If you made a mistake somewhere, try getting to a point where you can get *something* to converge. Start by disabling the optimization, and rerunning. If that is converged, then turn the optimization on and try again without reinitializing. It will use the previous successful run as the initial guess for this run, which helps.

Tada! Verify that the mole fraction of butanol is at least 98 mol% (or at least should be within 0.01% of 98 mol%, actually. Make sense?[7]).

You can also see the final value of the objective function by going to Convergence | Convergence | [$OLVER something] | Results. The solver corresponding to the SQP optimizer should have an Objective function value line, with the result. This is your objective function. You can go to the manipulated variables and iterations tabs to see those results as well. Mine are shown in Figure 6.9. You can see that it started guessing 6.4, then it tried 6.00776, then 6.05689, and then settled around 6.06109. In most problems that converge, you will see a few large steps that comes close to the final solution, followed by a long list of small, tiny steps as it perfects the solution.

TOM'S TIP: It is important to know that when you are using Aspen Plus in sequential modular mode (as you are doing now), it is not really possible to know if the optimizer actually has found the true, best solution (which is called the global optimum). If you get a successful optimization result, you only know that this is a local optimum.[8] This means that there are no other meaningfully better solutions at other values nearby, but there could be better solutions at places that have not been explored farther away. A lot of times though, the solution is in fact the global optimum, and even if it is not, it is still very good. Just be sure to evaluate it for quality before you decide to move on.

Q9) What is the resulting objective function value (in $/hr)? You can also check by looking in the Summary tab of the Model Analysis Tools | Optimization | [Your Optimization name, e.g., O-1] | Results section.

Iteration	Constraint / * Tear stream / #Tear-variable with CMAX	CMAX	OBJECT-IVE FUNC TION	KUHN-TUCKER ERROR	LAGRANG-IAN FUNCTION	CONSTRA-INT 1 C-1 TOL= .1000-03	CONSTRA-INT 2 C-2 TOL= .1000-03	CONSTRA-INT 3 C-3 TOL= .1000-03	VARY 1 DC2 COL-SPEC MOLE-BR
0			62.8618				0.0154064	0.0066685	0.014519	6.4
1	C-3	0	62.4213	0.0712016	62.3524	-0.00304405	0.00643777	0.014519	6.00776
2	C-1	30.4916	62.4888	0.0148495	62.4828	-0.0002489...	0.00647392	0.014519	6.05689
3	C-1	2.46331	62.4944	0.00119357	62.4941	-1.62973e-...	0.00647691	0.014519	6.06109
4	C-1	0.139552	62.4944	6.75915e-05	62.4942	-1.62366e-...	0.00647691	0.014519	6.06109

Figure 6.9 The iterations tab of the solver results form for this example.

---

[7]The idea here is that it costs more money to get higher purity. The cheapest way to produce butanol of at least 98% purity is to produce it exactly at 98% purity to avoid unnecessary expenses associated with purifying it further.

[8]Well, really, you only know that it is *probably* a local optimum.

Ok! Let's add in the final two variables. In the optimization Vary tab, hit the drop-down on the variable number tab, and select New. Add the manipulated variables to the optimization: the boilup ratio (MOLE-BR) of DC1 which can go from 0.5 to 10 and the reflux ratio (MOLE-RR) of DC3 which can also go from 0.5 to 10. Rerun (don't reinitialize because you want to use your previous result as the initial guess for this next run).

 TOM'S TIP: If you are having difficulty getting your optimizer to work, try any of the following:

- Check for stupid mistakes.
- Reinitialize the simulation and rerun.
- Remove the constraints and run the optimizer (don't delete the constraint, just move it from Selected to Available in the Objectives & Constraints tab of the optimization form). Then add back in one of the constraints and run. Then add the next and run, then the last and run (assuming each run worked each time).
- Leave in the constraints, but deactivate two of the three Vary variables in the Optimization | Input | Vary tab, and run. If you get results, then turn one back on and rerun, and then the last one and rerun.
- Deactivate the optimizer. Manually change some of the variables a little bit and keep an eye on the costs. Find some combinations with good values, and then reactivate the optimizer again. The optimizer will start from these new guesses, which should be closer to the true optimum and thus have a higher chance of converging.
- Disconnect the recycle (stream 7). Put in a temporary stream 7b that matches stream 7 and is fed into SC1 at the right spot. Then run the optimizer. 7 and 7b won't be exactly the same once the optimizer starts to change things, but they won't be all that different either. This will at least eliminate the possibility of convergence issues due to recycle, allowing you to just focus on the optimizer. Once you can get the optimizer to converge, you can reconnect the recycle (deleting stream 7b) and rerun the optimizer.
- Try reducing the bounds on the Vary variables to a small range, running, and then increasing them one at a time.
- If it is a single block that is giving you problems rather than the optimizer as a whole, ensure that all of your blocks are configured in a way that they can handle any possible input that might be visited without errors.
- You can try alternative algorithms for the optimizer. Instead of SQP, BOBYQA also works well. Change it in Convergence | Options | Defaults | Default Methods. Note that "Complex" is actually best only for simple systems. You can try it but that one is usually better as an initial guess generator for the others.

 TOM'S TIP: The SQP optimizer in Aspen Plus is quite handy, but sometimes you need something stronger, or quicker. If SQP is not doing it for you, especially with regard to run-times, consult Bonus Tutorial 2 for advanced methods in which you can use a technique known as parallelized particle swarm optimization by using Python.

Q10) What is the final resulting objective function value (in \$/hr)? It should be a huge improvement.

♫ Music break[9]

---

[9]Recommended listening: *Sling the Decks* by The Crystal Method.

Chemical Reactor Models

Objectives

- Use the kinetics-based `RBatch` and `RPlug` reactor models
- Use the specification-based `RYield` and `RStoich` models
- Use the equilibrium-based `REquil` and `RGibbs` models
- Use the Data Fit/Regression features to create models from experimental results in Aspen Plus

Prerequisite Knowledge

If you are not familiar with the difference between stirred tank reactors and plug flow reactors, review these videos on batch reactors,[1] continuous stirred tank reactors,[2] and plug flow reactors.[3] If you are not familiar with the concept of reaction kinetics and how that affects reaction yield, or how it is different from chemical equilibrium, review this video on the difference between reactor kinetics and equilibrium,[4] and this video on some basics of first- and second-order reactions.[5] See also this video on reversible reactions.[6] In addition, it will be useful to understand chemical equilibrium and common ways of using it to compute reaction extents, such as in this video.[7] Each video is about 5 minutes long.

---

[1] https://www.youtube.com/watch?v=_s5csM17Bxg. Peer-reviewed material produced by the University of Colorado, Boulder.

[2] https://www.youtube.com/watch?v=8jO6CWJXF3I. Peer-reviewed material produced by the University of Colorado, Boulder.

[3] https://www.youtube.com/watch?v=AOxqN18sA04. Peer-reviewed material produced by the University of Colorado, Boulder.

[4] https://www.youtube.com/watch?v=uJXOCpDhuSQ. Peer-reviewed material produced by the University of Colorado, Boulder.

[5] https://www.youtube.com/watch?v=toNzhxKKku4. Peer-reviewed material produced by the University of Colorado, Boulder.

[6] https://www.youtube.com/watch?v=kI9yO9_ss7s. Peer-reviewed material produced by the University of Colorado, Boulder.

[7] https://www.youtube.com/watch?v=-RDRYZqxrfs. Peer-reviewed material produced by the University of Colorado, Boulder.

This tutorial also makes use of regression in order to determine the reaction kinetics parameters. If you do not know what regression is, see this video showing how to use the least squares method in the context of reactor kinetics.[8]

Why This Is Useful for Problem Solving

For any given reaction kinetics, the choice, size, and operating conditions of the reactor will affect the conversion of reactants, the product composition, and the downstream product separation methods. Thus, reactions and reactors are at the heart of most chemical engineering processes, and learning how to represent them in a flowsheet is important.

As a design engineer, you should be able to use the Aspen Plus kinetic reactor models such as RBatch and RPlug (this tutorial doesn't cover RCSTR) to represent real kinetic and reactor data. You should be able to enter in the appropriate kinetic information and other information about a reactor related to size, length, temperature, or pressure. This allows you to simulate a reactor. In addition, you should be able to interpret the results enough to make sure it worked, and understand how one model can actually represent multiple pieces of equipment, such as in the RBatch case. This is incredibly useful for not only designing the reactor but also designing the system as a whole. For example, with the ability to simulate a reactor using its kinetics, you can make determinations about how separation and recycling unreacted reagents would affect the performance of the system, or make decisions about how to handle the thermal management of the reactor, perhaps by integrating it with other parts of the system.

Besides understanding how to enter the necessary model parameters, you should understand the basic relationships between the parameters and the outputs of the model. For example, you should understand how changing the reactor length (RPlug), batch time (RBatch), and other parameters affect the stream outputs. You should understand how equilibrium is approached but not necessarily achieved in kinetics-based models (and in real life), and that the outputs may not even be close to equilibrium at all. For example, you should know what would happen to the reactor output if the reactor size or residence time is increased.

It can also be very useful to understand how to use the less rigorous models that assume either a certain stoichiometric conversion, chemical equilibrium, or an approach to equilibrium. These models are much easier to use and require little experimental data, which can be very hard to find. Although these models are not able to incorporate physical reactor size into the simulation, they can be useful for quickly estimating the outputs of the reactor, the associated flash conditions, and heating or cooling requirements.

Finally, regression is a very helpful tool for creating a kinetic model from experimental data. It is very useful to know what it does, why it is there, what it produces, and how to use it. It is entirely possible that on a project you will need to create a kinetic model from batch data. This is because in many cases, kinetic information is scarce, since it varies from catalyst to catalyst and is often kept as a trade secret. However, it is possible to do certain experiments to help determine the kinetic parameters of a given model. This is done commonly enough such that Aspen Plus contains a feature to help you with this step.

---

[8]https://www.youtube.com/watch?v=yVkpq20OtcE. Peer-reviewed material produced by the University of Colorado, Boulder.

Tutorial

PART 1: RBATCH

The RBatch model in Aspen Plus provides a way to model batch reactors in continuous processes. A cyclic unit operation, such as a batch reactor, is usually integrated into a continuous process by means of holding tanks. Although not obvious from the icon, the RBatch model actually consists of a batch reactor and several tanks together, as shown in Figure 7.1, for a liquid-phase batch reactor.

TOM'S TIP: Although RBatch remains in the model library, it is being replaced by BatchOp, a general way of modeling batch reactions and batch systems more broadly. This is covered in Bonus Tutorial 3. The recently developed Batch Processing tools in Aspen Plus try to walk a fine line between the steady-state flowsheets to which we are accustomed in Aspen Plus and the rather general and complex Aspen Plus Dynamics or Aspen Custom Modeler software which can handle general batch processes in detail. In my opinion, RBatch will probably be preferable to BatchOp more often for most users, so it remains in this tutorial.

The holding tanks serve as a buffer between the upstream and downstream continuous processes. The continuous feed always feeds into Tank 1, never stopping. However, Tank 1 is drained only periodically, usually at the start of each batch reactor run. The batch reactor runs in batches, of course, typically with a fixed run time, or at least some known average. When the batch is finished, the products are drained into Tank 2. However, the contents of Tank 2 are continuously drained to the downstream process. A sample trajectory of key process variables might look as shown in Figure 7.2.

Aspen Plus assumes that there is no reaction during the feed and drain stages. This is not accurate, but it is a conservative assumption, and not so far off either when the feed and drain steps are quick. From the Aspen Plus perspective, they use the terms shown in Figure 7.3.

Let's look at an example by simulating the batch reaction of allyl alcohol and acetone to produce *n*-propyl-propionate, as shown in Figure 7.4.

Let's assume this reaction follows the following simple power-law kinetics:

$$-r_{AA} = k \exp\left(\frac{-E}{RT}\right) C_{AA} \, C_{ACE}$$

where k is the pre-exponential factor, E is the activation energy (let's use 6×10^7 J/kmol for this tutorial), and C_{AA} and C_{ACE} are the molar concentrations (a.k.a. the molarity) of allyl alcohol and acetone, respectively. The

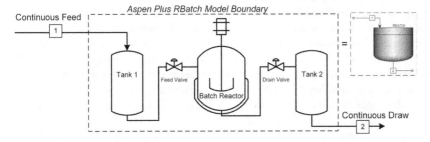

Figure 7.1 The RBatch model in Aspen Plus for a liquid-phase reaction.

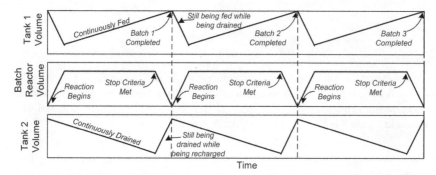

Figure 7.2 Sample volume holdup trajectories in the three units of Figure 7.1 as they often occur in real life. The feed to Tank 1 and draw from Tank 2 are always continuous.

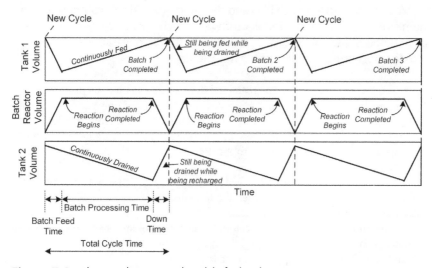

Figure 7.3 The RBatch conceptual model of a batch reactor in a continuous process.

reaction occurs without a catalyst in the liquid phase. *n*-propyl-propionate is a sweet-smelling food additive and is also used as a reagent for propionate derivatives, which make yummy artificial flavors and perfumes (e.g., ethyl propionate = "fruit punch").

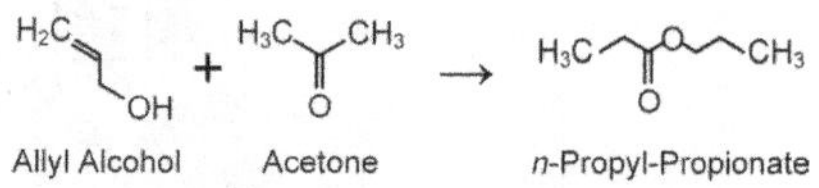

Since allyl alcohol is more expensive and acetone is easy to separate from the propionate (by distillation), acetone is used in excess to ensure maximum conversion of the allyl alcohol.

Figure 7.4 The reaction of interest for the first few parts of this tutorial.

Simulate a continuous process to produce *n*-propyl-propionate from 200 g/sec of allyl alcohol and 280 g/sec of acetone using an integrated homogeneous liquid-phase batch reactor system. The feeds are at 30°C and 1 bar, and the reactor should have a cooling system that maintains a constant temperature of 30°C. Assume no pressure drop and use the NRTL-RK model (*be sure your binary coefficients draw from* VLE-RK, *not* VLE-IG). Each batch cycle is considered complete when the reaction has achieved 98% conversion of the limiting reagent. For this simulation, the pre-exponential factor k is unknown. Our best guess is that it is about equal to 1.5×10^9 m³/kmol-min. Use this value for now.

Aspen Plus will integrate the dynamic mass and energy balance differential equations contained in the model. This means that you will get trajectories of the heat duty and molar holdups for each of the components as a function of time. You can see the results at the end of the batch if you go to the block's Results tab.

You can see the trajectories at any time by going to the Profiles tab of the block after it is run.

You have to first define the reaction in the Simulation | Reactions | Reactions tab. Make a new power law kinetic reaction corresponding to the given reaction kinetics, shown in Figure 7.5. Note that *the coefficients are negative for the reagents and positive for the products.* The exponents of the concentration variables are defined when you edit the reaction itself. The exponents of C_{AA} and C_{ACE} are both equal to 1 since they appear to the first order in the power-law expression on the previous page. The exponent of the propionate product is 0, since it doesn't show up (there is no reverse reaction), as shown in Figure 7.6.

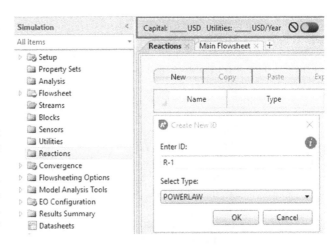

Figure 7.5 Defining a new kinetic reaction.

When you type in the kinetic information in the Kinetic tab, you'll quickly see that the units of k and C_i are not labeled in the form. This is quite the pain. Go to the Aspen Plus Help File and search for "Units for

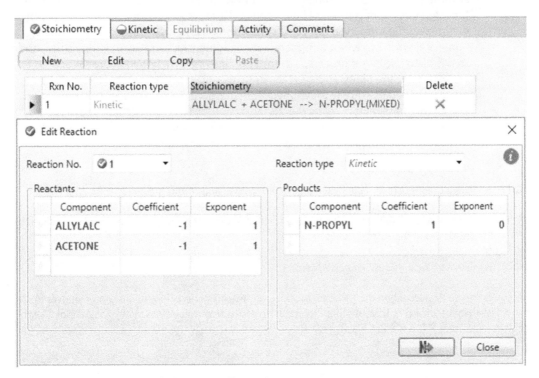

Figure 7.6 Defining the stoichiometric coefficients for a reaction.

Pre-Exponential Factor." (You can also click on the input box for k and hit F1. Then there's a link for the page there.) Since our kinetic rate law uses molarity, you should be able to figure out what units Aspen Plus expects you to type in for k. Note you'll have to convert the units given above to the form Aspen Plus expects.

To set up the RBatch block, first locate the RBatch unit under the Reactors tab in the Model Palette.

Specification Tab: Specify a constant reactor temperature of 30°C.

Reactions Tab: Be sure to add the reaction to the selected reaction set.

Stop Criteria Tab: Add a criterion for 98% conversion rate for the limiting reagent. Select Reactor for Location and Conversion for Variable type with a Stop value of 0.98. Then think about what the limiting reagent is for the component, and whether the criterion is approached from above or below the stop value during the batch reaction. Selecting Approach from Below means the conversion is approached via the forward reaction, while Approach from Above means the conversion is approached via the backward reaction. (Think carefully which is the one for your case!)

Operation Times Tab: Use a batch feed time and batch down time (safety time) of 60 seconds each. Use a maximum time of 2000 seconds and a time interval between profile points of 10 seconds.

TOM'S TIP: Dynamic simulations (i.e., something that changes with time) are usually best managed by numerical integration algorithms which control the size of the timestep used between points. Aspen Plus uses a method which will automatically reduce the timestep size during times of rapid change, but use larger timestep sizes during times of little change. This provides a good balance between minimizing the numerical error and simulation time (small timestep sizes reduce numerical error, but when things are not changing much, the error is very small, and so it is ok to use larger timesteps then). The "time interval between profile points setting" is not the timestep size. It is merely the intervals of time at which results are recorded in the profile for your viewing pleasure. So, in our case, the integration algorithm may simulate many more timesteps than what we actually see in the profile—it just simply doesn't store them or show them to us. This is done because saving too many data points can be inconvenient and unnecessary.

Q1)   How long does it take in seconds to achieve 98% conversion of allyl alcohol?

Q2)   How much of the propionate in kg is there in the reactor at the end of 200 seconds? (Hint: look in the profile tab of the RBatch block.)

Q3)   Is the reaction exothermic or endothermic?

If you go to the Profiles tab in the RBatch block result, the mysterious Plot menu should appear in the ribbon. Use the plot wizard to make a plot of the molar composition trajectories. They should look like the plot shown in Figure 7.7.

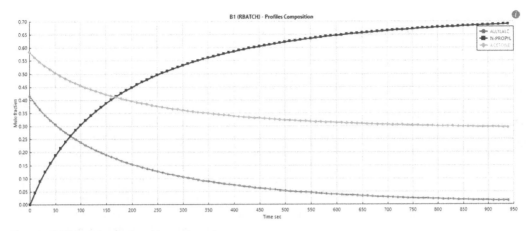

Figure 7.7 Reactor composition trajectories.

PART 2: DATA REGRESSION

Feeling uneasy about the model results, you order the development of an experiment to find a better estimate of the kinetics. A lab technician took 5 g of allyl alcohol and 7 g of acetone in separate beakers. Then
the technician placed each beaker in a bath of warm water with a heating control system that maintained the
temperature of the bath at 30°C. The technician left the beakers in the baths until the contents of each beaker
also reached 30°C. Once that was finished, the technician poured the contents of the alcohol beaker into the
acetone beaker, starting the reaction. The technician collected a sample of the liquid every 2 minutes and
determined the composition of each sample through titration, recording the data. The technician was not
able to determine the acetone concentration measurement reliably, but the results for allyl alcohol and
n-propyl-propionate were quite reasonable. The experiment was repeated many times, and the standard
deviation of the error in measurement time at each point of the measurement was 0.1 minute. The average
values of the concentration at each sample time are as shown in Table 7.1.

Table 7.1 Measured Concentrations of Allyl Alcohol and n-Propyl-Propionate Reaction
Broth Over Time

Sample Time (min)	Mole Fraction of Allyl Alcohol	Mole Fraction of n-Propyl-Propionate
2	0.2268	0.3513
4	0.1359	0.5320
6	0.0941	0.5724
8	0.0588	0.6632
10	0.0424	0.6517
12	0.0292	0.7017

In Aspen Plus, you can use the Regression functionality to try to find a better expression for the rate law. In this case, the pre-exponential coefficient needs to be changed. The problem is summarized as follows: find the pre-exponential coefficient k which best fits the simulation model used with RBatch to the experimental data.

To do this, Aspen Plus uses regression techniques to find the parameter that causes the model to best fit the data.

You can enter the data in the Simulation | Model Analysis Tools | Data Fit | Data Set tab. Create a new data set. The data you have from the lab are called profile data. On the Define tab of this form, select the appropriate model for which the data should be matched. Then identify the measured variables on this tab (note that you will have to choose and enter a variable name for the measured variables). On the Data tab, the measured variables names should appear in the column headings. Now you can enter the actual data.[9]

Then, once you have defined the data, you define the parameters to vary (k). Do this by going to the Model Analysis Tools | Regression and create a new regression. On the Specifications tab, list the data sets that are relevant to this analysis (i.e., the only one you've made so far). On the Vary tab, enter the parameter that will be varied. Note that k is the React-Var type. Pick a range that makes sense, say maybe ±20% of the original estimate (in the correct unit). Don't go too far outside the initial value because Aspen Plus really doesn't handle it well. If you pick one too big or too small, then Aspen Plus will throw an error and give up.

When you run the simulation, Aspen Plus will iterate on k and try to find the best possible value and use that as the final result.

Q4) What is the new value of the pre-exponential coefficient in m^3/kmol-min?

Q5) How long does it take to achieve 98% conversion of allyl alcohol using the k obtained from experimental data?

Q6) How much of the propionate in kg is there in the reactor at the end of 200 seconds using the k obtained from experimental data?

PART 3: EXPANDING TO PLUG FLOW REACTORS

At this point, you have successfully taken experimental data from a laboratory batch reactor and found a suitable kinetic model. The nice thing about kinetic equations is that they apply to other types of reactors. So let's simulate a plug flow reactor and see if we can get a similar result.

Use Aspen Plus to determine how big of a liquid plug flow reactor will be needed to achieve 98.5% conversion of allyl alcohol and the same inlet conditions. Use the experimentally determined power-law pre-exponential coefficient. Keep the diameter of the reactor at 40 cm and operate isothermally at 32°C.

The RPlug model can be used for this analysis. Make sure your valid phases are set to Liquid-Only.

Note, you may find it helpful to reinitialize and rerun if you notice that results are not changing like they should.

Q7) What is the length of the PFR that achieves 98.5% conversion of allyl alcohol? (Hint: consider using a Design Spec—review Tutorial 3 to see how to set it up.)

🎵 Music break[10]

---

[9]If you have the e-book version of this text, try copy-pasting it from the table into the Aspen Plus form.
[10]Recommended listening: *Kveldssong for deg og meg* by Odd Nordstoga.

PART 4: SPECIFICATION WITH RYIELD

In the previous parts, you worked with kinetics-based models, which are pretty advanced. In order to use those models, you needed lots of information, such as rate law kinetics and size. In this part, you will briefly work with two models that are very simple and do not require kinetic information or sizing information at all.

In this section, we will work with the reaction of lactic acid with ethanol to form ethyl lactate and water, as shown in Figure 7.8.

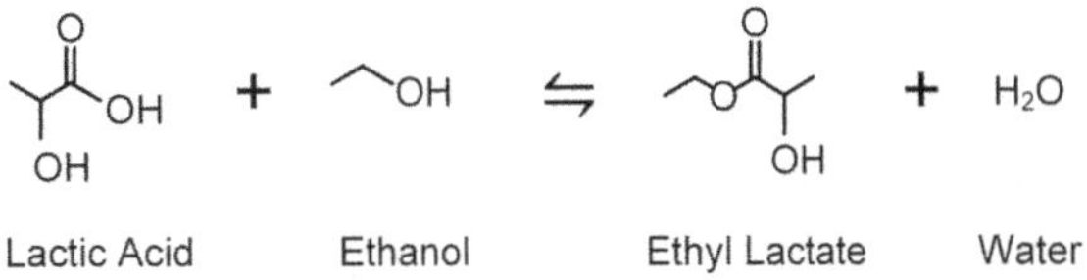

The `RYield` reactor model is incredibly simple. In fact, you literally tell it what the products of the reaction are, and it obliges by assigning the products to the output, even if your

Figure 7.8   The reaction of interest for Parts 4–6.

numbers do not make any sense. Let's do an example. Suppose we have 100 kmol/hr of lactic acid reacting with 100 kmol/hr of ethanol at 200°C and 1 atm. Suppose we desire that there will be 80% conversion of ethyl lactate in this reactor.[11] All we have to do in `RYield` is enter the flash conditions of the reactor (let's say it is adiabatic with no pressure drop to keep it simple), and then in the Yield tab, specify the component yield, which is what you want to come out of the reactor.

The tricky part is that the way you define the yield is a little strange. Instead of defining the absolute yield (as in the moles or mass of each chemical of output), you define the yield basis. For example, the mole basis yield of a chemical is the number of moles of that chemical that leave in the outlet per total *mass* of the feed. Similarly, the mass basis yield is the mass of the chemical found in the outlet per total mass of feed. You can choose which basis you would like to use for each chemical based on whichever is more convenient for you (I almost always prefer to work in moles whenever possible).

Let's do a simple example. For the ethyl lactate example, if I know that I have exactly 100 kmol/hr of each reagent, and I know the stoichiometry of the feed, then I can basically calculate what the outputs will be on paper if there is 80% conversion of ethyl lactate. Very simply, this means that 80 kmol/hr of both lactic acid and ethanol will be reacted away. We know from mass balances that there should be 80 kmol/hr of water and ethyl lactate each leaving the reactor, together with 20 kmol/hr of the two reagents each.

So, now we want to use `RYield` to make this happen. Set up a simulation with the given feed conditions using UNIQ-RK. We need to figure out the mole basis yield to type into `RYield`. We can do this in many different ways. One way is to use the molecular weights of the chemicals to figure out the total mass of the feed, and then since we know the individual component molar flow rates we want from the outlet, simply divide those outlet flow rates by that total mass flow rate. For example, you can find the molecular weights in the Properties tab by clicking the Retrieve Parameters button in the Home ribbon and looking at the MW row of the results. Or, you can be lazy about it and just type random garbage into the `RYield` model and run the simulation. Then look at the results for your feed stream to find the total mass.

Either way, I computed a basis yield of 0.00146899 kmol/kg for lactic acid. Type that into your `RYield` model as a mole basis yield. Note that there is no indication of units, but it uses the default units for mole and mass in your selected units set, which for MET and SI are kmol and kg and for ENG are lbmol and lb. In all of these sets, you still get the same number either way. So now, type in the remaining numbers into the `RYield` and run.

---

[11]Whether that is actually possible or not, well, RYield doesn't care!

Q8) What is the mole basis yield for ethyl lactate in this scenario?

Q9) What is the outlet temperature of the reactor?

One very important thing to remember is that `RYield` will only satisfy total mass balances. It does not actually satisfy the mass balance of each individual chemical, and as such it does not satisfy the first law of thermodynamics. This is because, by design, `RYield` will do its best to do exactly what you tell it to, regardless of how bad your instructions are. So go back and do something really dumb and change one of the numbers for your yield, maybe even setting one of them to zero. Now you know that it is impossible, but run it and watch what happens.

First, you get a warning. A quick check of the warning in the control panel shows the following (noting yours may be a little different):

```
*   WARNING
    SPECIFIED YIELDS HAVE BEEN NORMALIZED BY A FACTOR OF (0.867676)
    TO MAINTAIN AN OVERALL MATERIAL BALANCE.

*   WARNING
    THE FOLLOWING ELEMENTS ARE NOT IN ATOM BALANCE:
    C    H    O
```

Basically, `RYield` is doing two things. First, it is telling you that, hey, the molar basis that you entered doesn't make sense because if you calculate the outputs based on what you typed in, you get a mass yield that is less than the total input mass. So the warning is telling you that it went ahead and scaled the molar bases that you gave down (or up in my case by dividing them all by 0.87 or so) such that the total outlet mass flow rate is still equal to the inlet mass flow rate (go ahead and check).

The second warning is telling you, hey, the basis yields that you gave cause the atoms themselves to be imbalanced. For example, in my case, I would have more or less carbon, nitrogen, and oxygen atoms (which, in fact, are the only kinds of atoms I have in this simulation) in the outputs than the inputs (even with the scale up). Remember, you did not actually type any stoichiometry or define a reaction, so Aspen is trying to tell you that, well, you probably made a mistake because what you typed in is physically impossible. In practice, you may get this error even when you've essentially done everything correctly, because of issues related to significant figures, differences in molecular weight values that occur in the 6th digit, etc., so this message can be hard to put a lid on.

So why would you use `RYield` at all? It may seem really strange at first because you are basically forced to do all of the calculations and logic by hand and type it in, so the only information you are really getting out of the simulation is the heat duty calculation to compute its relationship with temperature. One thing to note is you can create a Calculator block to automatically overwrite the basis yield parameters for you based on the inputs (see Tutorial 9). That way, the block can be used in a situation in which the composition of the feed might vary from run to run (such as when inside a convergence loop). But even that seems like a lot of work.

Instead, there are two very convenient uses for this block. The first is when you have experimental data for a reaction that may be very complex. Consider if you have a reaction with many possible chemical outputs, which might be common especially for biological reactions. In many cases, you may be able to measure the contents of the reaction broth but have almost no idea what the reaction pathway was that obtained it. And, because experimental data is noisy and contains measurement error, it is unlikely that the atom balance holds exactly. Therefore, it is very convenient just to type in your reaction yield in a moles per kg of reaction product basis and just put that directly into `RYield`. Sure, you might get an atom balance warning, but as long as you are cognizant of what you are doing, you can keep this error in mind when analyzing the results of your simulation. By the way, you can turn up the control panel diagnostics by going to the Block Options

| Diagnostics tab for the `RYield` block and cranking the On Screen message level up to 5.[12] Then you can see the details of the mole balance to see how far off it is.

The second convenient use is when you are connecting this model to a much more complex reactor model. Suppose you have made your own special reactor model, say, in a Calculator block (see Tutorial 9), or in an external Microsoft Excel flowsheet (which you will also learn in Tutorial 9). You can use an `RYield` in which the complex model computes the basis yields and simply overrides that information in the `RYield` block. In that way, the `RYield` acts as a stand-in for the more complex, external model.

PART 5: SPECIFICATION WITH RSTOIC

The `RStoic` model is similar to `RYield` in that you simply specify the reaction conversion, except with this block you are required to provide the reaction stoichiometry. Go ahead and make an `RStoic` block and feed the same lactic acid and ethanol mixture into it as in Part 4. Keep the feed and flash conditions the same (adiabatic and no pressure drop). In the Reactions tab, you have to specify the reaction, namely one mole of lactic acid and one mole of ethanol react to form one mole of ethyl lactate and one mole of water. You can do this by clicking on NEW in the Reactions tab and then entering the corresponding information for reactions and products. The coefficient of a component is the number of moles you need of that chemical in the stoichiometry equation, and a negative sign means it is a reagent instead of a product. Go ahead and enter this information.

You then have to specify the products being generated. In this case, you can choose either a fractional conversion (a number between 0 and 1) or the molar extent of the reaction (which is the number of moles reacted divided by its stoichiometric coefficient). Again, simulate an 80% conversion.

Q10) What is the outlet temperature of the reactor?

The convenience over `RYield` in this situation is obvious since you have to do less math personally, and mole balances are always held. Moreover, as long as you are using fractional conversion instead of extent of conversion, you will never have a problem with limiting reagents. Try it with 80% fractional conversion, and change one of your feed chemicals to have only 10 kmol/hr and leave the other at 100 kmol/hr and run it.

Q11) What is the flow rate of water in the outlet?

Finally, it is useful to note that Aspen Plus is assuming that it is actually physically possible to obtain the reaction conversion you typed in. For example, this is actually a reversible reaction, and so it is limited by equilibrium. Is it even possible to achieve 80% conversion at this temperature, or did you just violate the second law of thermodynamics? Again, Aspen Plus will dutifully do the math with what you have given it, so remember, garbage-in, garbage-out!

PART 6: EQUILIBRIUM REACTIONS WITH REQUIL AND RGIBBS

The `REquil` block is used to model a reversible reaction system assuming that it achieves (or nearly achieves) chemical and phase equilibrium. The way it works is that the user enters the stoichiometric reaction equations, and using this, Aspen Plus will compute the equilibrium constants directly from the Gibbs free energy of reaction at the temperature of the reaction conditions. Using the equilibrium coefficient

---

[12]When things get really bad, I turn it up to 11.

combined with mass balances, energy balances, and a flash calculation, Aspen Plus can then calculate the outputs of the reaction. The mathematical details are best left for another day.

Let's try and see how the ethyl lactate system example works. Again, use the same 200 kmol/hr feed (containing 100 kmol/hr each of the two reagents) at 200°C and 1 atm; feed it to an REquil block where the flash conditions are again adiabatic and zero pressure drop. In the Reactions tab, define the reaction in much the same way as in RStoic. Note here that you can define an extent of reaction just like in RStoic, but you can also type an approach temperature instead. For now, leave the definition as having an approach temperature of zero. Now one quick catch: REquil requires you to have separate liquid and vapor outlet ports, so you need two outlet streams in this case. Note that the liquid stream should be completely empty because everything should be in the vapor phase in this system. This may seem strange, but it is just a model. As long as you know that an empty stream would never really be there, then there is no problem.

Q12) What is the extent of conversion of this reaction at equilibrium under these conditions?

The extent of conversion should actually be a lot lower than 80%. What does this mean? It means that my results of the RYield and RStoic examples above are basically complete garbage for the equimolar feed examples, and you never really knew that until now. Sure, Aspen Plus dutifully computed numbers for me, but now I know that the 80% conversion is thermodynamically impossible. Equilibrium is the absolute most I can ever achieve under these circumstances! So, this is an important lesson in the principles of garbage-in, garbage-out! Aspen Plus is not magic; it will only do what you tell it to (at best).

Even worse, the conversion computed here is the absolute best conversion that is thermodynamically possible, which can rarely be achieved in practice, especially when a lot of catalyst is needed or very large reactors. Fortunately, you can use REquil to approximate sub-equilibrium conditions, meaning that they approach equilibrium conditions but never actually get there. The reaction would be slightly less than the true equilibrium, which is more realistic.

To do this in practice, you can use an approach temperature. Essentially, what happens is that you intentionally use the equilibrium constant at the wrong temperature, one that is close to the actual temperature but off by about 10°C or so (this number is purely heuristic, you can choose other numbers). In this way, when you compute the yield at the actual temperature using the intentionally wrong equilibrium coefficient, you get a little lower yield than you otherwise would. In this way, we can approximate a more realistic situation which approaches equilibrium but never actually quite achieves it.

In REquil, you can achieve this by typing an approach temperature into the corresponding box on the reaction stoichiometry definition form. Aspen Plus defines the number you type as the number of degrees above the system temperature that you want to use for computing the new (intentionally slightly wrong) equilibrium coefficient. So in your case, since this is an endothermic reaction, we want to use a temperature that is a little bit lower than the actual temperature because conversion is generally lower at lower temperatures for endothermic reactions. In case you are confused about whether to type a positive or negative number for this system, just pick one and try it. If you get better conversion than the true equilibrium, this is thermodynamically impossible, and so you know this was the wrong one to pick!

Q13) What is the new extent of conversion with a −10°C approach temperature?

Lastly, there is one more equilibrium-based reactor model that is very convenient and interesting, RGibbs. This model can compute the chemical equilibrium conditions of the reaction without even being told the reaction equation at all! Without getting into the details very much, the second law of thermodynamics tells us that chemical equilibrium will eventually be achieved given an infinite amount of reaction time, and, that this chemical equilibrium will occur when the product mixture reaches its lowest possible Gibbs free energy state (in the absence of outside influences).

So what the RGibbs block does is solve an optimization problem that tries to find the exact reactor outlet mixture which has the lowest possible Gibbs free energy. It does this by a complex algorithm which essentially guesses the composition of the product mixture, computes its Gibbs free energy, and repeats this again and again until it decides that it has found the outlet mixture with the lowest possible Gibbs free energy. While it does this, however, it also ensures that the first law of thermodynamics always holds, so it makes sure that all of the atoms themselves balance (in other words, the total carbon in the reagents equals the total carbon in the products, etc.), the energy balances, and the flash conditions hold. It does not use any reaction equation information at all, which is really helpful because, in practice, the reaction equations could be incredibly complex and even unknown.

Try it yourself using the same feed conditions again as the other test cases. The only things you have to tell it are the flash conditions (again, use adiabatic and zero pressure drop) and which chemicals to consider in the outputs. By default, RGibbs will consider all chemicals in your chemicals list to be chemicals that could exist in the output when guessing-and-checking. However, if you know that some chemicals simply will not be products or should otherwise not participate, you can define a subset of your products to consider.

Q14) What is the extent of conversion of this reaction as predicted by RGibbs?

Note that your output should be exactly the same as in the first REquil case, which is amazing considering we did not even tell it what reactions there were!

However, like all models, you must use this block with caution. First of all, remember that this will only consider chemicals that exist in your model. So if you are missing important chemicals from your list because you do not know much about the chemistry of the system, it will dutifully report an output mixture that might be totally meaningless.

Second, be sure to ask yourself if true chemical equilibrium is really what you want to model. For example, consider a case in which you have one set of reactions that are very fast (perhaps with the benefit of a catalyst) and another set of reactions which are very slow. In practice, a real reactor might be designed such that it is only long enough such that the fast set of reactions approach equilibria, but the slow set of reactions do not because they are not catalyzed or simply very slow. In that case, RGibbs would be a terrible choice of a model, because RGibbs does not care about the speed of the reaction—it considers equilibrium after an infinite amount of time. If you used RGibbs, it would report that the slow reaction has reached equilibrium, when that would be physically unlikely in practice. In this case, you could consider either using REquil and specifically only modeling the fast reaction set, or using RGibbs and removing any unique products that might be in the second reaction set to prevent them from being considered, depending on the situation.

As an example, consider the reaction of methane with oxygen (using plenty of excess air) to produce carbon dioxide and water. In practice, this reaction does not even need a catalyst at a high temperature because methane will readily burn under these conditions, effectively achieving equilibrium very quickly.

However, suppose you had an air-deprived environment such that you did not have enough oxygen to combust all of the methane according to stoichiometry in the flame. In practice, there would still be some combustion, but this would leave lots of methane remaining leaving the furnace. The carbon monoxide produced is higher, but it is still relatively small comparatively. However, were you to model this with an RGibbs block, it would predict surprisingly large amounts of CO leaving the flame, which would be unrealistic. However, given infinite time, the CO would indeed form because the methane would eventually react with the steam, to form carbon monoxide and hydrogen gas (which is called the steam reforming reaction), and similarly, the carbon dioxide would also react with the hydrogen gas to form carbon monoxide and water (known as the reverse water gas shift reaction). These reactions are slow at normal furnace temperature without a catalyst, which is why they only proceed to a small degree in practice. But given infinite reaction time, sure, they would eventually react, which is why RGibbs would give that result.

♫ Music break[13]

---

[13]Recommended listening: *My Friends* by Red Hot Chili Peppers.

Rate-Based Distillation Models

Objectives

- Get more experience with the RadFrac model
- Get a deeper understanding of equilibrium-based models and distillation in general
- Calculate column diameters
- Use rate-based models in RadFrac

Prerequisite Knowledge

This requires a reasonable understanding of distillation itself in order to understand how it is being modeled. This includes understanding concepts like: how a distillation column uses volatility differences in chemicals to separate more volatile from less volatile components; how ordinary binary distillation columns generally have two products (the distillate and bottoms); that there is a temperature gradient through the column, with the colder part of the column being at the top driven by the condenser and the hotter part of the column being at the bottom; that the more volatile component (usually having the lower normal boiling point) is collected in the distillate and the less volatile component is collected in the bottoms; that when more than two chemicals are fed to the column, you generally still only collect two products using conventional distillation, meaning that at least one product stream will be a mixture of two or more chemicals; that the number of stages, the feed stage, the reflux ratio (RR), and the boilup ratio all contribute to the performance of the column; that additional streams called side streams can be collected from distillation columns, but, it is often very difficult or expensive to design a column where these side streams meet high purity in most cases; that the vapor-liquid splits on the trays approach but do not necessarily reach phase equilibria (often characterized by a "tray efficiency"); that without the presence of an azeotrope, *in theory* it should be possible to achieve any desired product purities of volatile chemicals in the distillate and bottoms streams of a binary distillation column at some combination of number of stages above and below the feed, RR, and boilup ratio, even if the costs and sizes are absurdly high and the energy required is in extreme quantities and temperatures; and other such properties.

If you still do not understand distillation, then I suggest you watch these videos from the learncheme.com website:

- Binary Distillation with Multiple Feeds
- Binary Distillation with Nonoptimal Feed
- Binary Distillation with Open Steam Heating
- Binary Distillation with Side Stream Product
- Binary Flash Distillation Example
- Distillation-Murphree Efficiency
- Distillation-Side Stream Feed
- Distillation using Partial Condenser Part 1
- Distillation using Partial Condenser Part 2

Why This Is Useful for Problem Solving

Distillation accounts for a very large proportion of the energy expended in the chemical industry, and so it is an important part of our profession. It is also very complex, especially for systems of many chemicals, with many possible ways to design and operate not just the distillation columns themselves, but the collection of distillation columns that perform multiple, difficult separations. Fortunately, we know a great deal about the theory of distillation and how it links to common chemical engineering concepts such as mass balances, energy balances, and phase equilibria for which data are readily available in many cases. We can even incorporate very specific information down to the size, number, and spacing of holes on the tray (and all the various types of trays or packing that could be used), and even use rate-based mass-transfer kinetics to understand how mass transfer will occur without even having to assume phase equilibrium is reached. It is rather remarkable, really, and the advantage of the modern chemical process simulator is that it can solve the system of thousands of equations for us so we can focus on the problem of design.

As such, if you know how to use even the basic features of RadFrac, you can get a lot of mileage out of it when it comes to designing a good distillation column or even a system of many distillation columns working to separate out mixtures of many chemicals into their individual components. Or, you can use it to understand how existing columns might respond to changes in feed conditions, and how the operators should change the operating conditions in order to respond to those changes appropriately. You can also use it to see how the same column can be operated differently in order to obtain different purity objectives.

Tutorial

PART 1: SIZING INFORMATION

Design a distillation column that will separate a feed of 100 kmol/hr of 50 mol% water and 50 mol% methanol at 25°C and 1.4 bar, into 99 mol% methanol and 99% water. Use PSRK as the property method. The following procedure is recommended to design the column:

- Use a DSTWU model to obtain a number of stages (N) versus RR profile for the separation and pick a suitable N.
- Use DSTWU to estimate the RR, distillate-to-feed (D:F) ratio, and feed stage at that N.
- Using the conditions obtained from DSTWU remodel the separation using the more rigorous RadFrac (equilibrium-based mode) model.

- Review Tutorial 5 if you have forgotten how to do these things.
- Use the design spec/vary feature within `RadFrac` (Blockname | Specifications | Design Specifications) to adjust the RR and D:F ratio so that the separation meets the product purity targets in the distillation and bottoms. It is just like the Design Spec | Vary feature for the flowsheet, except that it is custom designed to work within a single `Rad-Frac` unit (with higher success rates).

Furthermore, assume that the condenser is 1 bar and there is a 0.02 bar per stage pressure drop (pressure increases going down) in the `RadFrac` model, as shown in Figure 8.1.

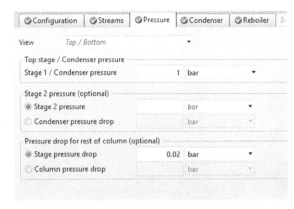

Figure 8.1 Setting the pressure drop in a column.

When you are done, check the distillate and bottoms temperatures to make sure they make sense. If you get a lower temperature in your reboiler than the condenser, then you probably specified the tray pressures incorrectly. Note that there is more than one right way to design this column. When you are done, check the liquid tray composition profiles and temperature profile to see if your column is over or under-designed.[1]

Mine is shown in Figure 8.2. The water liquid mole fraction hits the 99% on the bottom stage and 1% on the top stage, and the methanol liquid mole fraction does the opposite. That is exactly what the design specs were, so that is good. Notice also that there are no flat regions where the stages do not matter, so there are no stages to cut out. If the columns were over-designed, you would see large regions in which the mole fraction profiles change only miniscule amounts from stage to stage, and thus could be removed. Also, there is

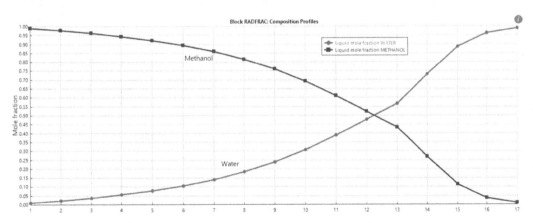

Figure 8.2 An example of liquid mole fraction profiles for a well-designed column.

[1]An under-designed column means either you are not getting the purities you want, or you could really benefit from adding some stages because so much work is being placed on the condenser and reboiler due to high reflux/boilup rates. An over-designed column is the opposite: either you are getting way higher purities than you need, or you have many more stages than you really need and could save some money by getting rid of a few with minimal impact on energy use.

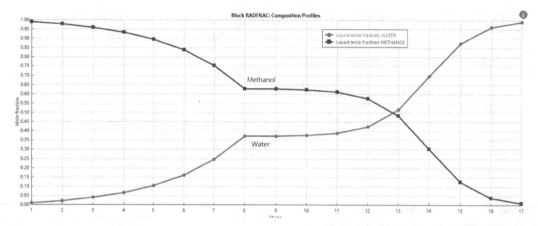

Figure 8.3 Example of mole fraction trajectories of a less-well designed column. The column still meets purity objectives, but the kink indicates that the feed stage is not optimally placed, and the flat regions indicate that some stages contribute little to the column performance and could probably be eliminated with minimal increases in energy consumption.

no "bump" associated with putting your feed into a suboptimal stage. So this is a good design. You may have another equally valid design with a different number of stages.

Figure 8.3 shows an example of that same column when I put the feed into a suboptimal stage (in this case stage 8) with the design specs enabled to ensure that both the distillate and bottoms still meet their desired product purities. The column still meets its objectives, but you can see the bump in the profiles at stage 8. This one requires way more energy than the previous example, and in fact the column should be shorter by two or three stages below the feed since not much change is happening in stages 8–10, and so removing those stages would cause a negligible increase in energy while saving a meaningful amount of capital costs.

Figure 8.4 shows an example schematic for a distillation column using sieve trays with four trays above the feed and three below it. RadFrac can model the column more rigorously by considering the details of the trays or packing and how they are designed. For example, you can choose between several kinds of tray or packing models:

Sieve Tray
- Cheap and easy to clean
- Requires good liquid/vapor flow rate balance to prevent flooding and weeping

Bubble Cap Tray
- Handles wider load ranges than sieve trays
- Consider using when sieve trays cannot do the job

Tunnel Cap Tray
- An alternative to bubble cap trays and used for the same purpose

Structured Packing
- Cheaper but not usually as efficient as a tray (more space required)
- Usually used for small-diameter columns (typically 2 ft. and smaller), but can handle wide variations in the balance of liquid and vapor loads

- Can integrate well with reactive distillation because structures can be designed to hold and support catalysts in particular ways that perfect the balance between catalyst liquid interfaces and contact times with the mass transfer kinetics of the separation

Unstructured (Random) Packing

- Loose materials that are dumped into the column (e.g., Berl saddles, Pall rings)
- Used in similar situations to structured packing, but ideally with even lower cost
- Typically lower efficiency than structured packing

In Aspen Plus, you can use the Column Internals folder within RadFrac to determine the column diameter (i.e., tray diameter) for different sections of the column. The theoretical diameter of the trays in the column is strongly dependent on the internal flows of vapor and liquid in the column. Aspen Plus uses these details to figure out what diameters should be used for the trays in different sections of the column.

Let's use a 25-stage column as an example (change the number of stages in your column). Remember, the trays are on stages 2–24 since stage 1 is the (total) condenser and stage 25 is the reboiler. To make sure we are on the same page, feed to stage 12 (above-stage). In your RadFrac model, go to Column Internals and add a new internals folder (mine is called INT-1). If you get a message about the simulation missing hydraulic data, click Generate. If you click Cancel, that's ok. You will just need to run the simulation again later. In [internals folder] | Sections, select Based on Flows from the Auto Section drop-down button, as shown in Figure 8.5. This will automatically create column sections for you by grouping stages together based on similar internal flow rates.

Aspen Plus divides the column into sections and automatically calculates the diameter of these different sections. Although the numbers that Aspen Plus calculates are called the "diameter," it is really a *minimum* diameter that is required to ensure that flooding is avoided within a certain safety factor. Check that the tray spacing of the column sections is 2 ft. (0.6096 m). The only other common standard option used in industry is 1.5 ft. (0.4572 m), but often the 1.5 ft. option is too close together and may cause flooding. Anything

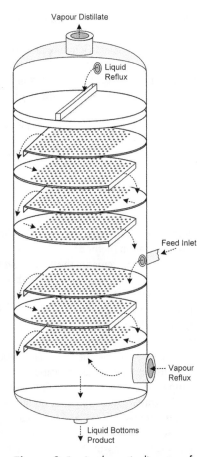

Figure 8.4 A schematic diagram of a distillation column.

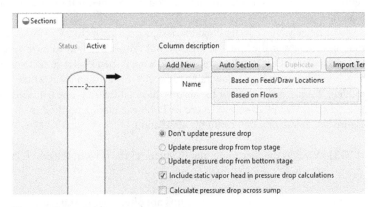

Figure 8.5 Adding a column section.

else except those two are usually custom orders, and way more expensive than just buying off-the-shelf tray stacks with 2 ft. spacing.

If you had clicked Generate in response to the warning message when adding the column internals, then you should see some numbers in the Diameter column of the Column Internals table, indicating what the minimum diameter should be. If you had clicked Cancel, run the simulation again and the numbers should be generated and appear on this form.

Q1) How many column sections does the column have?

Q2) What is the minimum diameter of the trays in the first section of the column?

You are welcome to play with other criteria such as grouping by where feed and side draws are located, or defining your own sections.

We can do more advanced stuff for our column design. For example, in Sections | CS-1 | Design Parameters there are several changes we can make to the Sizing criterion, Hydraulic plots/Limits, Design factors, and Calculation methods.

The two key items are the % Jet flood for design and the Jet flood calculation method, as shown in Figure 8.6. The `Fair72` method is the most commonly used way of computing the flooding velocity. Basically, it is an equation that takes as input the tray diameter, the tray spacing, the liquid and vapor compositions, the liquid and vapor surface tensions, and the liquid and vapor flow rates, and computes what the flow rate of the liquid has to be for the tray to start flooding. So above this flow rate, the tray will flood (and fail), but below it, the tray will function. Bigger diameters mean the flooding velocity will also be bigger (i.e., they can handle a higher capacity because it takes more flow to flood the tray). So what we are asking

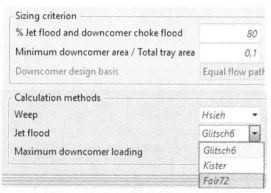

Figure 8.6 Changing the parameters of the flooding calculation that is used to compute column diameters.

Aspen Plus to do is to solve this equation backward to predict the diameter at which flooding will occur, and return a value for each column section. You would then want to make sure your actual diameter is bigger than this.

Well actually, the "80% Jet flood for design" is a slop factor. We don't want to pick the diameter to be such that we are right at the flooding velocity, but rather, we want to have a safety factor (we want to always be lower than 80% of the flooding velocity). So what Aspen Plus is going to do is find the diameter of my column which will have the liquid flow rates to be 20% lower than what would actually flood (as predicted by the model). We can make changes to this but the default 80% is a reasonable value to use.

Select the `Fair72` method as the jet flood calculation method in all your column sections, as shown in Figure 8.6. Reinitialize and rerun the simulation.

Q3) What is the new minimum diameter of the trays in the first section of the column using the Fair72 method?

In North America, trays and columns are often sold in standard diameters in 6 in. increments. Usually, the benefits of custom-sizing a diameter to a very particular size do not outweigh the high costs of custom manufacturing. So, usually diameters are rounded up to the nearest half foot. My results from Q2 and Q3 show that both flooding calculation methods agree: I should actually use a rounded up column diameter of 2.5 ft.

Also, you may have noticed from your calculations that Aspen Plus computes different diameters for different column sections. This is because the flow profiles in different sections of the column can be quite different in different sections. However, in your case, although the minimum diameters are different, when you round up to the nearest 6 in. diameter, both sections should have the same diameter in the final design.

This is a common result, but it is not always the case. For example, consider a column that separates A (which is lighter and more volatile) from B (which is heavier and less volatile), but the feed stream contains, say 10 mol% A and 90 mol% B. This imbalance means that the tray liquid flow rates below the feed would be much larger than above it, since so much more B would be present in the bottom half of the column. In these cases, it may be worth it to design a dual-diameter column. If you would like, see what happens if your feed was 10% methanol and 90% water. In my case, I required a 1 ft. diameter for the section above the feed and a 1.5 ft. diameter below it. So the distillation column would look similar to Figure 8.7, as is nicely illustrated on the column internals form. Even so, it can sometimes be more cost-effective and easier to maintain if both sections have the same diameter, especially for high-pressure applications where minimizing joints and welds can be important. If this is the case, you might be able to just make the whole column 1.5 ft. in diameter.

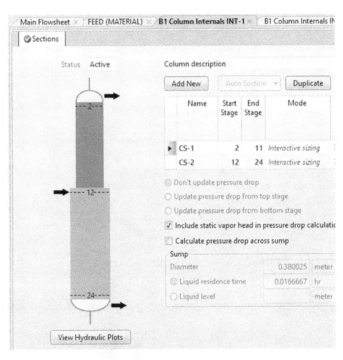

Figure 8.7 A dual-diameter column makes sense for the case where the feed is 10 mol% methanol, 90% water.

Now in order to answer the question of what tray spacing to use, rerun your simulation using a tray spacing of 1.5 ft. (0.4572 m) instead of 2 ft. (the only other option).

Q4) What is the new minimum diameter of the trays in the first section of the column?

In my case, my minimum diameter increased. So I could choose now between a shorter but wider column, or a taller but thinner column (both having the same number of trays). That tradeoff would be very case dependent. If your minimum diameter (when rounded up) stays the same, you should go with the smaller tray spacing in most cases.

♫ Music break[2]

---

[2]Recommended listening: *Selfless Cold and Composed* by Ben Folds.

PART 2: RATE-BASED SIMULATIONS

Now that you have a working equilibrium-based model, we can go one step further in accuracy by doing rate-based simulations. This is a mass-transfer-based, kinetically-driven, complex model that does not assume phase equilibria. This is more accurate because sometimes trays don't have enough residence time to sufficiently approach phase equilibria. However, in order to use rate-based calculations, the model requires more detailed information about the trays themselves.

First, switch your calculation type over to rate-based, as shown in Figure 8.8. This will remove the assumption of chemical equilibrium, and instead use rate-based mass-transfer kinetics, which is complicated, but kept under the hood. Next, go back to Column Internals | [internals folder], and in the Mode column of the Sections window, change all your column sections from `Interactive   sizing` to `Rating`. This means that instead of asking Aspen Plus to determine a minimum tray diameter based on the equilibrium assumption, the problem is reversed: you are now telling it to compute what the results will be for a given tray diameter that you specified, based on the mass transfer kinetics. Don't forget to change the tray spacing back to 2 ft. Next, go to the RadFrac Blockname | Rate-Based Modeling | Rate-based Setup | Sections tab and activate Rate-based calculation for all your column sections.

Now if you go to the column section folders

Figure 8.8 Switching to rate-based mode. Note: older versions of the software required specific licenses for rate-based mode and had less obvious ways of activating it. If you are using older versions and this drop-box is not available, consult your user guide.

of your column and look at the Geometry form, you will notice that you have a lot of column section design options including Section type (Trayed or Packed), tray type, tray dimensions, etc. You can get very specific such as to the diameter of the holes and the number of holes on the tray, as shown in Figure 8.9. You can also

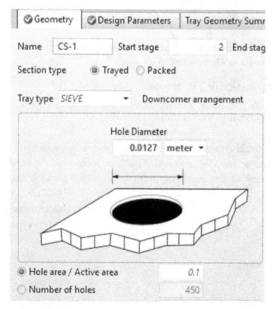

Figure 8.9 You can modify tray details such as packing/try type, hole diameter, and active hole area.

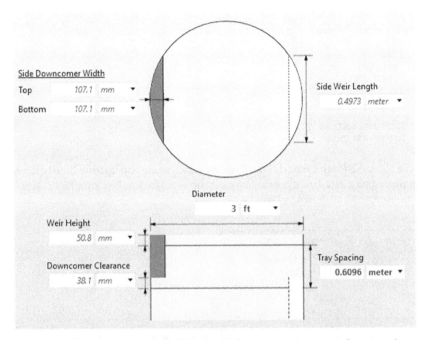

Figure 8.10 You can also modify weir and downcomer dimensions for use in the model.

modify details about the weir and downcomer dimensions, as shown in Figure 8.10. Just use the default values for this tutorial, but it is useful to know that you can change this for future applications.

Let's leave everything as they are. Don't mess with anything unless you know what you are doing. Now run the simulation!

Did it work? If not, check the Tom's Tips for ideas to try to get it to work. Once it works, you get all sorts of useful results. For example, look in RadFrac Blockname | Column Internals | [internals folder] | Column Hydraulic Results and you can see, for example, the actual pressure drop for each section. So our 0.02 bar estimate was not that bad (conservative really), whereas it is mostly about 0.005 bar on every stage.

TOM'S TIP: If it doesn't converge, check the control panel to see what happened. One common problem is that `RadFrac` needed more iterations because the default of 25 is often not enough. There are two places to change these. The first is in the Blockname | Convergence | Convergence as we learned in previous tutorials. However, this only affects certain parts of the solution procedure. If this is the problem, you will see the problem crop up in this section of the control panel output:

```
Block: B1        Model: RADFRAC

   Convergence iterations:
      OL    ML    IL      Err/Tol
       1     1     5       239.41
```

There is a second set of algorithms that is specific to rate-based mode. If it runs out of iterations in that step, the "Convergence Iterations" would actually converge first, and then you would see something like:

```
RateSep convergence iterations: B1
      Iter      Err/Tol
         0       800.39
              (...)
        50       757.27
**   ERROR
     RATESEP CALCULATIONS FAILED TO CONVERGE
        ITERATION LIMIT REACHED
```

Note that RATESEP is the old name for this model—you'll see the legacy name pop up from time to time. In this case, you can try increasing your rate-based convergence iterations in Blockname | Rate-Based Modeling | Rate Based Setup | Convergence, as shown in Figure 8.11.

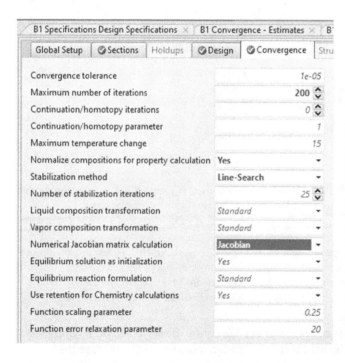

Figure 8.11 Since the model has so many equations, you may find a need to increase your maximum convergence iterations or turn on numerical Jacobians in order to achieve convergence. Jacobians are essentially a way of computing how the model equations change with regard to the model variables at the current solution guess, which the solver uses to generate new and better guesses for each iteration. Calculating them numerically (instead of analytically, which is the default) can be useful when the equations are not behaving "nicely." But that was probably more than you needed to know.

TOM'S TIP: If increasing the iterations does not help, try turning on numerical Jacobian calculations under RadFrac Blockname | Rate-based Setup | Convergence tab. This will slow the simulation down a bit, but you may not even notice. See Figure 8.11. This often helps me in tight circumstances. It is wise to reinitialize your model after each convergence failure if the convergence failure was not due to having too-few iterations. This is because it will use your previous run as the initial guess for the next run, and in a convergence failure, those values are usually so bad that they cause convergence failure... hence don't use them.

TOM'S TIP: If you are still having problems, go back and simplify the problem as much as possible. Reconcile your block by entering in the RR and distillate to feed ratio that you got from the design spec in equilibrium mode. Turn off the Design Specs | Vary entries (for now) and switch back to equilibrium mode. Then reinitialize and run it. If that works, switch back to rate-based again and run again, which uses the equilibrium results as the initial guess for the rate-based ones. If successful, the rate-based results should be very close to your final answer. Then turn the Design Specs | Vary entries back on again. Hopefully, that should do it.

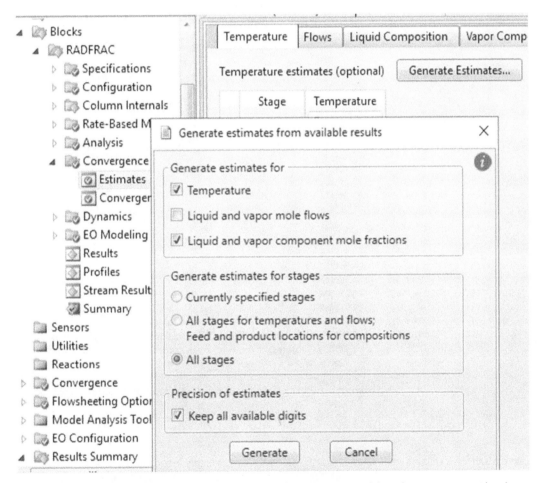

Figure 8.12   The generating estimates feature essentially stores your model results as estimates within the block. Estimates are used for initial guesses the next time the model is reinitialized and rerun. Good estimates help lead to rapid and reliable model convergence. Although you can enter your own estimates (and sometimes you may have to), it is particularly convenient to use the equilibrium mode to create estimates for rate-based mode.

TOM'S TIP: If you still cannot achieve convergence, try using the Estimates feature. Go back and reinitialize and resimulate this in Equilibrium mode and get it to work there. Then use the Generate Estimates button in the RadFrac `Blockname | Convergence | Estimates` form. Have it generate estimates for the intensive properties (temperature and mole fractions) on all the stages (see Figure 8.12 for setup). Then you can see that it will copy all of your temperature and mole composition results into the estimates section (see Figure 8.13). These will be used as initial guesses when you run it in rate-based mode and will make it much more likely to converge (and more quickly too!).

Stage	WATER	METHANOL
1	0.0100001	0.99
2	0.0225045	0.977495
3	0.037339	0.962661
4	0.0550487	0.944951
5	0.0763454	0.923655
6	0.102161	0.897839
7	0.133714	0.866286
8	0.17254	0.82746
9	0.220468	0.779532
10	0.279301	0.720699
11	0.34976	0.65024
12	0.429158	0.570842
13	0.429249	0.570751
14	0.429379	0.570621
15	0.429631	0.570369
16	0.430244	0.569756
17	0.431928	0.568072
18	0.43674	0.56326
19	0.450512	0.549488
20	0.488648	0.511352
21	0.582545	0.417455
22	0.744525	0.255465

Tab headers: Temperature | Flows | Liquid Composition | V...

Figure 8.13 The results of generating estimates for my example show that the model results from my previous successful run have been copied into the estimates form.

Even if you got a Results Available message, check the control panel. Did you see something like this?

```
INFORMATION WHILE GENERATING REPORT FOR UNIT OPERATIONS BLOCK: "B1"
      (MODEL: "RADFRAC")
        120.77%  JET FLOOD IN COLUMN B1        , OPTION INT-1    ,
      SECTION CS-1      EXCEEDS 80%.
INFORMATION WHILE GENERATING REPORT FOR UNIT OPERATIONS BLOCK: "B1"
      (MODEL: "RADFRAC")
      SECTION CS-1      IN COLUMN B1        OPTION INT-1
      IS ABOVE 100% JET FLOOD. HOWEVER, THE DOWNCOMERS
      HAVE SPARE CAPACITY. TRY DECREASING DOWNCOMER WIDTHS.
INFORMATION WHILE GENERATING REPORT FOR UNIT OPERATIONS BLOCK: "B1"
      (MODEL: "RADFRAC")
        110.19%  JET FLOOD IN COLUMN B1        , OPTION INT-1    ,
      SECTION CS-2      EXCEEDS 80%.
```

```
INFORMATION WHILE GENERATING REPORT FOR UNIT OPERATIONS BLOCK: "B1"
     (MODEL: "RADFRAC")
     SECTION CS-2     IN COLUMN B1        OPTION INT-1
     IS ABOVE 100% JET FLOOD. HOWEVER, THE DOWNCOMERS
     HAVE SPARE CAPACITY. TRY DECREASING DOWNCOMER WIDTHS.
```

This tells you that a 2.5 ft. tray diameter is too small, because your flooding rates are about your design limit (80% of maximum capacity), and in fact are actually above the actual limit too. So this means that the column will likely flood with a 2.5 ft. diameter. This is important to catch because it means that the equilibrium model undersizes the column and is likely to fail if constructed in practice. The message above suggests that the downcomers could be increased to avoid this problem. You could try that, but in my case, that did not help. Instead, I just increased the diameter to 3.0 ft. and that solved the problem. But this shows how important it is to consider rate-based mode when it comes to distillation tower design.

Q5) What is the section pressure drop for column section 1 in bar?

In the Results form of the column sections folders you can also see the pressure drop per tray in the By Tray tab. You can see that pressure drop is no longer assumed, but calculated! A selection of my tray results for column section 1 (CS-1) is shown in Figure 8.14.

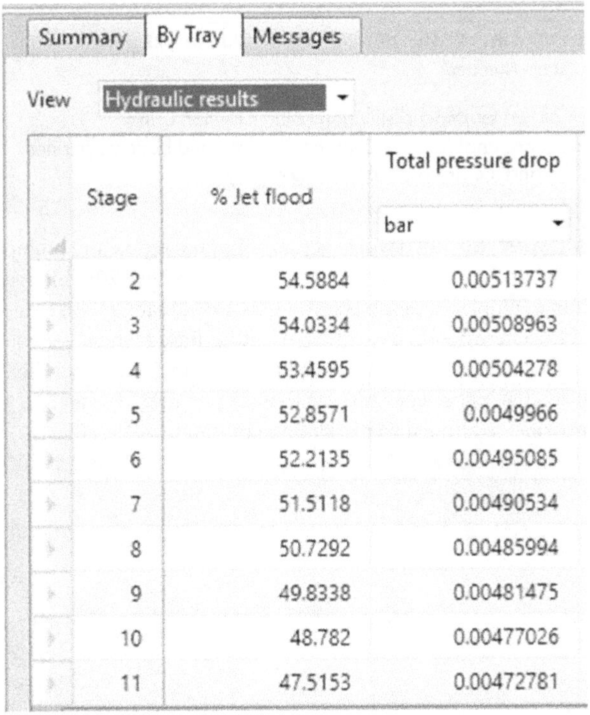

Stage	% Jet flood	Total pressure drop bar
2	54.5884	0.00513737
3	54.0334	0.00508963
4	53.4595	0.00504278
5	52.8571	0.0049966
6	52.2135	0.00495085
7	51.5118	0.00490534
8	50.7292	0.00485994
9	49.8338	0.00481475
10	48.782	0.00477026
11	47.5153	0.00472781

Figure 8.14 The rate-based mode can estimate the pressure drop on each stage so you do not have to assume it anymore.

Now, check your other results to see how things have changed. Look, for example, at your new reflux and boilup ratios. If you set up the design spec/vary like I did, then it will adjust the ratios to meet your purity objectives automatically. (See Figure 8.15 for a comparison between the mole-fraction trajectories for my "bad" design example.) So in my case, once all of the assumptions about equilibrium were taken away by

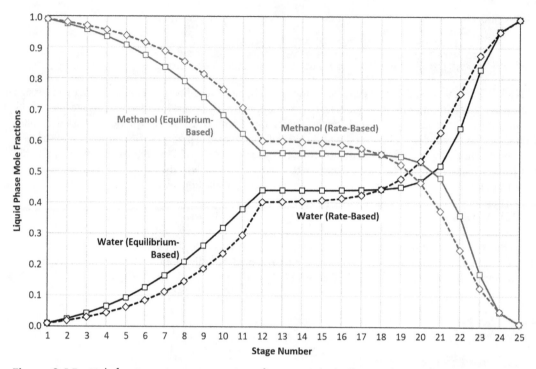

Figure 8.15 Mole fraction trajectory comparison of two example distillation columns when using equilibrium and rate-based modes, where design specs are enabled such that the distillate and bottoms purities are the same in both cases. You can see that stages 13 and 14 are probably unnecessary.

using the rate-based mode, the separation was a lot worse. So much worse actually that the reflux and boilup ratios needed to increase such that the condenser and reboiler duties were approximately 20% higher. In other words, if you had only done this column in equilibrium mode assuming that you reach phase equilibria, you would have mistakenly underestimated the energy costs by a lot! So use Rate-Based whenever you can, because you don't want to get in trouble when you design a column using Equilibrium mode, and then once you actually build it you realize that it requires 20% higher energy costs than expected to operate!

There you go. This is the most accurate way possible to simulate a distillation column in Aspen Plus.

♫ Music break[3]

---

[3]Recommended listening: *Zero* by Smashing Pumpkins.

Custom Models and External Control

Objectives

- Use the Calculator block to create custom models of units not included in Aspen Plus
- Use the Calculator block to compute intermediate values and generate useful text output to the control panel
- Use a Sep block combined with a Calculator block to make a custom hydrogen membrane module
- Use Aspen Simulation Workbook which allows external control over the simulation from Microsoft Excel

Prerequisite Knowledge

This tutorial requires very basic computer programming skills. Most engineers or engineering students develop these skills informally even if they do not receive any formal training. In this tutorial, I try to assume as little knowledge as possible. Specifically, the only concepts that are required are as follows:

- An understanding of what variables are as they are used in computer programs (that they store numbers and can be used inside of equations)
- An understanding that computer programs execute in a logical order (the first command in the program is executed first, and when that is finished, then the second command is executed next)
- An understanding that computer code can be grouped into functions
- An understanding that Aspen Plus, by default, executes blocks in a certain sequence (and often in loops), such that each block is essentially its own function

Even if you are new to computer programming, you should still be able to complete this tutorial. If you are familiar with or even an expert at computer programming, then this tutorial will make it clear how you can put your existing knowledge to immediate use inside Aspen Plus. Although Aspen Plus uses Fortran 77 syntax, having programming experience in any procedural programming language (like VisualBasic, C/C++, Python, Matlab) should be sufficient.

The second key prerequisite for this tutorial is a basic understanding of Microsoft Excel, such as how cells work, and how formulas can be entered and computed within those cells. In fact, even if you have never used

Microsoft Excel before, but have used a competing spreadsheet software such as Google Sheets or LibreOffice Calc, you should still be able to complete this tutorial. However, the Aspen Plus link feature only works with Microsoft Excel.

Why This Is Useful for Problem Solving

The Calculator block is another advanced tool you have at your disposal, alongside Sensitivity Analysis, Design Spec, and Optimization. It is incredibly powerful if you know how to use it. It's great for doing things such as connecting little bits of information around the flowsheet, calculating initial guesses to be used for tear streams, generating output files or text to the control panel to very quickly get at what you care about in a flowsheet, or even going as far as creating your own complex custom models.

Tutorial

PART 1: BASICS

In this tutorial, you will use the Calculator feature of Aspen Plus. The Calculator feature is one of the most powerful tools in your modeling arsenal. It gives you the ability to perform complex calculations, build user models, or otherwise make your job considerably easier. However, it also exposes the somewhat ancient underbelly of Aspen Plus and reveals clues to how the program has developed over the decades.

In the Calculator block, you can write Fortran 77 code (as in the year 1977) which is terribly inconvenient by today's standards, but it is what it is. We're going to do some very basic things, and the format is very similar to the Design Specs and Optimization blocks.

One very common use is to set the flow rates of streams relative to each other. For example, consider Figure 9.1 for the methane reforming reaction:

$$CH_4 + H_2O \rightarrow CO + 3H_2$$

Suppose we are building our model and we don't necessarily know what the flow rate of methane is going to be, but we do know that we want the molar flow rate of water to be 4.2 times the molar flow rate of methane. We can use a Calculator block to do this.

Start by setting up the flowsheet shown in Figure 9.1. In Aspen Plus, using the PSRK package, assume there is no pressure drop in the reactor, that the reaction reaches equilibrium (i.e., use REQUIL or RGIBBS), and that it is isothermal at 925°C. Type the known methane inlet rate (200 kmol/hr) into the input box, but type something wrong into the flow rate input box for steam (e.g., 10).

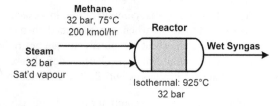

Figure 9.1 The flowsheet for Part 1.

Now, we're going to use a Calculator block to overwrite the value we just typed in. This first example seems a little contrived but it's just step one. The goal is that we're going to make a computer program which will execute when we run the simulation and set the steam rate to be 4.2 times the methane rate (by moles).

Make a new calculator folder from Flowsheeting Options | Calculator. In the Define tab, make two new variables, one is the molar flow rate of the steam, and one is the molar flow rate of the methane. This is just like the Design Spec and Optimization forms. However, we need to specify whether each variable is an `Import variable` or an `Export variable`. The difference is simple:

Import Variables

Import Variables are data that are read from the flowsheet. This is exactly like the variables that you type into the Define tabs of the Design Spec or Calculator block; all of those are Import variables.

Export Variables

Export Variables are data that are calculated by your Fortran computer program and then overwritten in the flowsheet. This is just like the variables that you type into the Vary tab of the Design Spec or Optimization block. The only difference is that now they are also in the define section, and you have to call them `Export variables`. See Figure 9.2.

Once you have defined your Import variables and Export variables, go to the Calculate tab. Here, you can enter executable Fortran statements into the box. Fortran is not nice, not like how Matlab is nice or any programming language made since the second Pierre Trudeau administration (1980–1984) is nice. What I mean is that the number of whitespaces makes a difference, and you are limited to a certain number of characters in the same line.

Here is what we are trying to do. We are trying to set the molar flow rate of steam (`FS`) to be equal to 4.2 times the molar flow rate of methane (`FM`), overwriting the incorrect value we typed in originally. To do this, you type in the exact following statement, where the underscores _ are six spaces:

$$\_\ \_\ \_\ \_\ \_\ \_\ \text{FS = 4.2 * FM}$$

The idea is that the result of the calculation on the right side of the equals sign is assigned to the variable on the left. This is just like in Matlab, Python, C++, PHP, Java, Basic, and even Excel.

The `FS` and `FM` are just my names for the flow rates that I made up, so you can use whatever names you want as long as they are the same that you defined as import/export variables. Watch out though, you also

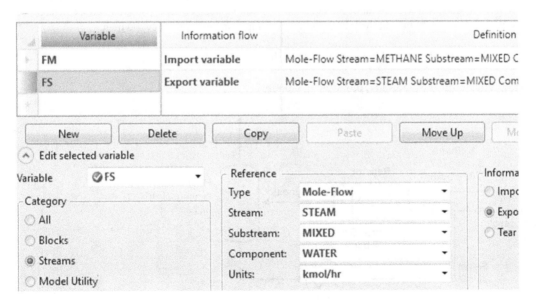

Figure 9.2 Defining import and export variables.

have to hit enter at the end of the line or else it doesn't always remember your input. This is not behavior which is typical of modern editors so this is often a source of unexpected bugs.

What happens if you don't put in the spaces beforehand? You get errors. The first six spaces are reserved for line numbers and comment indicators. We can make a comment by putting a little c in the first column, and then typing the rest. For example, I could type the following to make it easy for me to see this in the future and understand what it was I was trying to do:

```
C     This ensures that the flow rate
c     of steam is 4.2 times the flow
c     rate of methane. I needed to put
c     this on multiple lines because I
c     cannot have more than 72
c     characters in a line!

      FS = 4.2 * FM
```

Run the simulation. Go to the Input form for the steam (double left click). What is the flow rate you entered in the box? It should be whatever you entered to begin with. Then go to the Results tab for that stream. What is the actual flow rate used in the program? It should be 840 kmol/hr.

Q1) What is the flow rate of the syngas stream exiting the reactor in kmol/hr?

So that's a little contrived, but there are times when it helps. For example, what if we recovered and recycled the excess water? We could use a Calculator block to figure out how much new steam we need to add.

Consider the flowsheet of Figure 9.3. Suppose water is recovered from our wet Syngas by cooling it down in a flash drum (no pressure drop) to 105°C and collecting (mostly) liquid water. The liquid water is reheated to a saturated vapor and recycled to the reactor. Now, simulate this revised flowsheet. Again, make sure your simulation runs ok first without connecting the recycle stream.

Revise your Calculator block to ensure the total flow rate of steam entering the reactor (stream 1) is 4.2 times the flow rate of the methane. Think carefully about whether the recycle flow rate is an import or export variable, and thus which side of the equation it should appear. Note, if you are having convergence issues, the problem might be that the initial guess for the input steam is probably too high.

Q2) What is the flow rate of fresh steam in kmol/hr?

Q3) What is the flow rate of H_2 produced in kmol/hr?

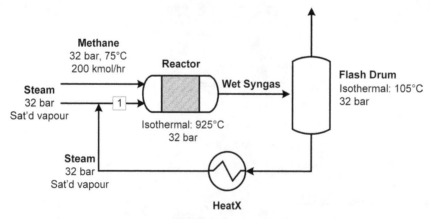

Figure 9.3   Methane reforming with water recovery and recycle.

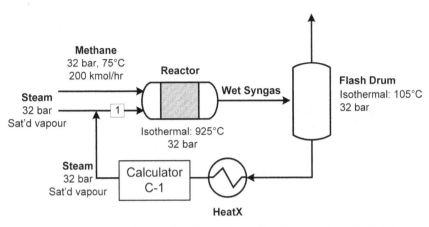

Figure 9.4 Methane reforming flowsheet with recycle, showing where the Calculator block occurs conceptually.

Why not just use a Design Spec? Well, we could actually. The problem is that a Design Spec is guess-and-check, but in this case, it's kind of a waste of time to guess-and-check when we can calculate the known amount directly. This ensures the total number of iterations is minimized, which can really speed things up. Take a look at the calculation sequence reported in the control panel (F7).

```
COMPUTATION ORDER FOR THE FLOWSHEET:
 $OLVER01 FLASH HEATER C-1 REACT
 (RETURN $OLVER01)
```

Here, C-1 is my Calculator block. There is only one Solver block due to the recycle stream and tear stream (practice: Which stream did Aspen Plus tear?[1]). You can see that Aspen Plus put C-1 after the heater but before the reactor. This means that Aspen Plus figured out that it needed the result of the flash calculation before it could determine the amount of fresh steam. Thus, it knew to run C-1 before the reactor.

Figure 9.4 shows what Aspen Plus is really doing conceptually. Aspen Plus places a Calculator block automatically in the best possible point in the calculation sequence, based on the import and export variables. If I took out the Calculator block and put in a Design Spec instead, this is what we end up with (one possibility):

```
COMPUTATION ORDER FOR THE FLOWSHEET:
 $OLVER01
 |$OLVER02 REACT FLASH HEATER MIXER
 | (RETURN $OLVER02)
 (RETURN $OLVER01)
```

And, the corresponding flowsheet would look like Figure 9.5. There are now two loops! So we need two tear streams and two solver loops. The Design Spec here changes the inlet flow rate of steam until the mixed stream (stream 1) is at the right ratio. This takes significantly more iterations to converge.

This is not as good because we now have two loops to deal with. This takes more time to solve and is less accurate because the Design Spec is only converged within a tolerance, rather than an exact calculation. You

---

[1]It tore the wet syngas stream, meaning that it used that as its starting point for its initial guess of the convergence loop.

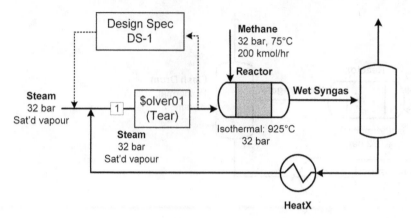

Figure 9.5 Methane reforming flowsheet with recycle, showing where a Design Spec and tear stream would be placed if a Design spec approach were used instead of a Calculator block.

can see that as things start to get complicated, using Calculator blocks instead of Design Specs wherever possible is a huge advantage.

🎵 Music break[2]

PART 2: CUSTOM MODELS WITH CALCULATOR BLOCKS

Another common situation is the creation of a model that doesn't exist in Aspen Plus. (For this one you definitely can't get away with using a Design Spec.) In this example, we'll look at the recovery of H_2 gas using a permeable, H_2-selective membrane (which is actually a common way of generating H_2). Here, the H_2 gas passes through the membrane (permeate) and the rest of the gas does not (the retentate). Of course, the yield and purity of the H_2 are never ideal. Figure 9.6 shows the change to the process.

Suppose we have developed a model[3] that can predict the yield of each species as a function of the partial pressure:

$$r_{CH_4} = 1 - \exp\left(\frac{-P_{CH4}}{1115\,\text{bar}}\right)$$

$$r_{H_2O} = 1 - \exp\left(\frac{-P_{H2O}}{160\,\text{bar}}\right)$$

$$r_{H_2} = 1 - \exp\left(\frac{-P_{H2}}{8.8\,\text{bar}}\right)$$

$$r_{CO} = 1 - \exp\left(\frac{-P_{CO}}{305\,\text{bar}}\right)$$

where r_i is the percent of species i which is recovered in the permeate, and P_i is the partial pressure (in bar) of species i in inlet gas.

---

[2]Recommended listening: *Hell* by Squirrel Nut Zippers.

[3]These are made-up numbers, but just go with it.

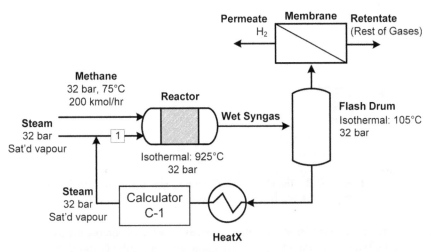

Figure 9.6 Updated process using a hydrogen membrane.

Since a membrane model does not exist in Aspen Plus, we can use a combination of tools to make one. The Sep block is a model which doesn't really do anything under the hood. It is used to model a separation unit where you already know the yield or the split fraction of each individual species. It is a lot like a splitter (FSplit) except that instead of splitting the whole stream, you can split individual components. The onus is on you to specify the split fractions correctly based on the model you want.

Our plan is to use a Calculator block which imports the necessary information from the vapor stream of the flash drum, uses the above equations, and then exports the split fractions to the Sep block.

Now add a Sep block representing the membrane process, as shown in Figure 9.6. The Sep block is found in the Separators section of the Model Palette. Note that you still need to give the Sep block some default values that will be overwritten by the Calculator block. Pick any split fractions for each chemical between 0 and 1, it doesn't matter (see Figure 9.7). You can pick either stream as the split fraction basis, but I suggest using the permeate for convenience just because the equations are written that way. The way Sep works is that whatever fraction you specify goes to one stream, the rest goes to the other stream (makes sense).

Specifications	Feed Flash	Outlet Flash	Utility	Comments

Outlet stream conditions

Outlet stream PERMEATE

Substream MIXED

Component ID	Specification	Basis	Value
METHANE	Split fraction		0
WATER	Split fraction		0
CO	Split fraction		0
H2	Split fraction		1

Figure 9.7 You need to specify default values in the Sep block, even though you are going to overwrite them with your Calculator block.

Now add a new Calculator block to compute the above equations. Note that you won't find "partial pressure" when you define variables. Then what should you define to calculate it? (Hint: What is the relationship between mole fraction, partial pressure, and total pressure?) For split fraction of membrane separation, use FLOW/FRAC of the Sep block as shown in Figure 9.8.

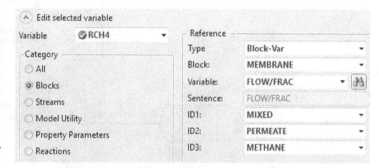

Figure 9.8   The FLOW/FRAC variable type corresponds to the separation factors on the Sep block that you want to change.

Note that the function EXP( ) can be used to compute the exponential in Fortran. Ultimately, it will look almost the same as an Excel formula: +, -, /, *, and ** can be used for addition, subtraction, division, multiplication, and power.

TOM'S TIP: After you have entered your Fortran code, either close the form or navigate to another tab, which serves to "save" the changes. Then reopen it again and check your code. If you forgot to hit enter anywhere (perhaps instead just tabbing away), your lines may not have been saved. It is quite old-fashioned and a common source of error.

Q4)   What is the resulting mole fraction of H_2 in the permeate?

If the whole system was at 5 bar pressure (I mean, the inlet streams, the reformer, heater, flash, and membrane are all at 5 bar, which admittedly is impossible but also is very convenient for now), answer these questions.

Q5)   What is the mole fraction of H_2 in the permeate?

Q6)   What is the total flow rate of the permeate in kmol/hr?

PART 3: MICROSOFT EXCEL AUTOMATION

You can actually connect Aspen Plus directly with Microsoft Excel. This is very useful in many cases. Sometimes, you just want a convenient way to get data from Aspen Plus to Excel, and this is a nice way of doing it. You can also use Excel to change things inside Aspen Plus. That means, if you know what you are doing, you can write your own custom model in Excel (perhaps with lots of equations!) and have it spit out a result to a block in Aspen Plus. Or, you can even write a program in Visual Basic for Applications (which is a programming language inside Excel) that runs Aspen Plus for you over and over, making changes as you go and recording the differences! We're not going to take it that far though, but it's good to know.

First, save and close all of your Aspen Plus simulations (you don't have to do this but it will make things less confusing later). Open a blank workbook in Excel and immediately save it to a new file. Look for the Aspen Simulation Workbook tab and click the Enable button. You should see a splash screen pop up for Aspen Simulation Workbook, and then when it's done loading, you should see the enable button change to a disable button, as shown in Figure 9.9.

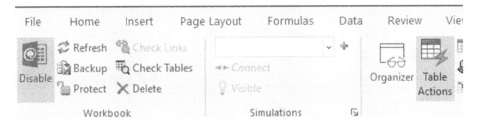

Figure 9.9 The Aspen Simulation Workbook tab in Microsoft Excel.

If you can't see the tab at all, in Excel, go to File | Options | Add-Ins and make sure that "Aspen Plus V12 64 bit Excel Calculator (ATL)" (or the 32 bit equivalent if that is what you use) and "Aspen Simulation Workbook V12.0" are both active. If you can't find them anywhere in your list, you'll have to select Excel Add-ins from the Manage drop-down, click Go, and then add them manually. Note that the Aspen Properties Excel Calculator is something different.

TOM'S TIP: You may have other issues if you have multiple versions installed, add-in protections, or certain security settings on your device. If so, there are some things you can try. First, try running the Aspen Excel Add-In Manager, which is a separate application installed in the Aspen Plus folder in your Start Menu. Make sure that the appropriate boxes are checked (see Figure 9.10 left). If you have older versions of Aspen Plus on your computer, make sure that the appropriate version that you are working with is enabled. Next, within Excel's File | Options | Add-Ins, if you see the items you want are inactive, you want to check the right boxes in the Manager for the COM Object Add-Ins. Depending on your version of Excel (for which we are using the desktop Excel App of Office 365 for this tutorial), you may find this at the bottom of the Add-Ins form, by choosing COM Add-Ins from the Manage dropdown box to the left of the Go button (see Figure 9.10 upper right). Finally, if the items you want are disabled (different than inactive), then select Disabled Items from the dropdown box, and then re-enable what you need on the next screen. In my example, that is `aswxladdinloader.dll` (see Figure 9.10 lower right), which means the Aspen Simulation Workbook Excel Add-In Loader. In many cases, you will have to close and reopen Excel for them to take effect because these add-ins are loaded upon startup, so you can surmise why you need the loader to be enabled.

Once it is enabled, you can connect to an Aspen Plus file by doing the following. First, click the Organizer button in Excel. In it, click Configuration | Simulations, then click the little green plus near the top to add a new simulation to the Excel workbook (see Figure 9.11).

Then, select the simulation workbook that you just finished for Part 2. This makes a link to it. Close the organizer. Now, back in the Aspen Simulation Workbook tab, click the Connect button under the name of your file to make it active. This basically loads the simulation in the background. You can't see it, but it's there in memory. Click the Visible box below it. After a brief pause, a window should pop up showing the file in the Aspen Plus application as you remember it. Now you have the power to run your simulation both from the main Aspen Plus program as normal and also through the arrow buttons in the Run toolbar of the Aspen Simulation Workbook tab in Excel.

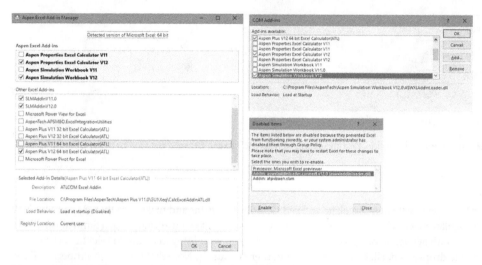

Figure 9.10   Different configuration settings for enabling the Aspen Simulation Workbook Add-In. *Left:* These are the correct settings for my machine in the Aspen Excel Add-In Manager (available as a standalone program in the Start Menu). This ensures that I can run V12 and not V11, since I have both installed together on my machine. *Upper right:* These are the COM Add-Ins that I need for my machine, in Microsoft Excel | File | Options | Add-Ins. *Lower right:* Some items may appear in the Disabled Items section of Excel's Add-Ins Manager. This may happen on machines with security settings, or if there was a problem using an add-in in a previous use of the program (in which case Excel blocks it from being loaded again). If Aspen Simulation Workbook mysteriously disappears from your ribbon in Excel, this is the first place to check.

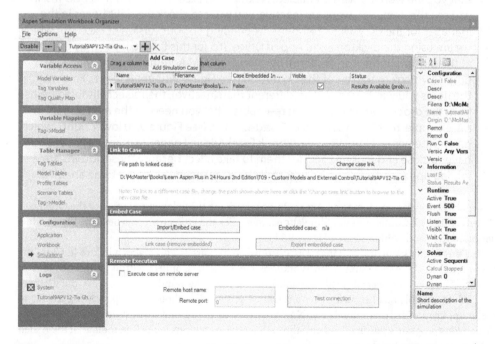

Figure 9.11   Connecting to an Aspen Plus file from within Aspen Simulation Workbook.

You can read variables from your simulation in Excel. Open the Organizer, and go to the Model Variables tab, and then click the binoculars on the ribbon. You can find the variables you need in the Simulations | *[Simname]* | appModel folder (you may have to expand some columns so you can actually see the names of the folders in the tight space of the form window). In there is a long list of stuff to which you can get access. You'll recognize the Streams and Blocks folders, for example. Drill down through the Streams folder until you find the MoleFlow of H_2 in the permeate of your membrane model. You'll find it in the *[StreamName]* | Output | Moleflow | Mixed | H2 (or your name for it) item (see Figure 9.12). Once you select it, hit Add Selected. In the Organizer window, there should be a column header called Status, which should indicate that this value is calculated by the model.

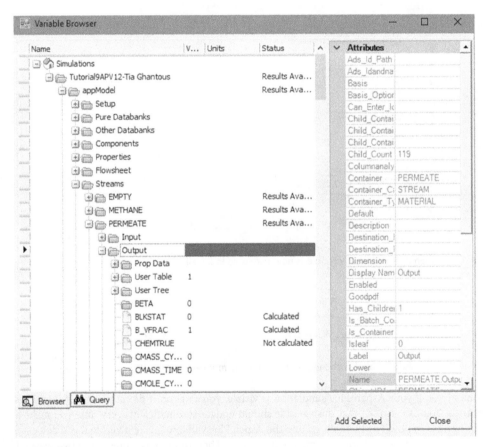

Figure 9.12   Finding variables to add to your Aspen Simulation Workbook.

That should make the variable appear in the list of Model Variables in the Organizer. Then, drag and drop this variable to a blank spot on the Excel worksheet to create a table (see Figure 9.13). This causes the Simulation Workbook Table Wizard to pop up. The table is basically an object in Excel that links directly to Aspen Plus. Just hit Finish (you can play around with the settings in the Table Wizard another time).

If you did it correctly, you should see some data show up, and if your simulation has been run, the number should appear! If it hasn't been run yet, you can run it either with the play button in the Aspen Simulation

	Value	Units	Status	Variable Name	Object ID1	Object ID2	Container	Container Type
PERMEATE	164.0742	kmol/hr	Calculated	MOLEFLO	MIXED	H2		PERMEATE MATERIAL

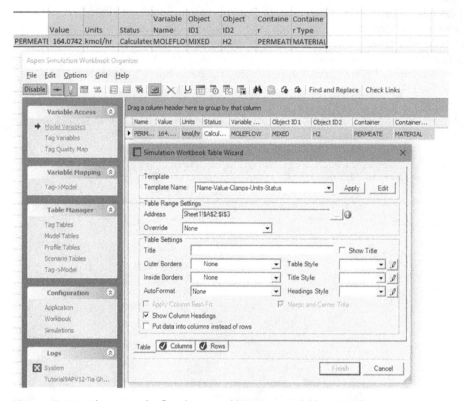

Figure 9.13 The Aspen Plus flowsheet variable is now available in Excel!

	Value	Units	Status	Variable Name	Object ID1	Object ID2	Container	Container Type
PERMEAT	180.4772	kmol/hr	Calculated	MOLEFLO	MIXED	H2		PERMEATE MATERIAL

Figure 9.14 The value updated after a change and rerun in the main Aspen Plus file.

Workbook Ribbon. Or, if your Aspen Plus simulation is visible, you can run it directly in the Aspen Plus window as normal. You can also see that this variable should update automatically every time you make a change to your simulation now. For example, go into the Aspen Plus window (make it visible if you don't see it) and change the flow rate of methane to some other number. Then run (choose one of the two ways). I changed mine from 200 to 220, and my number changed in Excel automatically (see Figure 9.14).

But, you can also make changes from inside Excel! Back in the Organizer, find another variable. This time find the flash drum temperature. It should be in the Blocks folder, under Flash | Input | Temp (because you have it as an Input degree of freedom specification at the moment). It should say Specified. That means you can change it. Again, click Add selected, and then drag and drop it into an empty excel area. In my case, the number 105°C shows up. Now, change this number to 160°C, click Run (the blue play button—right triangle—in the Aspen Simulation Workbook ribbon just under the Mode drop-down), and then watch the permeate H_2 flow rate change (I got 138.74 kmol/hr as my answer). You just ran Aspen Plus from Excel!

🎵 Music break[4]

---

[4]Recommended listening: *Tempus Fugit* by Miles Davis.

Capital Cost Estimation

Objectives

- Use the Aspen Capital Cost Estimator (formerly known as and still often referred to as Aspen Icarus) to generate cost estimates of chemical process equipment
- Use the features inside Aspen Plus which link your flowsheets directly to the cost estimator program to get cost estimates directly inside your simulation

Prerequisite Knowledge

This tutorial will show you how to estimate the capital costs of certain pieces of equipment, in this case, pumps, distillation columns, tanks, and heat exchangers, such as a kettle reboiler. So, you should have a basic familiarity with what those are, and completing the prior tutorials should be sufficient for this task. This tutorial only examines a small fraction of the cost models available and the features within Aspen Capital Cost Estimator, but it should be enough to give you an idea of how you can use it. For more, see AspenTech's User Guide which can be very helpful. There are also a few helpful video tutorials[1] on how to use the software.

Why This Is Useful for Problem Solving

Capital cost modeling is an important part of process engineering and plant design. Remember, the basic point of chemical engineering is to use a chemical reaction to create a commercial product from raw materials. A process systems engineer would then design a process to make that happen and then optimize to be the most profitable, almost always with the aid of a process simulator. So, we'll need this step to determine profitability. Cost matters!

---

[1]Dr. Patton at the Missouri University of Science and Technology has a good primer on the basics of using the economics feature within Aspen Plus: https://youtu.be/RSedanNo-10

To further illustrate the point, in chemical process design, we are often faced with trade-offs in which any number of possible designs could be "the best," but it really comes down to economics. For example, how many trays should you actually build in a distillation column? More trays usually mean lower energy costs but higher capital costs. Or, how large should my reactor be? A bigger volume usually leads to higher yields but with diminishing returns. Build it too small, and sure, the capital cost of that reactor is low, but my purification might be more difficult and my yield smaller, and so my overall cost per tonne of product actually goes up. How many stages should my compressor train have? What should my recycle ratio be? Is it better to compress my gas, or to condense it, pump it, and then vaporize it again? How much excess reactant should I use? Use a lot and my total product output might be higher, but then the downstream separation section might be much larger.

Over and over again, designers are faced with these questions, and although heuristics can help give us very good guesses, at the end of the day, it comes down to economics. The balance between capital costs and operating costs (including consumables, such as utilities, energy, and raw materials) very often determines our final design. It is the responsibility of the chemical process systems engineer to design a process that makes a good business case. And to do that, you need to have a good estimate of the cost. In this tutorial, you will learn how to estimate the capital costs portion of that.

Tutorial

PART 1: USING ASPEN CAPITAL COST ESTIMATOR AS A STAND-ALONE PRODUCT

Aspen Capital Cost Estimator is a beastly program, weighing in at a few gigabytes and containing an incredible amount of in-depth knowledge. Its purpose is to estimate the capital costs of common chemical process equipment. Costs are computed using a large database of detailed models of individual pieces of equipment, which is the most accurate method of estimation possible in the early stages of process design short of getting actual quotes. Other more traditional correlations are used to fill in the gaps in the data. Estimates are significantly detailed, which include labor costs to install (it varies depending on which part of the country/world you are in), what kind of ground you are putting it on (rocks? cement?), and how much paint you need for the outside. It also literally has a section called "nuts and bolts."

There are generally two ways to use the software. In this part, we will use the first way which is as a stand-alone product: Launch Aspen Capital Cost Estimator. It's not going to look anything like Aspen Plus. When it loads, it will ask you if you want to also load the Aspen Process Economic Analyzer (Yes).

First, you create Projects. A Project is basically a collection of pieces of equipment that are in your chemical plant. We'll start by creating a new project. If the default folder is no good for you, go to Tools | Options | Preferences | Locations and then Add your preferred directory to the list. Now, create a new project (File | New), pick a name, and put it in your new folder (see Figure 10.1).

Then on the next screen, select IP units. The default, IP, is inch-pound (also called "imperial"). Note that most American and Canadian companies still use IP for process equipment. For example, distillation columns are bought with diameters in standard sizes of 6-in. increments. If you want something that is 1 m in diameter (3 ft. 3.3 in.), that is a very expensive custom order.

Once you create the new project, you are immediately presented with a request to modify the "Input Units of Measure Specifications," as shown in Figure 10.2. Click on one, say Length and Area, and click Modify. This shows you the default measurements, as shown in Figure 10.3. You could, if you wanted, enter your own units here and a conversion. For example, if you want *pinky lengths* wherever inches are normally used, you could enter that here and put in the appropriate conversion amount. Let's not do this.

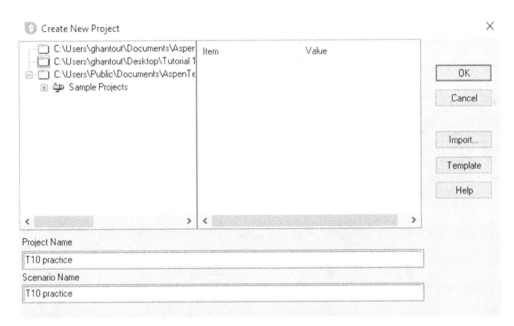

Figure 10.1   Creating a new project in Aspen Capital Cost Estimator.

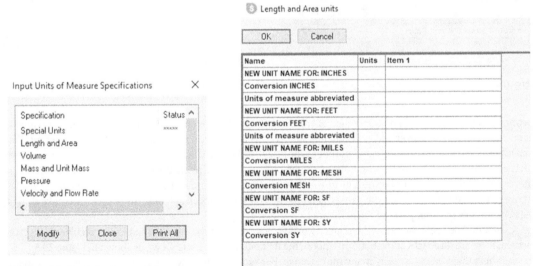

Figure 10.2   Selecting and changing the units of measure.

Figure 10.3   Modifying the default units for length and area.

After this (cancel and close), you are presented with the General Project data screen (see Figure 10.4). Here are defaults such as currency units, region, etc. We want to choose the United States as the Base country. In other words, all of their cost data are taken from American chemical plants and applications. However, suppose we are a Canadian company who will build this plant in Ontario, Canada, and thus prefer to

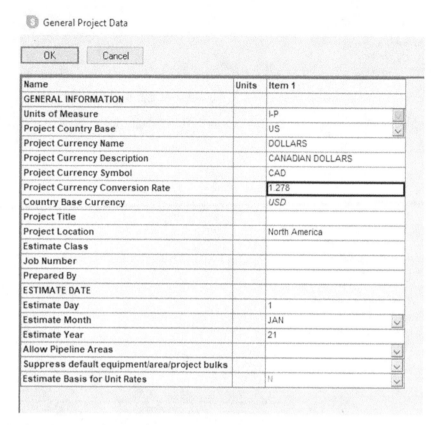

Figure 10.4 Changing the currency in the General Project Data form.

work in Canadian dollars. For convenience, you can change the currency description, symbol, and conversion rate. Enter in whatever today's exchange rate is or whatever you normally use for cost budgeting. For example, if you want to use the same number I did, 1.278 CAD = 1 USD, December 5, 2020, then type 1.278 in the box for Currency Conversion Rate, as shown in Figure 10.4. Update the description and other fields as necessary. At the bottom, enter the date at which you intend to purchase the plant (let's say January 1, 2021). It doesn't actually matter what the date is as far as the costs are concerned, but this is useful to make things easier to follow in other parts of the software.

Click OK. You are next shown the regular workspace screen. On the left column of your regular workspace screen, there are three tabs at the bottom. Choose the first tab (Project Basis View), as shown in Figure 10.5.

It is here that you can specify many more things. For example, go down to the Project Basis | Investment Analysis | Investment Parameters tab and double-click on it. Here we can change the key economic parameters like tax rate, desired rate of return, depreciation methods, etc. Change the tax rate from 40% (a typical U.S. amount is 35% federal + 5% state but it varies by state) to 26.5% (a typical amount is 15% Canada Federal + 11.5% Ontario Provincial) and click OK. Also, as we are assuming Canadian costs, we also need to bump our labor costs up. Double-click the Project Basis | Investment Analysis | Operating Unit Costs tab and bump operators from 20 to 40 $/hr and supervisors to 60 $/hr (again these are Canadian dollars). Also, set the electricity price to 15 ¢/kWh (0.15 $/kWh), which was the average Ontario's mid-peak price at the end of 2020, and click OK.

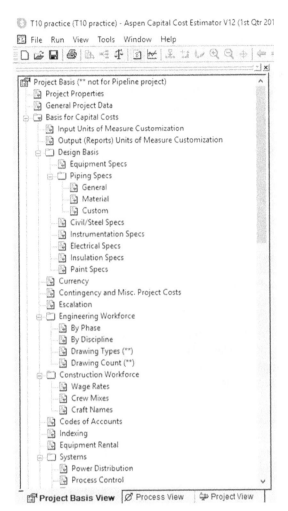

Figure 10.5 The Project Basis View.

Similarly, we can change the cost indexing, that is, how much more we have to pay than the base cost due to inflation and changes in the market. In the version used in this edition (V12), the base costs in the database are for the first fiscal quarter of 2019, and you can check yourself by looking in the title bar of the window of the program when you first open it. Because we left the Project Country Base as the United States, it will use its database of prices for things sold in the United States in the first quarter of 2019. If we wanted, the program also has databases for the United Kingdom, Japan, the European Union, or the Middle East as well. Let's assume that right now in the first quarter of 2021, Americans have to pay 5% more for equipment than they did in 2019, and Canadians have to pay 10% more than in the United States even adjusting for the exchange rate. This means that we are assuming that our 2021 Canadian costs are $1.1 \times 1.05 = 1.155$ (or 15.5%) more than the basis costs for the United States in 2019. Aspen Plus defines the base factor as 100 for the base case. So for a 15.5% increase in cost, we need to change the index for equipment to 115.5.

TOM'S TIP: When making adjustments for international projects, you can use the Purchasing Power Parity Index (PPPI) published by the Organisation for Economic Co-operation and Development (OECD), which gives you information about how the price of the same goods and services varies from country to country. According to this information, Canada's PPPI in 2020 indicated that, on average, things in general were slightly less expensive in Canada than in the United States (i.e., it would take fewer USD to buy something in Canada than in the United States, assuming that USD and CAD could be exchanged without penalty at the average bulk market rates for 2020). However, this neglects customs, duties, and shipping costs for goods manufactured in the United States and delivered to Canada, and thus the overall cost should usually be budgeted higher, and hence, the cost increase 10% assumption. Your own specific cases and applications will be different, and you can consider PPPI, customs, duties, and accessibility to materials when formulating your own assumptions.

TOM'S TIP: Although I had assumed things would cost 5% more than they had in 2019, you can make more data-driven assumptions using published cost indices. For example, *Chemical Engineering* magazine publishes the Chemical Engineering Plant Cost Index (CEPCI), which tracks costs of chemical plant equipment in various categories over time on a monthly basis. Similarly, IHS Markit regularly publishes cost indices over time for different applications in energy, such as oil and gas. A paid subscription to this information is often required.

Right-click on Project Basis | Basis for Capital Costs | Indexing and choose `Select`. You are picking between different index files. Just pick the default and click OK; it's too complex to go into this further.

Now right-click Basis for Capital Costs | Indexing item again, choose `Edit`, as shown in Figure 10.6. Select Material and hit Modify. Now you can see that "100" is the basis for all of these, so change Equipment to 115.5. Modify the rest and say that Piping should be 12% higher, Civil is 22% higher, Steel is only 5% higher, and all the rest are 15.5% higher,[2] as shown in Figure 10.7. Then click OK and Close.

Now that the base cost information is added, we can start adding and costing equipment to our plant. Switch to the Project View tab (third on the bottom right of the left column). It will show that you have a Main Area inside of your Main Project. Projects are like folders, you just group everything you are working on into one or more projects. Areas are geographical areas of your chemical plant, as in maybe the west wing of your factory, or some fenced-in place outside, etc. You assign pieces of equipment[3] to an Area.

On the right window pane, you should see the tab options for Projects, Libraries, Components, and Templates. Go to the Components tab. This is where all of the equipment models are located. Start by adding a `Centrifugal single or multi-stage pump`, as shown in Figure 10.8. You'll find it under Process equipment | Pumps | Pump-Centrifugal | Centrifugal single or multi-stage pump. To add it, drag and drop the icon into the whitespace in the middle column.

---

[2]By the way, did I mention that this software is detailed? You can even specify the number of coats of paint you put on each pipe...Project Basis | Basis for Capital Costs | Design Basis | Paint Specs...

[3]By the way, "equipments" is not a word. This surprises many English learners because you see it used so many times incorrectly. Instead, say "pieces of equipment."

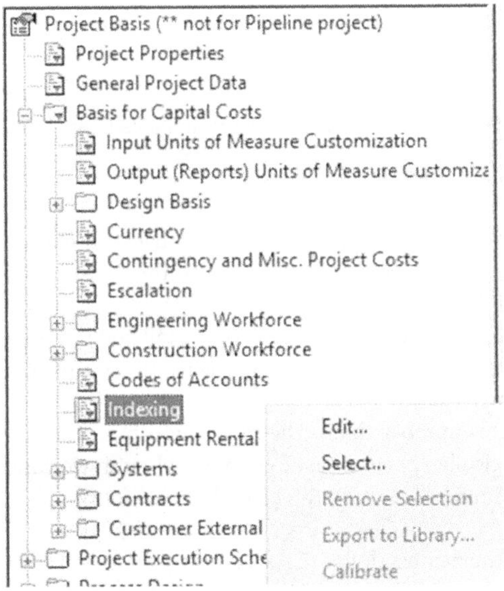

Figure 10.6 Getting access to custom cost indexes.

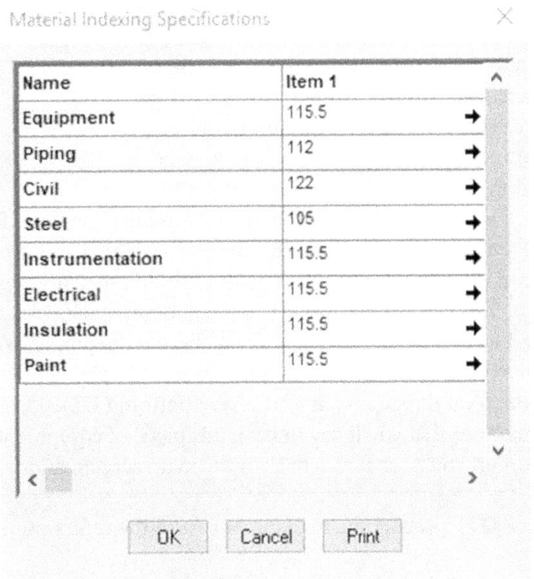

Figure 10.7 Editing the cost indexes.

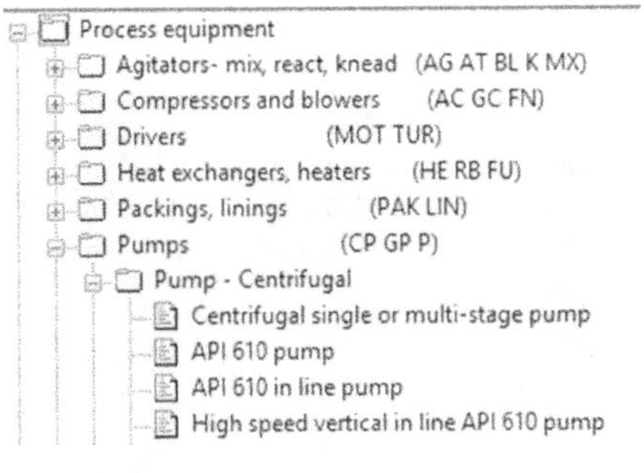

Figure 10.8 Adding a pump to your project.

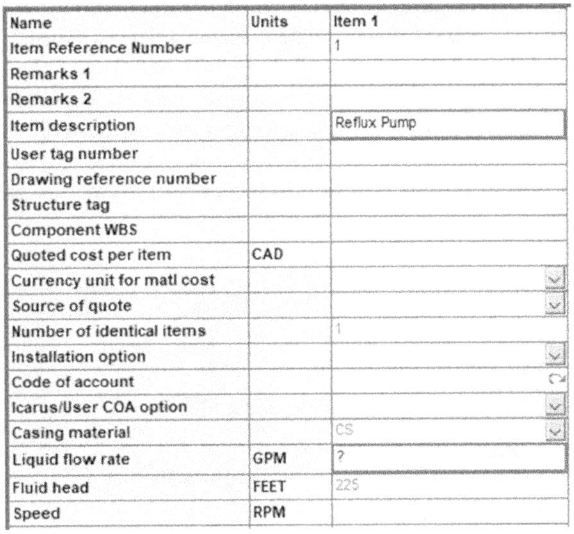

Figure 10.9 Editing the design parameters for the pump.

Give it a name such as Reflux Pump for the Item description. You are now presented with a form where you can fill in all sorts of information to ridiculous levels of detail, as shown in Figure 10.9.

The red boxes are items which must be entered before proceeding. The boxes with blue text are items which must be entered for Icarus to calculate the cost, but have a default option selected for you. The empty boxes are optional but can also be factored into the cost if you have that information available.

For this pump, change the casing material to stainless steel and update the flow rate, fluid head, and design gauge pressure according to the diagram on the next page. When ready, click OK. Your middle column on the main view should have something similar to Figure 10.10.

Item	User Tag Num...	Item Description	Model
■ 1		Reflux Pump	CP CENTRIF

Figure 10.10 The reflux pump appears in the item list.

Now, let's ask the program to compute the cost. Right-click on the pump in the item list, and choose Evaluate Item. ACCE will run something and produce an Item Report. Scroll down to the bottom, and see the equipment summary. You should see something similar to Figure 10.11.

You can see that while the actual pump itself costs $40,200 (CAD), it costs $2,422 to install and required 54 worker-hours[4] to do so. Then, there is the piping to connect it to the other parts of the plant, instruments such as flow meters, electrical wiring, and paint. The total material and installation labor cost, also known as the total direct cost, is at the very bottom ($75,100). It is this number that is the most important. It is the number that you'll pay to have this piece of equipment magically appear in your chemical plant in working order. You'll see it also back in the main screen, middle column, by selecting the List tab at the bottom.

Q1) Report the total direct cost of the reflux pump to the nearest dollar (CAD).

Similarly, add the remaining equipment, as shown in Figure 10.12: the condenser, reboiler, reflux drum, and distillation column. Use the specifications given in the figure, and leave anything else at their default values.

The trayed tower (DTW TRAYED) model should be used for distillation, which includes the trays but does not include the condenser, reboiler, or reflux pump. It is located at Process equipment | Towers, columns-trayed/packed | Tower-single diameter | Trayed tower. Change the Application to Distillation with kettle reboiler (DIS-RB).

For the condenser, you can use a Pre-engineered U-tube exchanger (DHE PRE ENGR). It is located at Process equipment | Heat exchangers, heaters | Heat exchanger | Pre-engineered (standard) U-tube exchanger.

The reflux drum is a vertical process vessel (DVT CYLINDER). It is located at Process equipment | Vessel-pressure, storage | Vessel-vertical tank | Vertical process vessel. In this case "height" is "tangent to tangent height."

```
                                                                 L/M
                    :---MATERIAL--:*** M A N P O W E R ***: RATIO  :
                    :    CAD      :    CAD      MANHOURS   :CAD/CAD :
EQUIPMENT&SETTING : 40200.      :   2422.          54    : 0.060 :
PIPING            : 18969.      :   6736.         156    : 0.355 :
CIVIL             :   818.      :   1401.          41    : 1.713 :
STRUCTURAL STEEL  :     0.      :      0.           0    : 0.000 :
INSTRUMENTATION   :   419.      :    124.           3    : 0.295 :
ELECTRICAL        :  1931.      :   2104.          49    : 1.089 :
INSULATION        :     0.      :      0.           0    : 0.000 :
PAINT             :     0.      :      0.           0    : 0.000 :
                  ------------------------------------------------
SUBTOTAL          : 62336.      :  12786.         303    : 0.205 :

INSTALLED DIRECT COST    75100.      INST'L COST/PE RATIO  1.868
========================================================================
```

Figure 10.11 An example Item Report for the pump.

---

[4] Hopefully, readers will forgive some of the legacy gendered language coded into the software.

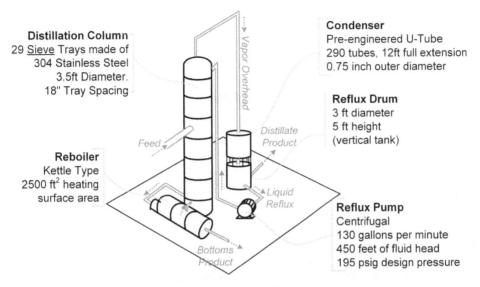

Distillation Column
29 Sieve Trays made of
304 Stainless Steel
3.5ft Diameter.
18" Tray Spacing

Condenser
Pre-engineered U-Tube
290 tubes, 12ft full extension
0.75 inch outer diameter

Reflux Drum
3 ft diameter
5 ft height
(vertical tank)

Reboiler
Kettle Type
2500 ft^2 heating
surface area

Reflux Pump
Centrifugal
130 gallons per minute
450 feet of fluid head
195 psig design pressure

Feed

Distillate Product

Liquid Reflux

Vapor Overhead

Bottoms Product

Figure 10.12 The distillation area of your chemical plant.

For the reboiler, use "Kettle type reboiler with floating head" (DRB KETTLE). It is located at Process equipment | Heat exchangers, heaters | Reboiler | Kettle type reboiler with floating head.

Q2) Report the total direct cost of the column (including trays) to the nearest dollar (CAD).

Q3) Report the total direct cost of the condenser to the nearest dollar (CAD).

Q4) Report the total direct cost of the reflux drum to the nearest dollar (CAD).

Q5) Report the total direct cost of the reboiler to the nearest dollar (CAD).

Then, once the individual pieces of equipment are added, you can run an economic analysis for the whole plant which uses them. This includes labor, operations, utilities, maintenance, loans, taxes, inflation, and investments. We will not go into this now. We will do one more thing though. Let's look into the depth of the calculations. When you have finished adding the equipment, click the Evaluate Project button in the toolbar and select Evaluate All Items, and let it do its magic (create a report). Note that you'll get an error message. It's okay for now as we are not designing a real plant and didn't go into a lot of details. Just click continue for the Scan Messages window, and close for the Capital Cost Errors window.

A new Report Editor window pops up in which Aspen gives you a suggested build-out plan for your plant containing this equipment (Mine is called CAP_REP.ccp—Report Editor). From the report we can see that Aspen is using vendor quotes from the first quarter of 2019, as shown in Figure 10.13.

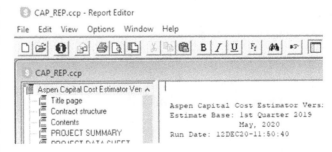

Figure 10.13 An example Capital Cost Report.

Q6) Go to the Project Schedule section (double-click on it) and determine how many construction workers you can expect to hire in week 5. (Each dot on the week column represents one person.)

🎜 Music break[5]

PART 2: INTEGRATED ECONOMICS IN ASPEN PLUS

Capital cost estimates can be directly integrated with Aspen Plus V12 in two ways. You can either export an Aspen Plus flowsheet into Aspen Capital Cost Estimator, or you can have capital costs predicted right in Aspen Plus itself. We will do the latter briefly here.

Figure 10.14 shows a very simple distillation of an 80/20 mixture of ethanol and butanol using an ordinary distillation column. Simulate the column in Aspen Plus using a `RadFrac` model for the distillation column and `NRTL-RK` for the property method. Run the simulation first and ensure that it converges correctly. Now let's make the Economics Active. If you haven't yet, go to the Economics ribbon and check the box for Economics Active (see Figure 10.15). Then, go to the Cost Options button on the ribbon (or Simulation | Setup | Costing Options). You'll see that you can enter in some of the basics that you could in Aspen Capital Cost Estimator. So go ahead and change the start of basic engineering to January 1, 2021. Although you have the ability to enter a currency symbol and conversion rate on the Currency tab, it does not get considered in the economic analysis. It will only work in USD (by default) while inside Aspen Plus, or you can select one of

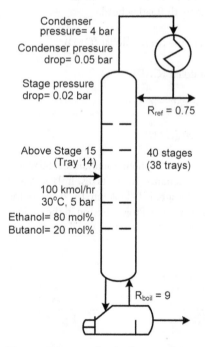

Figure 10.14 The distillation column used in this example.

---

[5]Recommended listening: *Kor e alle helter hen* by Jan Eggum.

the other built-in templates (European Union, Japan, China, the United Kingdom, or the Middle East) by changing the template on the Costing Options tab. To use currencies outside of those options, you should use the stand-alone Aspen Capital Cost Estimator application instead or make the conversion from USD to your units after the fact.

Figure 10.15 Activating the economics feature within Aspen Plus.

At this point, we need to map our simulation models to actual pieces of equipment. For example, our RadFrac model is just a set of equations which can represent many things (adsorption, distillation, extraction, rectification, stripping), so you have to map the simulation equations to a physical piece of equipment (or multiple pieces in this case) in the database.

So click Map in the Economics ribbon (you may need to rerun the simulation first). You'll get a Map Options prompt (see Figure 10.16). In this case, you want to use the Default basis, and you want to size the equipment and evaluate the cost. Sizing the equipment is an important step; it means that your simulation results are used to compute the sizes of the equipment (e.g., the length and diameter of the reflux drum of the distillation column).

You should see that Aspen Plus maps the column and supporting equipment collectively modeled in the RadFrac block to a Trayed Column (DTW TOWER), a condenser (DHE TEMA EXCH), a horizontal drum (DHT HORIZ DRUM), a centrifugal pump (DCP CENTRIF), two splitters (C), and a reboiler (DRB U TUBE). This is the result of the Standard configuration chosen by default (i.e., choosing the Default basis on the Map Options form). Switch to the Full – Split w/Circ. configuration. The mapping should then change to include more pumps, pre-coolers, etc. Let's change the reboiler to a different model. Select the DRB U TUBE item and change it to DRB KETTLE (Kettle type reboiler w/floating head) by selecting from the list, like was done in Part 1 (see Figure 10.17).

When you are done, click OK on the map preview page. You might get another prompt about custom sizing, if you checked that box by accident. Just leave it and click OK. You should see some familiar prompts.

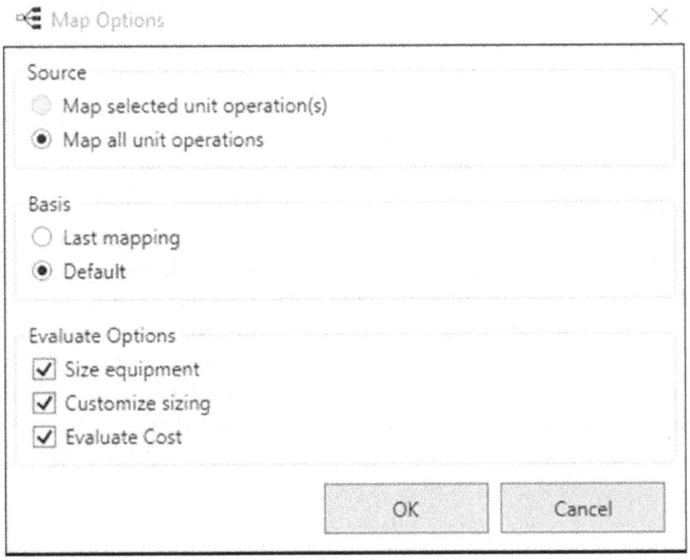

Figure 10.16 The Map Options form.

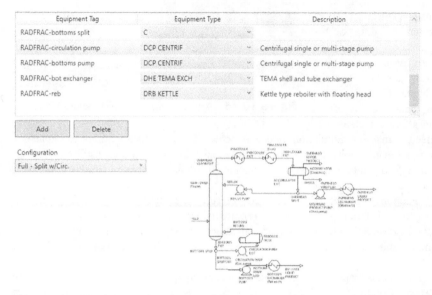

Equipment Tag	Equipment Type	Description
RADFRAC-bottoms split	C	
RADFRAC-circulation pump	DCP CENTRIF	Centrifugal single or multi-stage pump
RADFRAC-bottoms pump	DCP CENTRIF	Centrifugal single or multi-stage pump
RADFRAC-bot exchanger	DHE TEMA EXCH	TEMA shell and tube exchanger
RADFRAC-reb	DRB KETTLE	Kettle type reboiler with floating head

Add   Delete

Configuration
Full - Split w/Circ.

Figure 10.17 Changing the mapping of the RadFrac model to a new configuration.

If it works then you should see the items checked in the ribbon, shown in Figure 10.18.

Let's see the results! Hit View Equipment in the Economics ribbon. Explore the tabs, see what it comes up with and answer the following questions. Note that the Sizing step takes your simulation results and then does more calculations to determine how these translate into physical dimensions, heights, widths, etc.

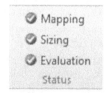

Figure 10.18 The three checkmarks in the ribbon indicate that the economics computation is complete.

Q7) What is the total installed cost of all of the equipment (in USD)?

Q8) The column in the simulation used 40 equilibrium stages. How many trays does this translate to (this incorporates inefficiencies, etc.) according to the result?

Q9) What is the column diameter it calculates?

Rerun the simulation using an inlet flow rate of 200 kmol/hr instead of 100 (doubling the capacity of the system). Then, when that is finished, hit Size in the Economics ribbon to resize everything and be sure to reevaluate the cost as well. Keep the "last mapping," which means that your reboiler configuration change from DRB U TUBE to DRB KETTLE is remembered from when you did it last time. Confirm that the Full – Split w/Circ. configuration option is still selected (if it isn't, reselect it). Notice from the result that the installed cost is significantly less than double even though we doubled the capacity. This is because of economies of scale, and fundamentally, why most chemical plants are so gigantic. The larger your plant, the more competitive your costs can be.

Q10) What is the new column diameter? Notice that the number of trays should remain the same.

---

[6]Recommended listening: *Peggy-O* by Tony Furtado.

♫ Music break[6]

Optimal Heat Exchanger Networks

Chinedu O. Okoli and Thomas A. Adams II

Objectives

- Use Aspen Energy Analyzer (AEA) to design heat exchanger networks (HENs)
- Learn to import thermal data of process and utility streams from Aspen Plus into AEA

Prerequisite Knowledge

You should now be familiar with heat exchanger basics, such as how heat duties and heat exchanger areas are calculated. You should also be able to differentiate between a hot stream and a cold stream, and understand what utilities are. You can review the prior tutorials related to heat exchangers (Tutorial 4) and utilities (Tutorial 6) to refresh your knowledge. You will also need to be able to model plug flow reactors (Tutorial 7) and equilibrium-based distillation (Tutorial 5). It might also be helpful to review the design specs and sensitivity features (Tutorial 3) for this tutorial as well.

Why This Is Useful for Problem Solving

After the design of a plant to meet quantity and quality specifications of a product, another important design phase is the design of a HEN. As the operating costs associated with utility usage can be a significant contributor to the cost of production, it is important to figure out ways to reduce these costs. A good HEN design seeks to accomplish this by utilizing heat integration techniques to improve energy recovery among process streams, and thus reduce the heating and cooling supply from utilities.

The concepts behind HEN design are very important in industrial practice, as many case studies have shown that energy savings of up to 30–50% in comparison to traditional practice are possible.

The different methods of HEN design aim to either minimize the utility usage of the process (maximize the heat recovery), or minimize the total cost of the heat exchangers and utility usage. Both methods could produce different results, and the choice of either one will depend on the overall design objectives. For example, if the cost of utilities is really high in comparison to the cost of heat exchangers then the objective of minimizing utility usage would be preferable. On the other hand, if the capital costs of heat exchangers are way higher than the costs of the utilities then it would be preferable to minimize the total cost of the heat exchangers. In addition, it may be better to minimize other aspects such as total annualized costs (TAC) (balancing both capital and energy while considering many business factors) or simply the number of heat exchangers that exist.

As a chemical engineer working on HEN designs, your knowledge and understanding of the concepts and methods that guide HEN designs will be critical for reducing process costs. The idea of minimizing the utility usage of a process is based on a concept called pinch analysis. The idea behind pinch analysis is to figure out where the most difficult heat exchange point (the point with the smallest temperature difference between the hot and cold streams, also known as the approach temperature) in the process exists and then start the HEN design from this point. This method is not covered in this tutorial since it is somewhat out of date, but you can look at these introductory videos,[1] if you are interested.

This tutorial will focus on the preferred method of minimizing the total cost of the HEN (this includes the capital cost of the heat exchangers and the operating costs of the utilities). It is an optimization-based approach, and you will learn how to use AEA to develop HEN designs.

Tutorial

BACKGROUND

There are three main steps in designing a HEN: data extraction, utility selection, and HEN design.

In AEA, data can be extracted from simulation files such as Aspen Plus or Aspen HYSYS, from Microsoft Excel, or entered manually. In Part 1 of the tutorial, we will be considering manual data entry, while in Part 2 we will look at automatic data extraction from Aspen Plus. In both parts of the tutorial, we will also be selecting utilities and doing the HEN design in AEA.

PART 1: GENERATING HENs USING MANUAL PROCESS STREAM DATA ENTRY IN AEA

Table 11.1 contains process stream data from which a HEN is to be built. As you recall, hot streams refer to process streams that require cooling, while cold streams refer to process streams that require heating. We will enter the data shown in Table 11.1 into AEA as a first step to building our HEN.

Load the AEA software. Once AEA loads, save a new file. Now go to Managers on your menu bar and click on Heat Integration Manager. A window pops up as shown in Figure 11.1.

Click on HI Project and Add. Next, right-click on HIP1 and select Add Scenario. A small window pops up and indicates the scenario should be named. I'll call mine `Example 1` (see Figure 11.2).

---

[1]Temperature interval method for heat exchanger networks. https://www.youtube.com/watch?v=7PpysQMD0WE. For designing a heat exchanger network, see https://www.youtube.com/watch?v=xZO2aSiakuw. This is peer-reviewed material produced by the University of Colorado, Boulder. On the page, navigate to the heat exchangers section for the video links.

Table 11.1	Process Stream Data			
Stream	**T_{in} (°C)**	**T_{out} (°C)**	**Mass-Heat Capacity (kJ/°C-hr)**	**Enthalpy (kJ/hr)**
Hot 1	320	220	17,280	
Hot 2	260	120		1,512,000
Hot 3	140	139	1,260,000	
Cold 1	120	220	11,520	
Cold 2	70	240	5,760	
Cold 3	170	220	29,520	
Cold 4	235	236	1,260,000	

Figure 11.1   The heat integration manager.

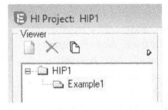

Figure 11.2   My heat integration project has one scenario in it.

Select your scenario (Example 1 in my case) and then the Data tab at the bottom left of your window, as shown in Figure 11.3. In the Data window, select Process Streams. A table view appears on the bottom right of the window in which you can enter your process stream data from Table 11.1.

In the form shown, enter the information given in Table 11.1. Start with Hot 1. Under the name column, click on **New** shown with the blue text. Type Hot 1 as the name of the new process stream.

Also, enter the inlet temperature (Inlet T), outlet temperature (Outlet T), and mass heat capacity[2] (MCp) data from the table. Notice that the enthalpy and heat transfer coefficient (HTC) fields become populated, as shown in Figure 11.4. The enthalpy is calculated based on the temperature and heat capacity information you provided, while the HTC is set at a default AEA value. Also, notice the downward-pointing red arrow in the Hot 1 row. AEA uses it to indicate that the stream is a hot stream whose temperature will be going down. For a cold stream, you will see a blue arrow pointing upward.

---

[2]The mass heat capacity is typically computed as the flow rate of the stream times its heat capacity assuming a constant heat capacity. Since heat capacity usually changes with temperature, the heat capacity number used for this calculation is usually either the average heat capacity (halfway between the heat capacities at the two temperature extremes), or even more accurately, an integral average heat capacity. The "enthalpy," as used in this table, is really the change in enthalpy of the stream: the mass flow rate times the integral of heat capacity over the temperature range. For a constant heat capacity, this is equal to the mass heat capacity times the change in temperature.

Q1) What is the enthalpy (or technically, the enthalpy change) of Hot 1 in kJ/hr? Oh, it's right there.

Now enter Hot 2 in the same way as Hot 1, but enter the enthalpy instead of the heat capacity. AEA will calculate the heat capacity for us. In AEA, you can enter either the heat capacity or enthalpy data (whichever is available).

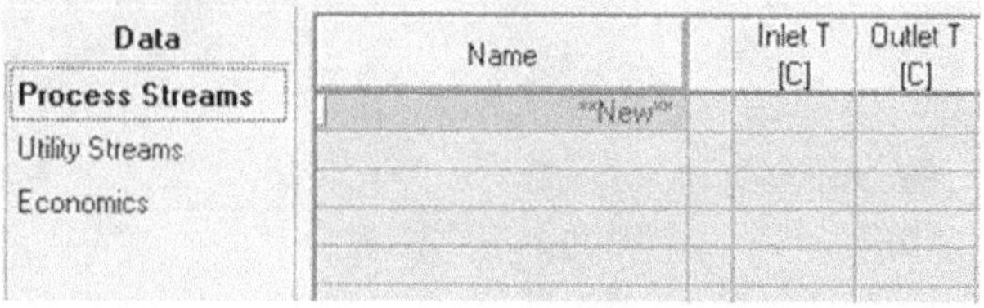

Q2) What is the mass heat capacity of Hot 2 in kJ/°C-hr?

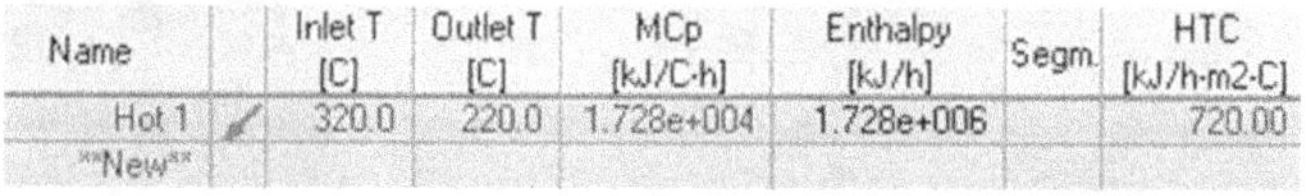

Ok, that was easy. In the same way, enter the remaining information shown in Table 11.1.

The next step is to choose your utility streams and enter the required information for them. When designing a HEN, utilities are required to supply any additional cooling or heating demands that cannot be met by matching hot and cold process streams together. Enter the utility information under Data | Utility Streams.

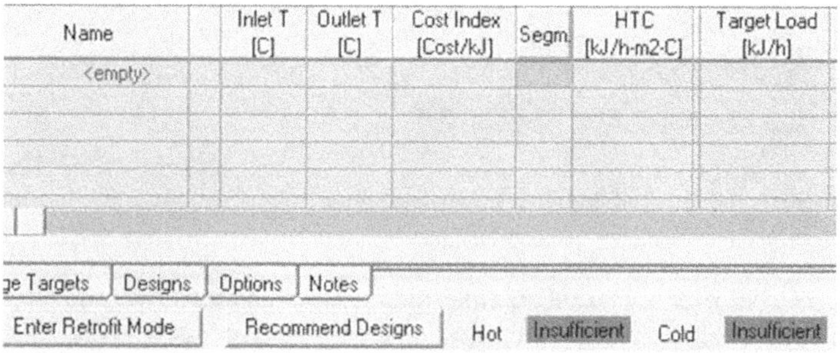

In the utility streams section, we can select the hot and cold utility streams. Take a look at the bottom of your AEA window and note that the Hot utility (Hot) and Cold utility (Cold) statuses are labeled Insufficient in

Figure 11.3 You can enter your own stream information in the Data tab manually.

Figure 11.4 Enter in the known information about a stream, and the software will fill in the missing columns when enough information becomes available. The red arrow indicates that this is a hot stream that needs to go down in temperature.

Figure 11.5 Add utilities to your scenario which the optimizer will use only when it cannot find enough heat or cooling available from the available process streams to meet all of the design objectives.

red (see Figure 11.5). This is because the process still requires external heating (for the cold process streams) and external cooling (for the hot process streams) in order to reach their specified temperature.

Now add a cold utility and see what happens. On your screen, under the name column, click on the drop-down for <empty> shown and select Cooling Water. Note that the cold utility status is now labeled as Sufficient in green, as shown in Figure 11.6. This is because you have selected cooling water, which is at a temperature cold enough to cool the hot process streams (second law of thermodynamics). So now it is physically possible for you to meet all of your temperature change objectives. In practice, be sure to select appropriate utilities that meet your temperature requirements but don't cost more than necessary (see Tutorial 6).

Add a hot utility. Select LP Steam (low-pressure steam) from the drop-down. Note that unlike the cooling water utility, LP steam is not hot enough to supply all the heat requirements of the process (see Figure 11.7). What do you think the reason is? Take a look at the inlet and outlet temperature of LP steam, and compare it to any of the cold process streams in the process stream data table. You will notice that LP steam can supply some of the heat for some parts of the cold process streams, say for Cold 2, but cannot supply the rest because its outlet and inlet temperatures are lower than the other cold streams. This means that hot utilities which are "hotter" than the cold streams are required.

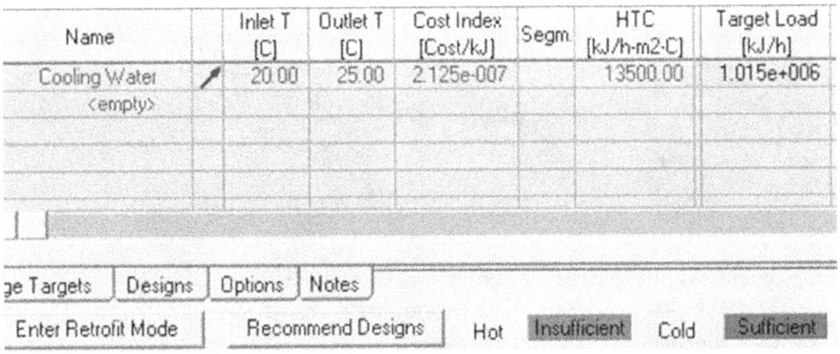

Name		Inlet T [C]	Outlet T [C]	Cost Index [Cost/kJ]	Segm.	HTC [kJ/h-m2-C]	Target Load [kJ/h]
Cooling Water	/	20.00	25.00	2.125e-007		13500.00	1.015e+006
<empty>							

ge Targets | Designs | Options | Notes

Enter Retrofit Mode | Recommend Designs | Hot Insufficient Cold Sufficient

Figure 11.6 The addition of the cold utility makes it possible to cool the Hot 1 stream that currently exists in the scenario. Because the cold utility is cold enough to use for Hot 1, the cold utilities are now "sufficient" to do the job.

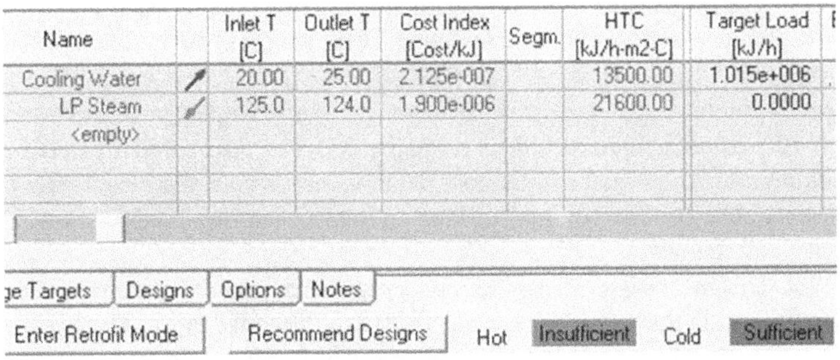

Name		Inlet T [C]	Outlet T [C]	Cost Index [Cost/kJ]	Segm.	HTC [kJ/h-m2-C]	Target Load [kJ/h]	
Cooling Water	/	20.00	25.00	2.125e-007		13500.00	1.015e+006	
LP Steam	/	125.0	124.0	1.900e-006		21600.00	0.0000	
<empty>								

ge Targets | Designs | Options | Notes

Enter Retrofit Mode | Recommend Designs | Hot Insufficient Cold Sufficient

Figure 11.7 This hot utility is not hot enough to heat up all process streams, so the available utilities are insufficient for our heating needs.

Ok, add HP Steam (high-pressure steam) as a hot utility. You will now notice that the Hot utility status at the bottom of your screen has now changed to Sufficient, with a green color. The temperatures of some of the hot process streams in Table 11.1 indicate that it is possible to generate LP steam, so add LP Steam Generation as a cold utility. Finally, your utility stream table should look as shown in Figure 11.8.

Finally, in the data tab, you can edit the economics information by making changes in Data | Economics, as shown in Figure 11.9. The information provided here is used to compute the TAC of the HEN with this equation:

$$TAC = Operating\ cost\ per\ year + (Annualization\ Factor \times Capital\ Cost)$$

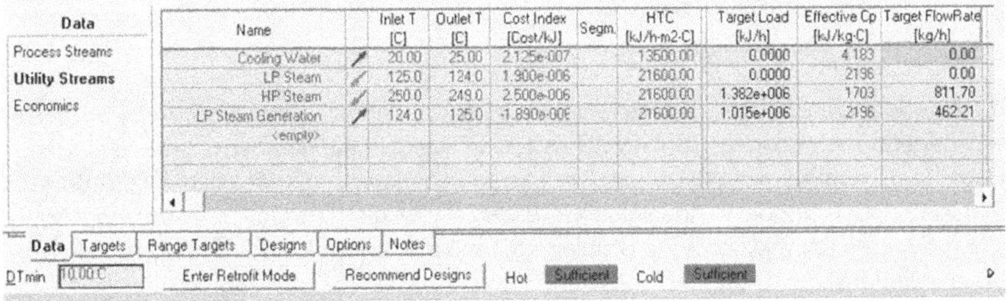

Data	Name		Inlet T [C]	Outlet T [C]	Cost Index [Cost/kJ]	Segm.	HTC [kJ/h-m2-C]	Target Load [kJ/h]	Effective Cp [kJ/kg-C]	Target FlowRate [kg/h]
Process Streams	Cooling Water	/	20.00	25.00	2.125e-007		13500.00	0.0000	4.183	0.00
Utility Streams	LP Steam	/	125.0	124.0	1.900e-006		21600.00	0.0000	2196	0.00
Economics	HP Steam	/	250.0	249.0	2.500e-006		21600.00	1.382e+006	1703	811.70
	LP Steam Generation	/	124.0	125.0	-1.890e-006		21600.00	1.015e+006	2196	462.21
	<empty>									

Data | Targets | Range Targets | Designs | Options | Notes

DTmin 10.00 C | Enter Retrofit Mode | Recommend Designs | Hot Sufficient Cold Sufficient

Figure 11.8 You can also generate steam from boiler feed water as a cooling utility.

Heat Exchanger Capital Cost Index Parameters				
Name	a	b	c	HT Config
DEFAULT	1.000e+04	800.0	0.8000	Heat Exchanger
New	....	....	....	

Capital Cost Index(Heat Exchanger) [Cost] =a+b(HeatExch Area/Shells)^c*Shells
Capital Cost Index(Fired Heater) [Cost] = a + b(Fired Heater Duty)^c
Capital Cost Target [Cost] =a(Min. for MER) +b(Area/Shells)^c*Shells

Annualization
Rate of Return (%): 10.0 ROR
Plant Life (years): 5.0 PL
Annualization Factor 0.2638 AF
(AF = [(ROR/100)*(1 + ROR/100)^PL]/[(1 + ROR/100)^PL -1])
Operating Cost
Hours of Operation: 8765.76 (hours/year)

Figure 11.9 You can change the parameters of the economic analysis, such as your target plant life and required rate of return. The capital cost index parameters correspond to the coefficients of a polynomial that represents capital cost as a function of heat transfer area.

The TAC is a useful way to compare different HEN designs, since it incorporates a balance of both capital (one-time) and energy (ongoing) costs. By default, the AEA designs are determined using an optimization algorithm in which the objective is to minimize TAC.

The Heat Exchanger Capital Cost Index Parameters are used to calculate the cost of the heat exchangers based on their attributes, such as heat exchanger area and number of shells. The Annualization Factor is calculated from the rate of return (ROR) and plant life (PL) time, while the capital cost is the total cost of all the heat exchangers in the HEN.

This information can be changed if you have data that you prefer to use. For example, you can have a longer lifetime for the plant such as 10 years or a higher ROR such as 15%. It is also possible that you have real plant data or vendor information about the actual capital cost parameters for your heat exchangers. Also, the hours of operation for the particular process you are working on might be known. Let's work with the AEA default values, so don't make any changes.

The next step is to design the HEN. This means we will ask AEA to try to match process streams to process streams, and process streams to utilities in the best way possible, that is, to minimize the TAC. Click on Recommend Designs at the bottom of your window. A window pops up called Recommend Near-optimal Designs, as shown in Figure 11.10.

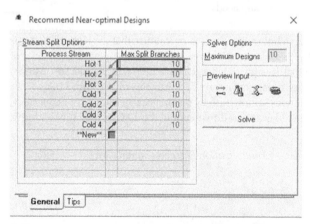

Figure 11.10 The recommended near-optimal designs feature uses an optimization approach to try to match your streams using heat exchangers. The maximum split branches option refers to the number of times a stream can be split into smaller pieces (e.g., it might want to use one very large heat source to heat lots of little streams by breaking it into pieces). If the solver is unable to find solutions, especially when the problems are large, try turning this number down to reduce the complexity of the problem and make it easier to solve (though possibly missing out on potentially better designs). Note also that these are "near" optimal designs. The best design reported may or may not be the true global optimum design. Even if it is not, it usually isn't very far off, and it almost always is way better than what you could have come up with on your own.

Check to see that in the Stream Split Options table the maximum split branches of all the process streams is set to 10. This value can be more or less, but leaving it at 10 allows AEA to have a good number of options for matching streams without making the problem too complex to solve. Leave the Maximum Designs under Solver Options as 10. Again, this value can be more or less but 10 is a good value to choose to start. Click Solve.

The AEA solver runs and generates 10 different designs which you can see in your scenario folder at the top left section of your window. If you go through the 10 designs, you will notice the green bars at the bottom of the HEN diagrams.

Green indicates that all heat exchanger matches are feasible, and the heat requirements (heating and cooling) of all the process streams are satisfied. If so, you will see text for that design that says "Infeasible HX: 0." Some of the designs that result may be infeasible, and in fact, 7 out of the 10 designs AEA generated were infeasible in this example. If so you will see something like "Infeasible HX: 1" or some other number under the HEN diagram, without the green background. Infeasible designs are those which either violate the second law of thermodynamics (because they have temperature crossover), or they don't technically violate the second law, but they have one or more heat exchangers in which the approach temperatures are so small that the costs and efficiency of the heat exchanger is likely to be very impractical.

TOM'S TIP: In practice, sometimes AEA cannot find any feasible designs. It may then report only designs with at least one infeasible heat exchanger match. This is more likely to happen when there is a phase change. Obviously, the infeasible exchangers cannot be built in real life, but the cost numbers of that infeasible system are at least somewhat useful because they provide an estimate of the lower-bound on cost (in other words, the real system should be more expensive than this and there will likely be nothing feasible that is cheaper than this). If AEA cannot find any feasible designs, try any of the following:

- Try rerunning using fewer maximum stream matches.
- Pick one of the better (but infeasible) HENs with a small number of infeasible heat exchangers. Manually delete the infeasible heat exchanger(s) and replace them with your own appropriate utility matches for the streams. Keep all the other feasible heat exchangers. Use Add Heat Exchanger button to manually add a heat exchanger connection to one of the utilities in the network, but it is tricky to figure out the right buttons to press. Right-click and hold the Add Heat Exchanger button (upper left of the HEN diagram), drag the icon to the stream that needs a new utility connection (should be a dashed line, which means the stream is unsatisfied and needs a heat exchanger), and release the right-click when the cursor turns to a bull's eye. Then you will get a single red circle somewhere on that line (look for it, it won't likely be where you actually released the click). Left-click and hold the new circle, and drag to the utility connection line and release. That will make the connection and create the heat exchanger. Then, left-click that new heat exchanger to bring up its details. Click the "tied" checkbox, and then enter in one more degree of freedom. I like to specify the duty, because you can get the exact unspecified duty from the Unspecified Streams View (second button from the right above the HEN diagram). If you did it correctly, it should recalculate everything automatically and give you the green bar with Infeasible HX: 0 message.
- Remove the offending streams one at a time and keep rerunning each time until you get a feasible HEN. Then you would manually design the HEN for the missing streams, usually by direct use of utilities, or make a separate HEN for the missing streams.
- Decrease the minimum approach temperature (ΔT_{min}), which can be edited on the bottom left side of the Design | Data form (see bottom left of Figure 11.8). For example, when I changed the ΔT_{min} to 2°C for

the current problem, all 10 designs were feasible. However, based on heuristics, you don't usually want a ΔT_{min} lower than 5°C, but on the other hand, it is just a heuristic. If you go lower than this, and you find heat exchanger matches which are feasible and the costs are reasonable, then you may have to use it.

In the process flow diagrams, all the hot process and utility streams are represented with red arrows, while the cold process and utility streams are represented with blue arrows. The gray heat exchangers are used to show heat exchange matches between a hot process stream and a cold process stream, the red heat exchangers represent matches between a hot utility stream and a cold process stream, while the blue heat exchangers represent matches between a cold utility stream and a hot process stream.

Review these designs and make a HEN selection based on the TAC. The best design (mathematically speaking at least) will be the one with the lowest TAC. But, other design options are given because maybe there are other factors that may weigh into your decision, such as physical proximity within the plant, ease of construction, maintenance, control issues, safety issues, etc.

Click on the Designs tab in your scenario folder, and take a look at the Total Cost Index column of the different designs. This is actually TAC in units of $/sec.

Q3)   What is the value of the lowest Total Cost Index?

Q4)   In your scenario folder, click on the HEN design which corresponds to your answer from Q3 to see its diagram. How many heat exchangers are on the HEN diagram, and how many are process-to-process heat exchangers?

Q5)   Go to the A_Design6 | Heat Exchangers tab of your scenario folder, and take a look at the heat duty (Load) of heat exchanger E-112. How much cooling water is used by this heat exchanger in kJ/hr? Note: If you can't see the numbers clearly, you can expand the column (the same way as you would do it in Excel). You can also double-click the blue icon for the heat exchanger that is connected to the cooling water stream, and see the information there.

Q6)   What is the total heat duty (kJ/hr) of the process-process stream heat exchangers of A_Design5? These are the heat exchangers with the light gray icons beside them.

♪ Music break[3]

PART 2: GENERATING HENs USING DATA IMPORTED FROM ASPEN PLUS

In this part, we will learn how to import process stream data from an Aspen Plus simulation into AEA, and then use the data to generate HENs.

Naphtha is an intermediate hydrocarbon stream that is obtained from the refining of crude oil in a petroleum refinery. Naphtha is usually catalytically reformed in the refining process into smaller molecules to produce a high octane blend for gasoline. However, the raw naphtha feed is rich in sulfur-containing compounds which have to be removed to avoid poisoning the naphtha upgrading catalytic units downstream of the petroleum refinery. The desulfurization of naphtha is done through a catalytic chemical process called hydrodesulfurization. Figure 11.11 shows a process for the hydrodesulfurization of naphtha. As the process is energy-intensive, it is important to recover as much energy as possible within the process through heat integration. This will help reduce the energy demands of the process.

---

[3]Recommended listening: *A Sky Full of Stars* by Coldplay.

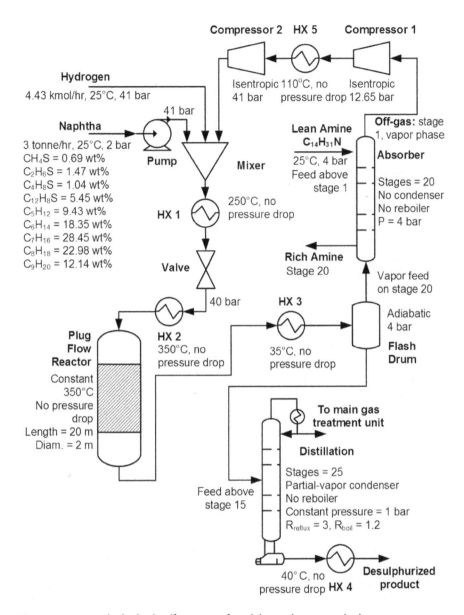

Figure 11.11 The hydrodesulfurization of naphtha. In the reactor, hydrogen gas reacts with various sulfur compounds to produce hydrocarbons and H_2S. The heavier chemicals are condensed out of the product, leaving lighter gases and H_2S in the vapor. The H_2S is removed from the vapor by absorption. A distillation column separates the heavier from the lighter hydrocarbons for different refinery uses.

Simulate the process in Figure 11.11 using Aspen Plus. Additional information for the process is provided below:

- Use the GRAYSON model as the property method for the process, except for the absorber for which the AMINE model should be used. You can change the property method in your absorber block by going to Specifications | Block Options.
- For the plug flow reactor use the reactions and power-law kinetics (mole fraction basis) shown below:

Reactions

1. $H_2 + CH_4S \rightarrow CH_4 + H_2S$

2. $2H_2 + C_2H_6S \rightarrow 2CH_4 + H_2S$

3. $2H_2 + C_4H_8S \rightarrow C_4H_{10} + H_2S$

4. $2H_2 + C_{12}H_8S \rightarrow C_{12}H_{10} + H_2S$

Rate expressions

1. $-r_1 = k_1 \exp\left(\dfrac{-E_1}{RT}\right) C_{H_2} C_{CH_4S} C_{CH_4}^{0.35} C_{H_2S}^{0.35}$

2. $-r_2 = k_2 \exp\left(\dfrac{-E_2}{RT}\right) C_{H_2} C_{C_2H_6S}$

3. $-r_3 = k_3 \exp\left(\dfrac{-E_3}{RT}\right) C_{H_2} C_{C_4H_8S}$

4. $-r_4 = k_4 \exp\left(\dfrac{-E_4}{RT}\right) C_{H_2} C_{C_4H_8S}$

Kinetic parameters

1. $k_1 = 0.2$ kmol/m^3-sec; $E_1 = 200$ J/kmol

2. $k_2 = 0.3$ kmol/m^3-sec; $E_2 = 600$ J/kmol

3. $k_3 = 0.4$ kmol/m^3-sec; $E_3 = 400$ J/kmol

4. $k_4 = 0.5$ kmol/m^3-sec; $E_4 = 200$ J/kmol

- Use a `RadFrac` model in equilibrium mode for both the absorber and distillation columns.
- In the absorber, vary the amine flow rate between 35 and 45 kmol/hr until 95 wt% of H_2S is recovered. Consider a Design Spec or Sensitivity block instead of doing it by hand.

TOM'S TIP: Build your simulation block by block, in increasing complexity. For example, after you add a new block, make sure the system converges correctly before adding another block.

TOM'S TIP: Connect the recycle loop (outlet of Compressor 2 to the Mixer) last.

Simulate the process. However, if you are unable to do it, an Aspen Plus file with the complete process simulation is provided on our website (see the Solutions section for a link).

Before you proceed, ensure your Aspen Plus simulation converges without errors. If the simulation converges with errors, AEA will not be able to transfer all process stream information and may fail to generate a HEN if information is incomplete.

Next, we will define utilities in the Aspen Plus simulation (see Tutorial 6). Add Cooling Water and LP Steam Generation as the cold utilities, and Fired Heat (1000) and HP Steam as the hot utilities, as shown in Figure 11.12. Leave the utilities at their default values. Click Yes if Aspen Plus prompts you to add water to your components.

Now turn on the Energy panel in the Simulation environment and allow it to run until completion (see Figure 11.13). It will calculate the available energy savings before we do a proper HEN design.

Q7) What are the available energy savings in Gcal/hr?

Next, switch to the Energy Analysis environment in Aspen Plus by clicking on Energy Analysis ribbon at the bottom left of your window. The Project 1 | Saving Potentials section shows a breakdown of the utilities consumption in the process, and also shows details of the heat exchangers in the process. Notice though that this HEN is not heat integrated (worst design possible), as all the heat exchanger matches are between process streams and utilities. As there are no heat exchanger matches between process streams, it means that there is still scope for improving the HEN by using heat integration.

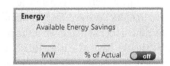

Figure 11.13 The Energy panel in Aspen Plus connects to AEA. The on/off switch is handy because if you are running lots of Aspen Plus simulations, you may often want to switch off the energy analyzer while you are working until you have settled on a design.

Name	Type	Status	Description
CW	WATER	Input Complete	Cooling Water, Inlet Temp=20 (
FIREHEAT	GENERAL	Input Complete	Fired Heater, Inlet Temp=1000
HP-STEAM	STEAM	Input Complete	High Pressure Steam, Inlet Tem
LP-GEN	STEAM	Input Complete	Low Pressure Steam Generatior

Figure 11.12 The utilities for the naphtha sweetening example.

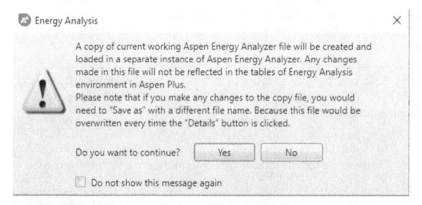

Figure 11.14 When you leave Aspen Plus to change the details of the HEN, those changes will not affect the Aspen Plus results and only appear in AEA.

Click on the Details icon in the Home ribbon to open the associated AEA file. An Energy Analysis window pops up (see Figure 11.14). Click Yes.

An AEA file opens, which you should save. The utility and process stream data from the Aspen Plus simulation file have now been imported to AEA. Click on Scenario 1 to access this Data, just like as in Part 1 of this tutorial.

In the Process Streams table, you will notice that AEA names the process streams by using the stream names of the corresponding half-heat exchanger (`Heater`) in Aspen Plus. The distillation column has two process streams, one for its reboiler (requires heating) and another for its condenser (requires cooling), while the plug flow reactor also has a process stream which requires cooling.

Just like in Part 1, it is possible to add and remove streams here, edit stream names, and also to adjust the temperatures, enthalpy, heat capacity, etc. of the different streams depending on what extra information or knowledge of the process you have. Similarly, you can also make adjustments for the Utility Streams and the Economics if you have to. Note that AEA uses the same utility streams and specifications that were inputted in your Aspen Plus simulation file.

Click Recommend Designs | Solve to generate improved HENs. Depending on how you set up the simulation, you may notice that after generating only a few designs (less than the default 10 specified), an AEA window comes up saying that the program could not generate the specified number of near-optimal designs, as shown in Figure 11.15. The designs that it does generate may be feasible or infeasible. Click OK to close the window if you do get this message.

If you are not able to generate feasible designs, try some of the tips given at the end of Part 1, particularly with regard to reducing the ΔT_{min}. Remember that this is a complicated problem, and so sometimes you have to spend some time understanding what the problems are. As long as you get at least one feasible solution that is not the base case (the base case uses utilities entirely and has no process heat integration), then it is unlikely that there is some amazing hidden HEN that is hiding out there that AEA cannot find. If you get no feasible solutions, then either your problem is completely impossible (maybe all of your cold streams are hotter than your hot streams, for example), or the constraints of the problem are too tight (e.g., ΔT_{min} is too small).

Using AEA is a quick and easy way to generate HENs for processes. Very little skill is required to learn how to use the software, and as a result it can be hard to appreciate how useful and advanced this tool is. Prior to AEA, it was incredibly difficult to generate high-quality HENs. More experienced users (read: older users) will remember the `MHEATX` tool inside Aspen Plus, which uses zone-based interval analyses

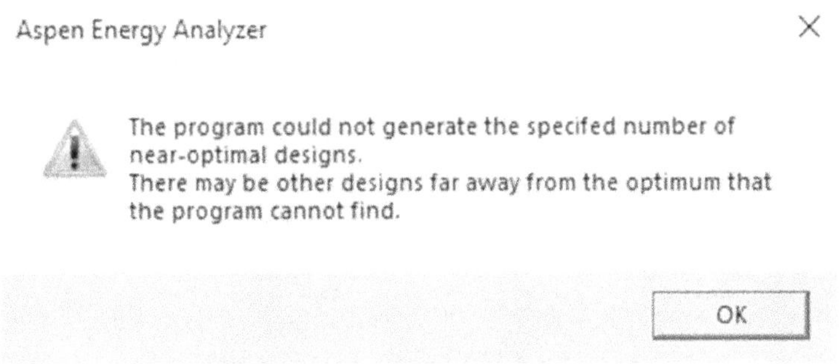

Aspen Energy Analyzer   ✕

⚠ The program could not generate the specifed number of
near-optimal designs.
There may be other designs far away from the optimum that
the program cannot find.

OK

Figure 11.15 Sometimes, AEA cannot find as many different combinations as you had hoped. This message indicates that there might be other really bad designs out there, but it isn't going to bother looking for them.

techniques that often result in very messy, impractical, and suboptimal HENs. The optimization-based approach is far superior in terms of HEN quality, and the graphical interface of the AEA program itself makes it far easier to interpret and use the resulting data. In our own work, we have found that AEA can synthesize good HENs for even very large chemical plant simulations with relatively little effort, something which would take months to do "by hand." As such, AEA effectively makes MHEATX defunct for HEN design purposes and represents a paradigm shift in chemical process simulation methodology.

However, for better HEN designs, more thought is required in selecting utilities and process stream conditions. Furthermore, to get guaranteed optimal designs, optimization-based formulations which can be solved using software such as GAMS might be required. However, in most cases, AEA is satisfactory.

🎵 Music break[4]

---

[4]Recommended listening: *Chasing Cars* (Armin van Buuren Remix) by Snow Patrol.

Electrolyte Chemistry and Solvent-Based Carbon Dioxide Capture

Objectives

- Learn how to use electrolyte chemistry in Aspen Plus
- Learn how to simulate complex CO_2 capture systems using absorption
- Use supplementary Aspen Plus data packages
- Resolve ID conflicts when importing data from other files
- Build a complex flowsheet that manages steady-state multiplicity
- Perform a solvent makeup calculation to close a complex recycle loop

Prerequisite Knowledge

This tutorial assumes you have completed the physical properties tutorials (Tutorials 1 and 2), problem-solving tools tutorials (Tutorials 3 and 6), heat exchangers tutorial (Tutorial 4), equilibrium-based distillation tutorial (Tutorial 5), and Calculator blocks portions of Tutorial 9, and have a good understanding of tear streams and convergence strategies for recycle systems (see Preface). Overall, this material is more advanced, so you will use much of what is in your bag of tricks that you have learned up to this point. Note: this may take more than 2 hours to complete depending on your preexisting knowledge and expertise, but if you have mastered the previous tutorials, I think you can do it in two. However, since this has been the most requested new material for the book, and it combines so much of what we have learned so far, I think it is ok to go over time on this one.

Why This Is Useful for Problem Solving

Electrolyte chemistry is the study of how electrolytes (such as hydronium and carbonate ions) interact in solution. It is important because when they are present, they can have strong impacts on the behavior of mixtures, particularly with regard to phase equilibria. If you are modeling systems with ions in solution, which are common in many applications such as CO_2 capture from power plants, natural gas sweetening, syngas cleanup, or other applications, you may find that the classic physical property models in Aspen Plus are unable to accurately predict important physical properties (especially phase equilibria). In those cases, you will likely find better performance with electrolyte-based models, which require special treatment within Aspen Plus.

Tutorial

PART 1: ELECTROLYTES

For certain liquid mixtures, the formation of electrolytes can be an important consideration when considering fluid properties. In particular, vapor-liquid equilibria (VLE) predictions can be inaccurate when predicting electrolyte formation. For example, in the simple mixture of CO_2 and H_2O, the CO_2 dissociates to form H_3O^+ (or H^+, as some chemists prefer to model), $CO_3^=$, and HCO_3^-. That's what makes it so tasty![1]

Aspen Plus can help you predict what electrolytes will form. For CO_2 in bulk water, for example, you can use the electrolyte wizard on the Component Specifications sheet (see Figure 12.1). Make a new simulation

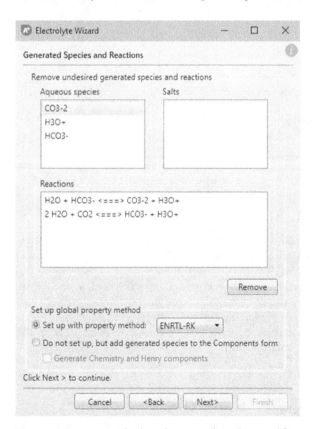

Figure 12.1   Example electrolyte wizard window used for this example.

---

[1] "...HCO_3^- and H^+... are the principal salient stimuli for detection of carbonation." Chandrashekar J. *et al.*, The taste of carbonation. *Science*, 2009, 326:443–445.

file in Aspen Plus V12 (use the Electrolytes with Metric Units template this time). Enter CO2 in the Component | Specifications form (water should already be in there, and if you did not use the Electrolytes template, you should add water in as well) and then click the Elec Wizard button. Select the default database on the first page (AVP120 Reactions) and leave the reference state as unsymmetric. Then, on the second page, make sure that both CO2 and H2O are selected as base components, that the hydronium ion is modeled (H_3O^+) instead of H^+, and that salt formation (only) is included. When you click next, you should see two reactions which are in their database involving the ions H_3O^+, $CO_3^=$, and HCO_2^-. Then, you should see the option to use the Electrolyte NRTL with Redlich-Kwong physical property package (ENRTL-RK). I generally prefer this to ELECNRTL which also uses Redlich-Kwong equations of state but handles mixtures of ions a little differently, and has an increased level of model consistency. Both should work similarly in most cases, however. Select ENRTL-RK and click Next. On the next page, keep the default setting using a True component approach. We will discuss the difference between True and Apparent components later.

After a final confirmation page where you are invited to review the information (and you click Finish), you can see that Aspen Plus added the three chemicals to the components form and has added the two equilibrium reactions to the chemistry section. Also, it has changed your physical property model to ENRTL-RK. You should see something similar to Figure 12.2 in the Methods | Specifications folder. You will need to click on the Components | Henry Comps folder, Parameters | Binary Interaction (check the HENRY-1 subfolder) and Parameters | Electrolyte Pair folders (check the five GME subfolders) of your Properties ribbon to have Aspen Plus finish the job and fill in these parameters.

Looking at the updated Properties | Methods | Specifications form, you will notice that the base method has changed. So have the Components | Henry Comps and Chemistry folders, which both now have folders called Global (you can change the name). For example, the Global chemistry specifications are in the Chemistry | GLOBAL section, as depicted in Figure 12.3.

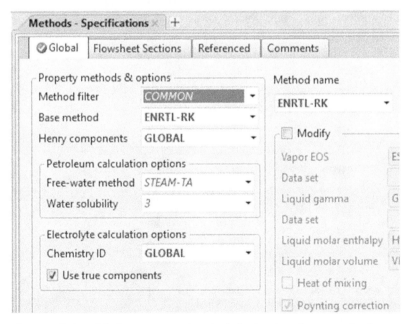

Figure 12.2 Global properties and methods for the electrolyte example after completing the electrolyte wizard.

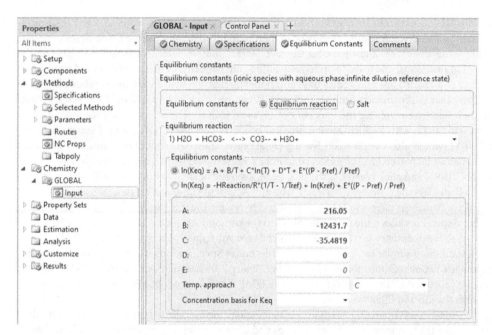

Figure 12.3 Example chemistry specifications for the electrolyte systems in this example.

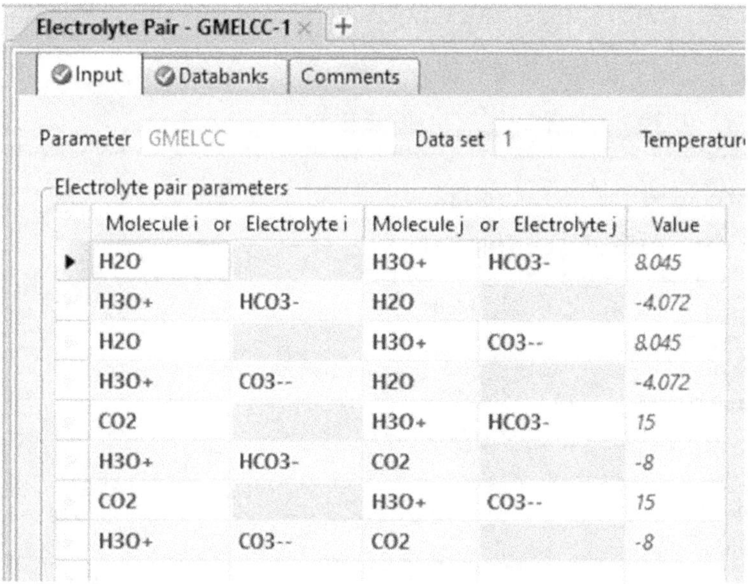

Figure 12.4 Electrolyte pairs modeled by Aspen Plus in this example. This information will be populated automatically when you first visit the form if it is not there already.

Finally, in Figure 12.4 are the electrolyte pairs that are modeled in ENRTL-RK. These are similar in function to the parameters you find in the Methods | Parameters | Binary Interaction | NRTL-1 subfolder except now these guide the ion interactions.

Ok. Now we have that settled, let's start a simulation using it. The way the ENRTL-RK model works is that it uses the electrolyte interactions to help predict more accurate VLE. So let's try it out. Using this property model you have created, perform a constant pressure adiabatic flash of an equimolar mixture of CO_2 and water at 40 bar and 35°C (choose any nonzero flow rate you want). For the inlet streams, just specify the CO_2 and H_2O components and leave the ions at zero flow or mole fraction (as explained later).

 TOM'S TIP: Check your control panel. If you get a warning about all your NRTL binary pair values being zero, go back to Properties | Methods | Parameters | Binary Interactions | NRTL-1 | Input tab and see if there is anything there. If not, then go to the Databanks tab and move the APV120 ENRTL-RK database over from Available to Selected. Then go back to the Input tab and the parameters should be there just like they are in Figure 12.5. Then rerun.

Notice that the electrolyte compositions are essentially zero in your final result. The interesting thing is that ENRTL-RK needs these components defined in Aspen Plus as a requirement of flash calculation convergence even though they have only trace quantities in the final result.

Q1)   What is the mole fraction of CO_2 in the liquid phase, as predicted by the ENRTL-RK model?

Now, do the flash again using a regular NRTL-RK model without the electrolyte chemistry. (Do it in a totally new flowsheet where you never specified electrolyte chemistry to make it easier on yourself.) Again, you may need to select the NISTV120 NRTL-RK databank in the Databanks tab of Methods | Parameters | Binary Interaction | NRTL-1 to retrieve binary parameters. It is usually easier to select this databank before you have chosen NRTL-RK as a selected method. Otherwise, you have to add in the chemicals yourself and then hope it automatically populates.

Q2)   What is the mole fraction of CO_2 in the liquid phase, as predicted by the NRTL-RK model? Do you have two phases or one phase?

Binary Interaction - NRTL-1 (T-DEPENDENT)	Control Panel	+

⊘Input	⊘Databanks	Comments

Parameter	NRTL		Help	Data set	1		Swap	Enter De

Temperature-dependent binary parameters

	Component i ∇	Component j ∇	Source ∇	Temp. Units ∇	AIJ ∇	AJI ∇
	H2O	CO2	APV120 ENRTL-RK	C	10.064	10.064

BIJ ∇	BJI ∇	CIJ ∇	DIJ ∇	EIJ ∇	EJI ∇	FIJ ∇	FJI ∇	TLOWER ∇	TUPPER ∇
-3268.14	-3268.14	0.2	0	0	0	0	0	0	200

Figure 12.5   Binary parameters present after selecting the ENRTL-RK database for this example.

Q3) Which is the more accurate model? Note that the experimental value for 35°C and 40 bar is 1.563 mol% CO_2 in the liquid phase.[2]

By themselves, electrolyte-based property models are pretty simple to use, but integrating them into flowsheets that also use nonelectrolyte models, or even just additional chemicals that are not a part of the electrolyte chemistry, can be a serious headache. This is why I encouraged you to use a separate flowsheet for the NRTL-RK model. Here are some tips in case you ever need to use both electrolyte and nonelectrolyte models in the same flowsheet.

To start with, it is helpful to understand the difference between True and Apparent components. (Apparent components means not checking the "Use True Components" box on physical property definition forms.) Almost all physical property models use "true" component approaches, meaning that each chemical present in a mixture, including ions, is considered when making physical property calculations such as phase equilibria. The problem, though, is that usually only the electrolyte models have data available for individual ions like hydronium or carbonate.

For example, suppose you have a flash drum with water and CO_2 in it and you are modeling with ENRTL-RK. The liquid output of that flash drum will contain trace amounts of ions in it, as you can see in your answer to Q1. Suppose that liquid is then sent to another block which uses PSRK or some other nonelectrolyte model. That block will try to access physical property parameters for those trace ions (which it does not have) thus potentially causing a solver failure due to missing parameters. One solution to this is to set each individual unit operation on a flowsheet that uses the electrolyte model to use "apparent" components (go to the blocks' Block Options form). This means that the ion concentrations will in fact be considered and computed during flash calculations as desired, except that when the results are reported, the ions are bundled back into their "apparent" components (water and CO_2) when reported in the stream. As such, the liquid output stream leaving the flash drum will have exactly 0% ions in it (not even a trace amount). This way, downstream units using nonelectrolyte property models do not see electrolytes at all, preventing lots of problems later. The second option is to uncheck the "Use true components" option on the Properties | Methods | Specifications form for the default method, if that method is an electrolyte method. In either case, the electrolytes are considered "under the hood," you just don't see them in the stream conditions.

There are some minor under the hood differences between True and Apparent component approaches, which can sometimes, but not often, give meaningfully different results. However, RGIBBS, REQUIL, and some of the shortcut distillation models like DSTWU, Distil, and BatchSep, must use Apparent components, and sometimes RCSTR or RPlug depending on reaction details. Also, certain special models, like for CO_2 capture, work only in True component mode, which we will do next. You can refer to the help documentation included with the software for the minutiae. In most cases, it does not matter which you choose, and so I recommend starting with Apparent unless you really need True.

It is more challenging, however, if you want to change property models between blocks, to switch from one that supports electrolytes to one that does not. For example, if a downstream unit does not require electrolyte considerations and if it is better modeled in some other fashion, you should use the Block Options to set the immediate upstream unit operation(s) to Apparent components such that no ions will be present in the stream feeding to the downstream unit. In fact, on the downstream block, you may need to right-click the Chemistry ID and hit clear to get rid of the chemistry specification when changing the property model, because the Chemistry ID drop-down box does not have a "none" option. An example is shown in Figure 12.6.

[2]Valtz *et al*. Vapour–liquid equilibria in the carbon dioxide–water system, measurement and modelling from 278.2 to 318.2 K. *Fluid Phase Equilibria*, 2004, 226:333–344.

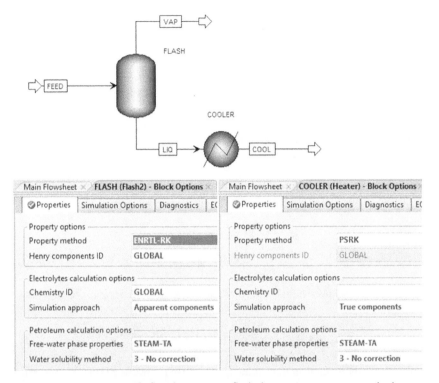

Figure 12.6 An example flowsheet using a flash drum using ENRTL-RK and a heater block (which cools the stream to 25°C) using PSRK. The correct properties settings for these blocks are shown.

PART 2: SOLVENT-BASED CO$_2$ CAPTURE—SETTING UP PROPERTIES

CO$_2$ capture systems have long been a part of chemical engineering. Historically, they are used to produce industrial CO$_2$ for many purposes such as beverages and enhanced oil recovery, to remove CO$_2$ as an unwanted byproduct of gasification reactions, or to remove the CO$_2$ that is naturally present in raw natural gas. More recently, it has risen to prominence for power generation where CO$_2$ can be captured from coal or natural gas power plant exhaust; for the production of synthetic gasoline, synthetic diesel, synthetic natural gas, dimethyl ether, or methanol; as a part of hydrogen production from fossil fuels; and other cases. In many cases, the CO$_2$ can be captured at certain purities and raised to supercritical pressures such that it can be pipelined either for sequestration in an underground reservoir for "permanent" storage, thus avoiding emissions to the atmosphere. Although the subject of what processes make the most sense for society is actively researched and debated,[3] CO$_2$ capture remains a prominent feature in many future "green" or "blue" energy projects.

Solvent-based CO$_2$ capture remains one of the most difficult processes to simulate successfully, largely due to convergence difficulties in individual process units, the presence of electrolyte and reactive chemistries, the complex interactions that result from recycle streams, and the phenomena of steady-state

---

[3]My research group examines this issue closely. For a big-picture look at the most promising possibilities and their potential, see Comparison of CO$_2$ capture approaches for fossil-based power generation: Review and meta-study. *Processes*, 2017, 5:44. An open access copy is available at http://psecommunity.org/LAPSE:2018.0134.

multiplicity that can sometimes occur in complex processes. It takes expert modeling skills to get it to work in most flowsheet software packages, including Aspen Plus, making this one of the most requested subjects for this book. However, if you've made it this far, you actually understand all of the important under-the-hood mechanisms in order to achieve it! This tutorial will get you started on one example, and is not at all intended to be an exhaustive coverage of CO_2 capture, process options, and solvent options, as your own needs will change from problem to problem.[4] Rather, the solution strategy demonstrated here is the most important takeaway.

In this example, we will consider the capture of CO_2 for the purposes of producing Blue Hydrogen. A Blue Hydrogen process is one in which a fossil fuel (usually natural gas) is used to produce high purity hydrogen gas (H_2) for use in fuel cells or as an industrial reagent, but CO_2 capture and sequestration technology is included in order to reduce greenhouse gas emissions. There are many varieties of Blue Hydrogen processes. We will use a simple one shown in Figure 12.7. In this process, natural gas is reformed by reaction with steam to produce syngas (a mixture of CO and H_2) as follows:

$$CH_4 + H_2O \leftrightarrows CO + 3H_2$$

Heavier hydrocarbons in the natural gas, like ethane and propane, also convert to CO and H_2 in a similar fashion.

In the next reactor, the carbon monoxide is shifted by further reaction with steam in water gas shift reactors as follows:

$$CO + H_2O \leftrightarrows CO_2 + H_2$$

At this point, almost all of the carbon atoms in the natural gas have been converted into CO_2, and most of the energy in the natural gas we started out with ends up as produced H_2. Then, we can use a CO_2 capture process to separate the CO_2 from H_2. The H_2 stream would contain some uncaptured CO_2, unreacted natural gas, and perhaps N_2 or other light gases, and would be sent for further purification. The captured CO_2 would be compressed, dewatered, and pipelined for sequestration.

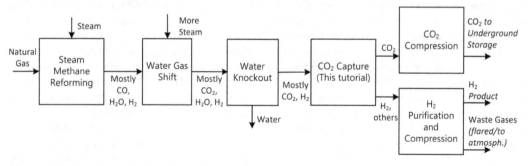

Figure 12.7 An example of a Blue Hydrogen process. It is similar to the upstream portion of a natural gas combined cycle (NGCC) process with pre-combustion capture.

---

[4]For more examples with other solvents, physical property models, and processes, as well as more details in general, check out my book chapter "Processes and simulations for solvent-based CO_2 capture and syngas cleanup." In: *Reactor and Process Design in Sustainable Energy Technology*. Elsevier; 2014, edited by Fan Shi.

For this tutorial, we will use methyl-diethanolamine (MDEA) as the primary solvent (or really, a mixture of MDEA and water). MDEA is quite selective for CO_2 over H_2 (meaning it will dissolve CO_2 in large amounts, but H_2 in very small amounts), so it is ideal for CO_2/H_2 separations. Like CO_2, MDEA forms electrolytes in water, resulting in the following electrolyte reaction system:

$$MDEA^+ + H_2O \rightleftarrows MDEA + H_3O^+ \tag{1}$$

$$HCO_3^- + H_2O \rightleftarrows H_3O^+ + CO_3^= \tag{2}$$

$$CO_2 + 2H_2O \rightleftarrows H_3O^+ + HCO_3^- \tag{3}$$

$$2H_2O \rightleftarrows H_3O^+ + OH^- \tag{4}$$

$$HCO_3^- \rightleftarrows CO_2 + OH^- \tag{5}$$

Thus, it makes a lot of sense to use an electrolyte physical property model in Aspen Plus. Because of the high demand for solvent-based CO_2 capture models, and the difficulty in finding and acquiring good physical property model parameters, several "Data Packages" have been provided with Aspen Plus V12 that are easy to miss if you don't know to look for them. We're going to start by setting up our physical property package using one of these data packages.

Start by creating a blank simulation. Start with the Electrolytes with Metric Units template (you can use one of the others but this is the most convenient for this case). Now, instead of adding the chemicals yourself and using the Electrolyte Wizard, we are going to use a package that is already set up for you. Go to File | Import and choose the Data Package called `KEMDEA.bkp`. I found it in the `C:\Program Files\ AspenTech\Aspen Plus V12.0\GUI\Datapkg\` directory on my computer, but your install path could be different. You may be able to locate it by going to the Resources Ribbon and clicking Examples, going up one directory to the `GUI` directory, and then going down into `Datapkg`. You can see that they have some other ones here for other solvents like monoethanolamine (MEA), diglycolamine (DGA), and diethanolamine (DEA). There are also packages in that folder for other kinds of electrolyte systems unrelated to CO_2 capture that you may find useful in other applications, such as ammonia in water, sour gas in brine, hydrogen chloride in water, sodium hydroxide in water, and sulfuric acid in water.

In any case, go ahead and import the file. Immediately, you will be prompted with a dialog that asks you what you want to name each chemical in your simulation. Type in whatever names you want into the Parameter Value boxes (you will have to double click each one or click Edit Value on each row). The names I chose are in Figure 12.8. Then click OK.

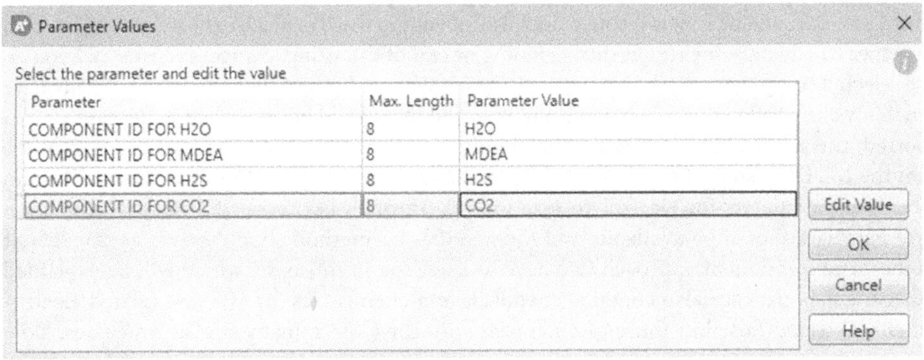

Figure 12.8 Choosing chemical names when importing chemicals from another file.

Figure 12.9 ID conflicts happen whenever you import objects that already exist with the same name in your current flowsheet. These objects can include chemicals, chemistries, reactions, unit sets, calculator blocks, streams, flowsheet blocks, and much more.

Now, depending on what template you started with, you may receive a Resolve ID Conflicts dialogue, as shown in Figure 12.9. This shows up whenever you are importing something from a file that has the same name as something that you already have in your current file. In my case, I used the Electrolytes template and so my simulation started with H2O as a chemical and ELECNRTL as the default property method. Both of those were included in the data package as well and it is asking me what I want to do about it. Click on each item that shows up and make the appropriate decision. Replace means that the object in your current flowsheet is replaced by the one you are importing. Merge means that the imported object is appended to the existing one, and has different meanings in different circumstances (use cautiously). Edit ID means that you will import the object in the file with a new ID that you give it, so you have both the new and the existing objects. Adding a Prefix or Suffix is a convenient way of editing the IDs of a large batch of objects that you select altogether by adding some text at the beginning or end of the string. Ignore means that you do not import the selected component at all. In my case, I chose to ignore both of them because I already have those specified in my current flowsheet. I could do a replace and that would be equivalent in my case.

Once imported, the physical properties section should already contain all the necessary electrolyte compounds for the MDEA system. The Chemistry folder should contain the definitions and equilibrium constant correlation parameters for electrolyte reactions (1) through (4) above. All Binary Interaction and Electrolyte Pair data should be available, and the ELECNRTL method should be set as the default method with the True component approach (we have to use True in this case, which will be explained later). Note that the data package also contains chemicals and chemistries for H_2S and related electrolytes. We will not be using those but those often appear with the CO_2 in many similar processes. Do a properties run just to make sure everything has been imported correctly. You should see no warnings or errors in the control panel.

Next, add the following chemicals, which we will need to describe our gas stream to separate: CO, H2, CH4, N2. If you get a dialogue about updating parameters on the "form," yes, you do want that. I like to reorder mine so the ions are on the bottom of the list but that's optional. Do another run to ensure that it worked.

(♫) Music break[5]

PART 3: SOLVENT-BASED CO$_2$ CAPTURE—ONCE-THROUGH SIMULATION

Now, let's take a look at the process. We'll use a simple one, shown in Figure 12.10. In this process, we have 100 tonne/hr of gas at 40°C and 27 bar from the upstream dewatering step, containing 74.5 mol% H$_2$, 18.9%

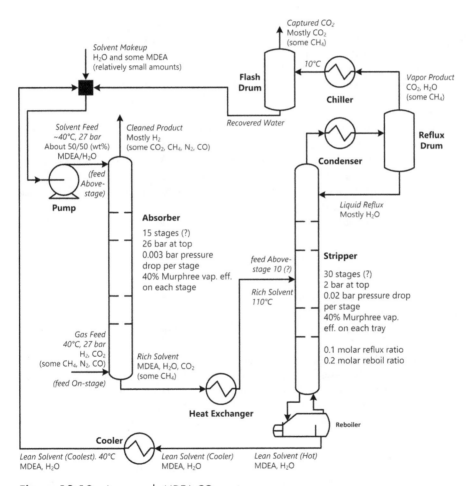

Figure 12.10 An example MDEA CO$_2$ capture process.

---

[5]Recommended listening: REM's *Electrolyte*. I know, how *apropos*. Is it possible I wrote this tutorial just so I could reference the song? No comment.

CO_2, and then small remainders like 5.8% CH_4 (unreacted natural gas left over from our original feed), 0.4% CO (unreacted gas left over from the water gas shift reactor), 0.2% H_2O, and 0.2% N_2 (natural gas can sometimes contain a little nitrogen). This gas feed will be sent to an absorption tower, in which lean solvent (a mixture of roughly 50 wt% MDEA and 50 wt% H_2O) is fed to the top. The CO_2 (and unfortunately also the CH_4) tends to dissolve into the solvent, but not the rest of the gases, which leave through the top. The absorber pressure should usually be high, because not only does it usually absorb better at higher pressures, but we usually want our gas product to be used at high pressures downstream as well. So typically you set the absorber pressure close to the pressure of the feed gas so you do not suffer large pressure losses.

The rich solvent (which contains the CO_2) is regenerated in a stripper, which is just a distillation column with a small reflux and a vapor distillate product. Most of the CO_2 and CH_4 are released from the solvent and exit through the distillate, while the solvent in the bottoms (MDEA and water) leaves with little CO_2 or CH_4 in it. The vapor distillate also contains a lot of water in it as well, and so this stream is chilled and flashed to condense out most of the water, leaving just the CO_2 and CH_4 gases to be collected. The recovered water is reunited with the solvent from the bottoms and then recycled to the absorber again. Because a small amount of water and MDEA will be lost through either of the two gas product streams, a small amount of makeup H_2O and MDEA needs to be added here. However, the presence of water is nice because it is more volatile than MDEA, meaning that almost no MDEA will be lost in the captured CO_2.

The biggest challenge is knowing where to start, but I recommend you start by modeling the absorber. Create the feed gas stream, the absorber, the product gas streams, and the lean solvent feed to the top of the column (start just after the pump), using the description above and the figure. You do not know the solvent rate yet (you get to choose it), nor it's exact makeup, so start by creating a guess of 50/50 MDEA and H_2O by weight with a total mass flow rate four times that of the gas feed (so use 400 tonne/hr). For reasons that will be better understood later, I recommend you do this by specifically specifying the total mass flow rate of the stream, and specify the mass fractions of H_2O and MDEA, as opposed to component mass flow rates. Note that we essentially have no idea what solvent rate we want when starting out. Our goal will be to just pick something that converges, and then change it later.

Absorbers are notoriously difficult beasts to work with, unless you can find the right settings and good initial guesses. Use `RadFrac` to model the Absorber by setting the condenser and reboiler types to `None` (an Absorber is basically the column part of the distillation column without the reboiler, condenser, and reflux drum). Note that without these parts, stage 1 will be the top tray and stage 15 will be the bottom tray. When you specify the stream feeds, the solvent feed should go to stage 1 using the default Above-stage convention (meaning that the liquid will literally be sprayed above the top stage and thus fall down onto the top tray). However, the gas feed needs to be fed with the On-stage convention at the bottom (stage 15). This is because the bottom tray needs a gas feed to work or have any purpose, and if you feed the gas above stage 15, well, it will just go up and miss it entirely.

In the efficiencies section, set the Murphree vapor efficiency to 0.4 on every tray (you can just enter 0.4 for stage 1 and stage 15 and it will also assign it to everything in between). In the convergence section, I recommend you set the maximum iterations to 200 as absorbers often need a lot of iterations to converge. Also, in the Advanced tab of the Convergence | Convergence folder for the block, set "Absorber" to "Yes." This will modify the convergence algorithm to use heuristics more favorable to absorbers. It is a good trick to know whenever modeling absorbers, and the average user would not know that unless someone tells you.

Finally, and this is important, we actually have to specify reactions in this column. The electrolyte chemistry are reactions, and it is important to understand that they take place at different reaction rates. In distillation columns, the liquid residence time on each tray is usually pretty small (3–9 seconds[6]) so like other kinds of reactions, electrolyte equilibrium cannot always be assumed. Therefore, we need to model the

---

[6]Chuang KT, Nandakumar K. Tray columns: cesign. In: *Encyclopedia of Separation Science*. Elsevier; 2000.

electrolyte reaction kinetics on the trays themselves. Fortunately, all that information is already contained in the Data Package that you imported, which you can find in the Reactions folder. There are two reaction sets in there; MDEA-ACID is for cases where you have H_2S (you do not), so you will just ignore it and instead use the MDEA-CO2 reaction set. A quick glance shows that reactions (1) through (5) are listed, three of which are equilibrium reactions, and two of which are kinetic, with the appropriate kinetic and equilibrium parameters already entered. Also, this is why you need to use the True components approach for this column, so that they are considered as terms in the reaction equations.

To add reactions to your absorber column, go to its Specifications | Reactions folder and add a line saying that MDEA-CO2 occurs on all stages (1 through 15). Because it is kinetic, you need to specify the residence times of the trays. Rather than working with tray geometries, we will indicate that each tray should have a liquid residence time of 5 seconds in the liquid phase (leave the vapor phase column blank). This way, Aspen Plus will compute the extent of each kinetic reaction based on this reaction time, which may or may not approach equilibrium. When you are all done, run the simulation. If it works, I strongly suggest you immediately use the Generate Estimates feature (for all stages with all details) inside the Absorber to record this state such that future simulations will converge much faster and more reliably.

TOM'S TIP: If you cannot get your absorber to converge, I suggest you modify the solvent flow rates or feed flow rates until it does, reinitializing before each run. The moment you get convergence, use the Generate Estimates feature, then change the solvent or feed rates to the desired amounts, which should work much better because the initial guesses will be better, thanks to the data stored in Generate Estimates. If you can still not get convergence, reinitialize everything, set the electrolyte convention to Apparent Components in the Absorber's Block Options, and remove the reaction and residence time information. That has a higher chance of convergence. Then once converged, Generate Estimates from there, then return to True components for the block and run again. If converged, Generate Estimates once more to lock in even better initial guesses. Then add the reaction back in again and run once more. If this approach still doesn't work, try adding your own temperature estimates for the column (with about 44°C at the top and as much as 100°C at the bottom…the electrolyte reactions are exothermic), and perhaps some liquid or vapor phase estimates (the top stage vapor flow should by about 91 mol% H_2, 6% methane, and a little bit of CO_2 and water, and you can guess at some liquid phase compositions for that stage; the bottom stage liquid should be about 80 mol% water, 7% MDEA, and the rest mostly CO_2 and you can guess a corresponding vapor concentration). Generally, I find that having just one or two guesses that are basically somewhere in the ballpark is enough. Once you have basically anything that has converged, and store the results using Generate Estimates, the simulation should be much more robust to future changes.

Now, check the composition of the product streams. The loaded solvent stream should contain most of the MDEA, H_2O, and much of the CO_2. The remainder of the gases will leave out the top. We want this stream to be mostly H_2, but it is not realistic that we will capture all of the CO_2. Let's shoot for 95% of the CO_2 in the gas feed ending up in the loaded solvent.

Q4) What solvent flow rate will result in 95% of the CO_2 in the gas feed ending up in the loaded solvent in the absorber? Hints: (1) Use a Design Spec, but start by guessing manually to get a sense of where the answer lies. That can make a great guess for the Design Spec. (2) The answer should be some-where between 300 and 875 tonne/hr of solvent. (3) Calculate the % recovery by using the CO_2 in the hydrogen product stream, not the CO_2 in the loaded solvent, since the CO_2 in liquid phase will dissociate into electrolytes and it is harder to compute. (4) The design spec should work if you are starting from a good initial guess. If it does not, you might have made a silly mistake; check for that before messing with convergence stuff.

Next, it is a good idea to check the column profiles of the absorber, and make sure they make sense. Make a custom plot of the resulting liquid phase mole fractions of the four main chemicals of interest: H_2, CO_2, MDEA, and H_2O. Mine is shown in Figure 12.11. The 15 stages for the absorber was a guessed number. We can look at this plot to determine if we should add or remove stages. When the lines are flat, it means that those stages are not doing much work, and they are probably unnecessary and could probably be removed. When the lines are steep, it means that the stages are doing a lot of work. Based on this plot, I do not see any flat regions, and so I would not want to remove any stages. But should I increase them? The MDEA, water, and CO_2 lines are quite steep throughout most of the column. If I add more, the stages I add might be able to do more useful separations work. Furthermore, because I am trying to find the solvent rate that achieves 95% capture, if I increase the stages, I may be able to reduce my solvent load significantly. Let's add five stages and see what happens. First, save your file! Then try to get a converged solution with 20 stages. First, disable the solvent flow rate design spec if you have one, increase the number of stages, and readjust feed locations if necessary. Don't forget to change the vapor efficiency, reaction stage, and residence time definitions. Add some stage 20 estimates (just copy the old stage 15 results onto a new stage 20). Run and Generate Estimates again if it works, then answer the next question.

Q5) What solvent flow rate will result in 95% of the CO_2 in the gas feed ending up in the loaded solvent in the absorber when using the 20 stage absorber?

So having more stages did little to help us in terms of the solvent. A look at the new composition profile plot (Figure 12.11, bottom) shows that there are now some large flat regions too. There seems to be little benefit for having more stages, so let us keep it at 15 stages. Note that this kind of analysis is really helpful for making really good base case designs for columns. In reality, you would want to do a formal systems optimization that balances capital and operating costs for the plant as a whole, considering the number of stages, solvent rate, design objectives, and other factors, when creating your final design. That's a lot of work, but this technique is a great way to size columns quickly and get some really good initial design estimates.

Now let's move onto the stripper. First, add a heater block to represent the heat exchanger that heats the loaded solvent to 110°C with a small pressure drop of 0.2 bar. The temperature was chosen because I know in advance that we are going to want a feed stage temperature inside the column on the order of 85°C, and that bringing this feed up to 110°C at this high pressure near 27 bar will result in roughly the appropriate temperature after it flashes. If you don't know your feed stage temperature in advance, you could skip this step, and then play with the temperatures later. However, experience has shown that this feed temperature actually has an important impact on convergence success rates in the stripper.

Then, create the Stripper using a `RadFrac` block. Start with 30 stages, with the feed above stage 10, a partial vapor condenser (because you want a vapor distillate product, not liquid), and a total reboiler. Again, set the Murphree stage efficiency on all the trays (stages 2 through 29, because 1 is the condenser and 30 is the reboiler) to 0.4, and place the reaction on all 30 stages. Set the residence time on each tray to 5 seconds and the residence time in the reboiler and condenser (really, the reflux drum) to 300 seconds each. Set the reflux and reboil ratios to 0.1 and 0.2 (molar), respectively.

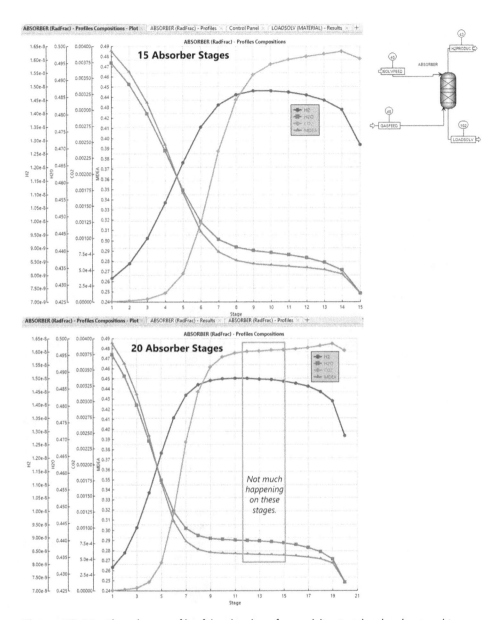

Figure 12.11 The column profile of the absorber after modeling just the absorber to achieve 95% removal of CO_2 (locally) from the gas stream. (inset) The corresponding flowsheet. (top) The case when using 15 absorber stages. (bottom) The case when using 20 absorber stages. Note that there are four separate y-axes here, one for each chemical. This helps you see the changes, but also recognize that H_2 is so small everywhere it doesn't matter much.

This column will have much lower pressure (we are going to assume about 2 bar), so set the condenser to that pressure with a 0.02 bar pressure drop per stage throughout. The optimal pressure is unknown: the closer the pressure gets toward atmospheric, the easier (and thus less expensive) our separation will be, because as you lower the pressure, CO_2 will more readily bubble out of the solvent. However, capturing CO_2

at lower pressures means that you will have higher compression costs downstream if you want to use it for sequestration purposes. The optimal pressure will again be the result of systems level optimization, which is too much work for this tutorial.

TOM'S TIP: Sometimes when looking at the resulting conditions of a stream, certain elements, like mole fractions, may be blank. If this happens, the only fix that works for me consistently is to save the file, close it, reopen it, and rerun.

Use the usual bag of tricks to get it to converge: increase the maximum number of convergence iterations, and generate some initial estimates for the stages. The stage temperatures are the most important for convergence. I suggest guessing 58, 83, 94, 100, and 132°C for stages 1, 2, 12, 25, and 30. This really shapes the feed profile well enough; it is not enough to just do a few stages, you have to give it enough information to describe most of the column. If that does not work, try adding mole fraction guesses, the stage 1 conditions should be about 0.998/0.002 water/CO_2 in the liquid phase and 0.2/0.75/0.05 water/CO_2/CH_4 in the vapor phase, and the reboiler conditions should be about 0.86/0.14 water/MDEA in the liquid and 0.93/0.07 water/ CO_2 in the vapor phase. You can also try removing the reaction from the stages, and perhaps also using Apparent components. Again, the point is to get *something* to converge, and then once converged, make judicious use of Generate Estimates to lock in converged solutions as you work toward your final goal. When finished, be sure to save the file as a backup, and then make a plot of the liquid stage compositions for water, CO_2, and MDEA. Mine is shown in Figure 12.12.

In this case, we can see a couple of regions in which very little is happening (the profiles are mostly flat), both above and below the feed. A quick check of the column products shows that the column is working well: the bottoms has very little CO_2 left, and the distillation is mostly CO_2 and water, with the unintentionally captured methane as well. We could approach this by eliminating these stages (above and/or below the feed), reducing the reflux and/or reboil ratios, or a combination of those things. The tradeoff here is again to save capital versus saving energy. Since the reflux and reboil ratios are both already small, let's remove the stages instead. To remove five stages above the feed, just move the feed stage up by five stages (from stage 10 to stage 5). Then, drop the total number of stages to 21 (thus removing five above the feed and four below the feed). Don't forget to adjust the estimates, reactions, efficiencies, and holdup stage definitions as well as they have now changed. Run it again, and check the composition profiles again (mine are in Figure 12.12, bottom).

From the plot, it looks like we can do essentially the same job without those nine extra stages. In fact, you could probably remove one more stage above the feed and probably a few below it as well. Alternatively, we could reduce the reflux and reboil ratios to save on energy and use these stages instead. Let's leave it at 21 stages for the sake of time, but in practice, you might spend some more time on these questions.

Q6) What are the mole fractions of water and CO_2 in the vapor distillate product?

The final bits are to add a cooler for the lean solvent (the bottoms product of the stripper) to bring it down to 40°C, and to add a flash drum that cools and flashes the vapor distillate at 10°C (I recommend a 0.2 bar pressure drop for each unit). The vapor distillate contains mainly CO_2/CH_4 and water, and this will condense almost all of that water leaving the gases behind as the final product. Then, connect up your recovered water with the lean solvent in a mixer, and the pump that brings it back to the absorber feed pressure of 27 bar. *Do not attempt to close the recycle loop or add a makeup solvent stream just yet!*

At this point in the process, you should check to make sure your cleaned gas and recovered CO_2 streams make sense. The gas product should basically be hydrogen with some leftover light gases like uncaptured CO_2, CO, N_2, and CH_4. There should also be some water present since this stream will now be humidified

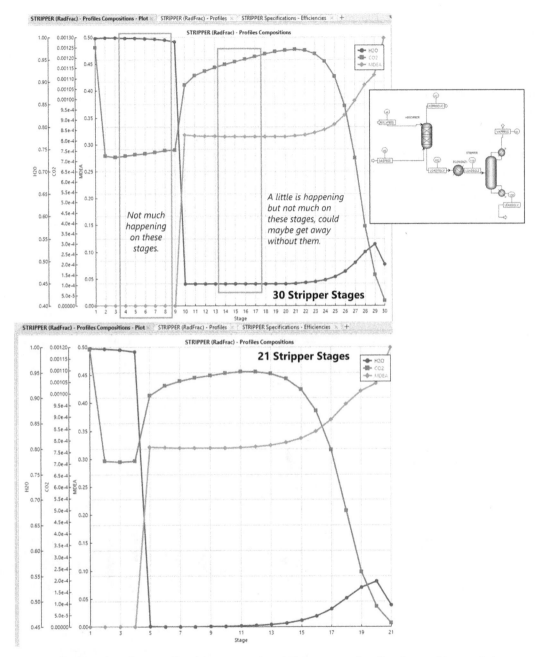

Figure 12.12 The column profile of the stripper. (inset) The corresponding flowsheet at this stage in its development. (top) The case when using 30 stripper stages. (bottom) The case when using 21 stripper stages.

through its contact with water. The captured CO_2 product stream should be about 94 mol% CO_2, with the rest light gases. The large amount of CH_4 present in both streams tells me that for my Blue Hydrogen process, I might want to do something to get rid of this methane downstream, or reconfigure my upstream process to consume more methane so that it does not show up here in the first place.

A side-by-side comparison of the mixer output should show that the lean solvent is very similar to, but not the same as, the solvent feed that we started with in the very beginning. It should have a little less water and MDEA, since some will be lost in the product streams, and some trapped-up impurities like small amounts of CO_2 and CH_4. For many purposes, you can stop here, because the key results of this simulation will tell you most of what you need. In cases where the solvent leaving the bottom of the stripper is almost completely lean (meaning devoid of CO_2), closing the loop may not result in too many differences. However, for other cases where there is a meaningful amount of CO_2 still in the stripper (which can often be the case because it is often cheaper overall to allow some CO_2 to remain), closing the loop is much more important.

My flowsheet looks like Figure 12.13. But the flowsheet is not truly complete until we close this loop. This is the final challenge!

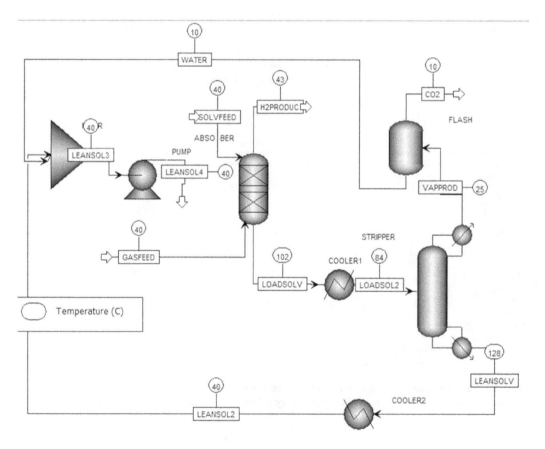

Figure 12.13 The flowsheet in the "once-through" configuration (no recycle or solvent makeup).

PART 4: SOLVENT-BASED CO$_2$ CAPTURE—CLOSING THE LOOP

In this case, closing the loop is complex because of two key challenges. The first is understanding what the flowsheet makeup should be. The second relates to choosing solvent scale relative to the gas feed. In a sequential modular framework, it is not straightforward to set either one.

Let's start with the makeup stream. If we look back at Figure 12.10 of our process as a whole, we can write the overall mass balance of each chemical species i. If you have forgotten how to take a mass balance, the basic principle is that at steady state and with no reaction, the flow rate of each chemical into our system must equal the flow rate of it out. This gives me the equation below, where the left-hand side are the streams that come into the flowsheet (the completed one in Figure 12.10, not what we have made so far), and the right-hand side are the streams that come out:

$$F_{\text{makeup}, i} + F_{\text{gasfeed}, i} = F_{\text{cleanproduct}, i} + F_{\text{capturedCO2}, i}$$

This means that we can use this equation to calculate what the flow rates of the water and MDEA should be in the makeup stream. This means we can write:

$$F_{\text{makeup}, i} = F_{\text{cleanproduct}, i} + F_{\text{capturedCO2}, i} - F_{\text{gasfeed}, i}$$

for i = MDEA and H$_2$O. Then, because we know what the three variables are on the right-hand side from our simulation output, we can calculate the makeup flow rates directly. We could do this by hand and type it into the makeup solvent stream which feeds to the mixer (and honestly that's not a bad place to start). However, when we close the loop and connect up our recycle stream, we know that we are going to create an iterative solver situation if we ever change anything at all on this flowsheet, and so it won't be correct. Instead, create a Calculator block which does the above calculation automatically. That way, it will automatically update whenever any changes are made to the system at all. On the other hand, if you do not do this, once you add the recycle stream, the flowsheet will **never** converge. This is because without this piece, your flowsheet will be unable to satisfy mass balances **by definition**.

As a tip, when creating the makeup solvent stream, I recommend you set up the stream such that you can specify the individual molar flow rates of water and MDEA directly, rather than mole fractions, so you can use the previous equation directly. You can pick any nonzero dummy value for these flows since they will be overwritten immediately, but I recommend using excellent initial first guesses by using mass balances you have just written—that way, the first iteration will be very close to the final result and this is the best way to avoid convergence issues. The stream can be at 40°C and 2 bar.

Finally, it's time for the recycle. Once you have ensured that your code is working and the simulation runs, it's time to connect the mixer output to the solvent feed. You can delete one of the streams and connect them or else connect them together with a dummy block that does nothing, like a mixer. However, before we run, we have two important things to do. The first is to choose a tear stream. Now that we have created the recycle loop, a tear stream needs to be specified so the solver knows where to start its initial guesses. The best place to do that is the solvent feed to the absorber itself, because that is where we know the most information in the loop, and for reasons you will see in a minute, that is the spot where we can have control over solvent flow rate because we have specified it in a particular way. So, after you have connected up the recycle, set the solvent feed stream to be the tear stream in the Convergence | Tear.

The second key thing is to disable the design spec from your absorber that changes the solvent flow rate until 95% capture is achieved. This will help avoid some convergence challenges for when you make that critical first run upon closing the recycle loop. Once that's done, go ahead and run. My final flowsheet is shown in Figure 12.14.

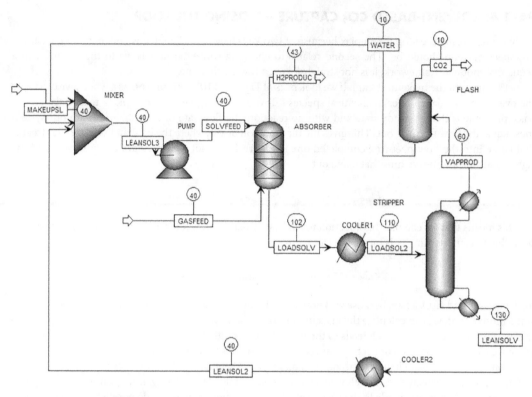

Figure 12.14 The flowsheet in the "recycle" configuration with solvent makeup.

If everything was done correctly, it should converge. If it does, congrats! Generate Estimates in both the stripper and absorber, and save the file. Check the outputs to make sure they make sense. Note that you may not actually be at 95% CO_2 capture as before, and that the solvent flow rate in the feed stream may now be different. In fact, in my case, my CO_2 capture rate has suddenly dropped to 90%. This is because the sequential modular solver will start with your initial guesses for the solvent flow rate, but they will be wrong, especially because other things will be caught up in there like CO_2. Therefore, the solvent flow rates will be overwritten with new values in the next convergence loop iteration, and again and again until convergence, meaning that all mass and energy balances are closed. However, it is likely that the final values for the solvent rates will be rather close to the ones you guessed to begin with.

This is an ingesting demonstration of the phenomena of *steady-state multiplicity*. Steady-state multiplicity means that there is more than one steady state for a given design and a given set of parameters. It's a lot like having a set of equations that have more than one solution. For example, consider the equation $x^2 - 1 = 0$. It has two solutions, −1 and 1. It is quite possible that the equations of our very complex Aspen Plus models can also have such properties, such as having a set of finite, discrete solutions, or an infinite number of solutions within some continuous range. Even more interestingly, if the model is a good representation of real phenomena, then it means the real phenomena could also have multiple steady states in a natural way.

In our case, we have a flowsheet with fixed settings for all of the blocks: absorber stage counts, stripper reflux and boilup, flash drum settings, and so on. Our gas feed input is fixed, and the makeup solvent is defined by our calculator block. What then determines the solvent flow rate looping around on the inside?

Is there only one possible flow rate that can result in system convergence and satisfaction of the mass and energy balance? No. There is actually an infinite number of possible steady states that can occur in this system, characterized by the internal solvent flow rate and its corresponding CO_2 capture rate. When you have steady-state multiplicity in any system of equations as we do, you can usually find different steady-state solutions just by using different initial guesses for the variables and then seeing where the numerical solvers take you. So if we change our tear stream guess, we will arrive at new steady states. Make a small change to the solvent flow rate in the tear stream and see what happens.

To get back up to 95% capture again, we can turn the Design Spec back on that adjusts solvent flow rate until 95% capture is achieved. This will only work, however, if the stream that the design spec will vary is the same as the tear stream. What will happen now is that the Design Spec is not actually going to change the solvent mass flow rate to the absorber. Instead, in each design spec iteration, it will change the solvent flow rate *initial guess*. After it changes the initial guess, the sequential modular solver will reconverge the flowsheet, which will result in a solvent flow rate that is different than the initial guess used by the Design Spec. What we are hoping for is that the Design Spec will eventually find an initial guess for the solvent flow rate that will just happen to result in the solver finding a steady state that has the desired characteristics.

Q7) What is the solvent flow rate that achieves 95% capture of the final system, with the loop fully closed?

Q8) How much makeup water and makeup solvent do you need in the final configuration (in terms of mass flow rate of each)?

The results show that the solvent flow rate needed to get 95% CO_2 capture is about 3–4% higher than our earlier estimates before we closed the recycle loop. This difference will get worse for stripper designs that do not purify the lean solvent as well.

If you have made it this far, congratulations, you are now a Power Aspen Plus User!

TOM'S TIP: For some other CO_2 capture solvent examples using Aspen Plus by the author, see http://psecommunity.org/LAPSE:2021.0100 for a Rectisol example (a low-temperature process using methanol and water the primary solvent) and also http://psecommunity.org/materials/calculationfiles for other examples with MDEA for syngas cleaning, MEA, and DGA for post-combustion carbon capture from natural gas or coal power plants), and MEA for pre-combustion carbon capture for integrated gasification combined cycles.

♫ Music break[7]

•

---

[7]Recommended listening: Armen van Buuren's *Be in the Moment*.

Solids Processing

Objectives

- Learn basics of solids handling in Aspen Plus
- Learn the basics of stream classes
- Use particle size distributions as a part of solids modeling

Prerequisite Knowledge

This tutorial assumes that you are familiar with Aspen Plus and can do basic simulations involving loops, select physical properties (Tutorials 1 and 2), and use equilibria-based rector models such as `RGibbs` or `REquil` (see Tutorial 7).

Why This Is Useful for Problem Solving

Many chemical process engineers are more comfortable working with fluids rather than solids because fluids can often be represented by elegant models, well-understood chemical structures, and whose behavior can be predicted by beautiful theories of thermodynamics. Solids, on the other hand, are sort of like the messy, unpredictable, and inconvenient older brother, whose bedroom you avoid because of its strange funk. In practice, when it comes to classical chemical process modeling, solids simply do not quite fit in. Aspen Plus was originally designed for fluid-based simulations, but AspenTech got serious about solids when they acquired SolidSim Engineering GmbH in 2012 and integrated their models into Aspen Plus. The models have been further improved and now are much more useful than they once were for problem-solving purposes. This short tutorial will highlight some of the basics so you can get started on solids modeling.

Tutorial

PART 1: SIMPLE SOLIDS

This is a quick overview of solids that in no way does justice to the capabilities of Aspen Plus. However, we will learn just enough of the basics in order to get you going so you can learn more on your own. In this example, we are going to model chemical looping combustion via iron oxide, which is a form of advanced power generation that is not yet fully commercial but shows promise as a potential future application.

Figure B1.1 illustrates a chemical looping combustion process for producing power and hydrogen from syngas. Syngas (a mixture of high-energy CO and H_2 gas along with CO_2, steam, and small amounts of methane and nitrogen left over from upstream syngas generation processes) is oxidized in an adiabatic reducer at 55 bar using Fe_2O_3 as the oxidant (O/Fe molar ratio of 1.5). A significant portion (but not all of it) is oxidized, producing heat and combustion products. The Fe_2O_3 is either fully reduced to FeO (O/Fe ratio of 1) or partially reduced to Fe_3O_4 (O/Fe ratio of 1.33) because there is an excess of Fe_2O_3 to ensure that the syngas is oxidized as much as possible. The combustion products are sent to Cyclone 1, where it is assumed that gas-solid separation is perfect. The hot, high-pressure gases are sent downstream to a heat recovery and steam generation plant (HRSG) to produce electric power. The solids are sent to an oxidizer, which are oxidized at 559°C via high-pressure steam (the reactor has a considerable cooling requirement, but fortunately at a high enough temperature to make steam). In this case, H_2O is the oxygen carrier producing a considerable amount of H_2 as a (valuable) waste product. Most of the FeO is therefore oxidized to Fe_3O_4. After another cyclone, the Fe_3O_4 is oxidized further into Fe_2O_3 using air in an adiabatic "combustor."

Now, in order to model this, we need to understand substreams. Aspen Plus uses substreams to differentiate between classes of chemicals. So far, by default, you have always used the MIXED substream in this book, which really means mixed liquid and vapor phases, with solids in liquid solution (so it can still be modeled as a liquid phase). If you want to use a solid, you have to add a new substream to your model. There are two options. CISOLID means conventional inert solid. This is for homogenous solids. In other words,

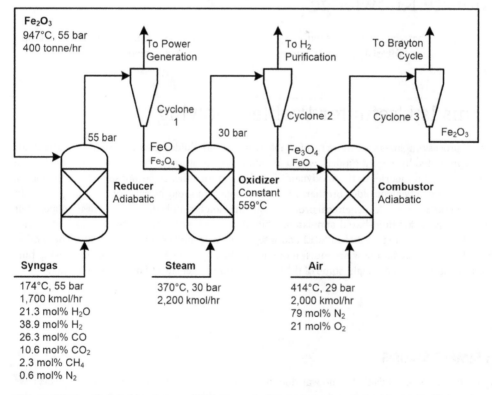

Figure B1.1 Chemical looping combustion process for producing power and hydrogen from syngas.

N2	Conventional	NITROGEN	N2
O2	Conventional	OXYGEN	O2
FE2O3	Solid	HEMATITE	FE2O3
FEO	Solid	FERROUS-OXIDE	FEO
FE3O4	Solid	MAGNETITE	FE3O4

Figure B1.2 Converting solids in simulation from type "conventional" to type "solid."

Fe_2O_3 and so forth would be modeled here. The other option is NC (nonconventional). This is how you model something that is heterogeneous, like coal, ground-up wood chips, or the mysterious substances that my children stick to the wall. For the iron oxides, we'll use CISOLID.

Let's start simulating with the Chemicals with Metric Units template. First, add in all the components you need, but in the Components | Specifications form, change the type from `Conventional` to `Solid` for the iron compounds, like in Figure B1.2.

Use the `PR-BM` physical property package. If you have a Required Input Incomplete display, it is probably because your Binary Interaction parameters have not been updated. Click on the Methods | Parameters | Binary Interaction folder to update them.

 TOM'S TIP: The next button (the blue N with the arrow in the title bar) takes you to the next form in which required information is missing. It is useful for trying to figure out what you forgot to enter when faced with the Required Input Incomplete message.

Next, we need to change the default stream class. Go to the Simulation | Setup | Stream Class form and change the default stream class for GLOBAL from CONVEN (which basically just means mixed vapor-liquid substreams only) to MIXCISLD. This allows both the MIXED (vapor-liquid) and CISOLID (homogeneous solids) substreams but does not model particle size distributions (we will worry about that later). Note that when you go to input the streams, the CI Solid tab is now enabled. You enter liquid-gas streams into the Mixed tab and the solid streams into the CI Solid tab, as demonstrated in Figure B1.3.

| ⊘ Mixed | ⊘ CI Solid | NC Solid | Flash Options | EO Options | Costing | Comments |

(ᴧ) Specifications

Flash Type	Temperature	▾ Pressure	▾	┌ Composition ───
State variables				Mole-Frac ▾
Temperature	174	C	▾	Component	Value
Pressure	55	bar	▾	H2O	21.3

Figure B1.3 Entering solids information to a material stream.

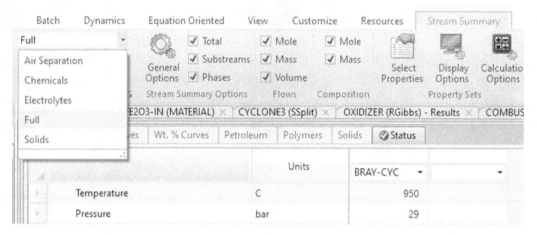

Figure B1.4 Selecting "Full" results to be shown in the stream summary tab.

Notice also that when you look at the results of the stream, the MIXED and CISOLID substreams are kept separate. There is also a Total Stream which is basically the combination of the other two. To see this most easily, you may want to switch your stream results format to Full, as shown in Figure B1.4. This way, you can clearly see whether the flows you are looking at are for MIXED, CISOLID, or Total Stream.

For the cyclones, Aspen Plus has a few simple models that we can try. For this section, let's use the Substream Splitter (SSplit) model in the Mixers/Splitters tab of the models library. See the Command Index for the icon, if you can't find it.

What this basically does is determine what portion of each substream goes to each of the various outlet streams. If you are assuming 100% gas-solid separation (which you may assume for this tutorial), you can simply specify that 100% of the MIXED substream goes one way and 100% of the CISOLID substream goes another. However, in so doing, we ignore any pressure drop losses from the cyclone, and we have to assume that perfect separation is actually feasible. Let's do that for now.

So, give it a shot and simulate this system. You can assume all of the reaction steps go to equilibrium. Another tip: do you really need that recycle loop for the iron oxide to complete your objectives? See if you can get away without it at first. That means, try to find a place where you can break the loop manually, because you know exactly the conditions of the stream, so you don't need to calculate it with a model. Also, a tonne (1000 kg) weighs more than a ton (2000 lb) by the way, don't confuse them!

Q1) What is the cooling duty required by the Oxidizer to three significant figures in MW? Answer as a positive quantity (give the absolute value).

Q2) What is temperature of the stream leaving the Combustor to three significant figures in °C?

Q3) What is the Fe_3O_4 content of solids leaving Cyclone 1, in terms of mass%? Answer to three significant figures.

🎵 Music break[1]

---

[1]Recommended listening: Kristin Husøy's *Pray for Me*.

PART 2: SIMPLE SOLIDS WITH PARTICLE SIZE DISTRIBUTIONS

Now, we're going to consider what happens once we consider solids with particle size distributions. Clearly, the iron oxide we are looking at is not a big sheet of metal; it is a collection of ground-up particles with some distribution in size. So, go back to the Simulation | Setup | Stream Class form and change your stream classes to `MIXCIPSD` (mixed vapor-liquid substreams plus conventional solids with a particle size distribution). This means that you will have to go back to your streams and redefine your solid inputs. You will also need to update the split fractions in your cyclones. When you update the Fe_2O_3 input stream, you will notice that you now need to define a particle size distribution at the right. You can choose between specifically identifying the weight fraction of each bin or specifying a distribution function with some standard deviation (like a Gaussian) and hitting the Calculate button and letting it fill in the bins for you. A bin is just a range of particle sizes. So for example, if you say that 10 wt% of your particles are in the 180–200 µm bin, it means that 10% of your total mass exists in the form of particles which have effective diameters in somewhere between 180 and 200 µm.

First, let's make the bins. Fe_2O_3 particles in this example are normally distributed and have mean particle size of 2 mm with a standard distribution of 0.3 mm. Looking at the size distribution table, you will notice that the default bins for the particle size distribution are really inconvenient, since they are sized between 0 and 200 µm (or "mu" in Aspen Plus). Choose a distribution function instead then type the appropriate numbers into the Distribution function section and click Calculate (yes, you want to normalize), as shown in Figure B1.5. If you look at the bins, you'll see that almost everything fits into the 180–200 µm bin because they are way too small. So, change the bins! Looking at Figure B1.6, it seems like bins from 1 to 3 mm in 0.1 mm increments are sufficient. We can also have 1 large bin from 0 to 1 mm and 1 large bin from 3 to 4 mm, since that represents the "tails" of the distribution with very little in it.

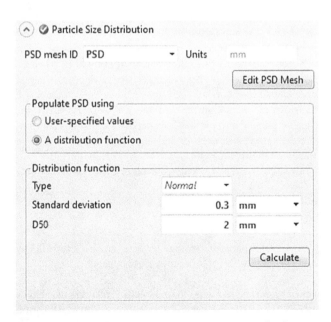

Figure B1.5 Defining the particle size distribution for this example.

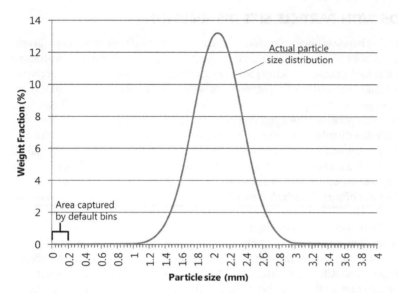

Figure B1.6 The particle size distribution for this example.

Set that up by clicking the Edit PSD Mesh button and editing the mesh at the right. You can do this quickly by using a combination of the Equidistant and User PSD mesh types to customize the mesh. First, select and set up the Equidistant PSD mesh type with 22 intervals from 0.9 to 3.1 mm. Hit the Create PSD mesh button to complete it. Next, then switch to the User PSD mesh type (yes you want to keep your intervals) and adjust the beginning interval (change 0.9 mm to 0 mm) and the end interval (change 3.1 mm to 4 mm). An image of the completed window is given in Figure B1.7.

Now go back and recalculate the normal distribution. You should get more reasonable-looking bins. For example, particles between 1.9 and 2 mm in size account for 13.1 wt% of the total. So does 2–2.1 mm in size.

Ok! Let's do a realistic cyclone simulation now. Instead of the SSplit block, use a rigorous cyclone model (Cyclone) for Cyclone 1, available in the Solids Separators tab. Select Cyclone as the model and use Simulation mode. Select the Muschelknautz Calculation method and the Lapple-GP Type. Let's not worry about what these mean; it's basically the model to predict how particles are separated from the gas phase. Leave the Efficiency correlation parameters at their defaults and select 1 cyclone with 0.5 m of diameter. The final set of specifications for the block should look like Figure B1.8.

Now that you have been able to specify Cyclone 1, go ahead and update your other cyclones in the same way. Run the simulation!

Q4) What is the pressure drop of the first cyclone, in bar?

Q5) Is our previous assumption of perfect gas/liquid separation reasonable? You can check the PSD by going to the Results tab for a stream and clicking the PSD plot button that appears.

Now, go back and redo your particle size distribution using the same methodology (including number of bins) as we did before except use an average particle size of 0.02 mm with a standard deviation of 0.005 mm instead. These would be rather fine particles. Rerun the simulation and answer these questions.

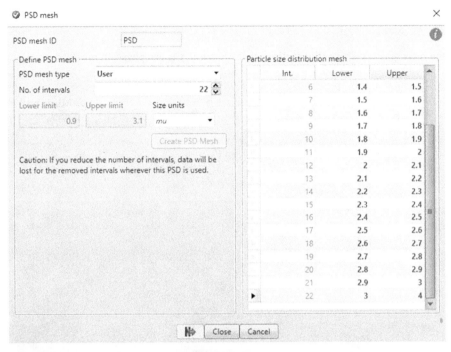

Figure B1.7   Changing the particle size distribution mesh.

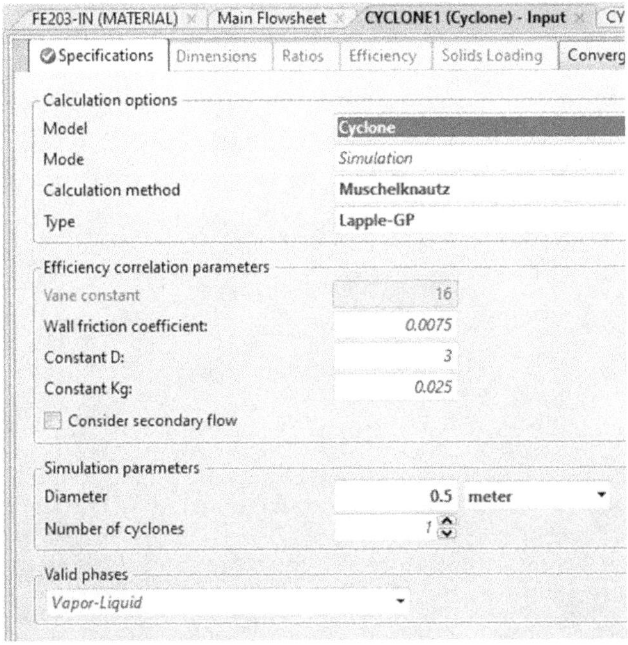

Figure B1.8   Completed specification block for the cyclone in this example.

Q6) What is the separation efficiency of the first cyclone? That is, the percentage of solids that ends up in the solid stream?

Q7) What is the mean particle size of the particles that remain in the gas phase, in mm? Use the PSD plots to help.

(♫) Music break[2]

---

[2]Recommended listening: *In the Mood* by Glenn Miller.

Parallel Computing Tools: Reduced Order Models, Optimization, and High-Performance Computing

Objectives

- Use Aspen Multi-Case to run many simulations in parallel on computers with multiple CPU cores
- Build reduced models using the results from many Aspen Plus runs
- Use Python to automate Aspen Plus simulations
- Use Python to run thousands of Aspen Plus simulations using parallel computing, while handling a variety of errors in a robust fashion
- Use particle swarm optimization to optimize Aspen Plus flowsheets on desktops and high-performance computers using parallel computing

Prerequisite Knowledge

This tutorial assumes you have enough knowledge of the basics of Aspen Plus, and that you have access to a computer with four or more CPU cores (nearly any modern desktop or laptop has at least four by default). For the second part of the tutorial, you should have a good understanding of the concept of optimization, as discussed in Tutorial 6. It is also helpful to have rudimentary programming experience in any language [especially Python, but any major programming language will do, or even Visual Basic for Applications (VBA), which comes with Microsoft Word or Microsoft Excel]. However, this tutorial is written so that most Aspen Plus users can take advantage of these features without knowing any programming languages. Code files are available for download and most users will be able to readily adapt them to their own needs with minimal changes.

This is likely to take closer to 4 hours, but it is a bonus tutorial of a frequently requested advanced topic, so enjoy!

Why This Is Useful for Problem Solving

Like many other commercial process modeling software packages, Aspen Plus is built around a traditional "single, sequential" computer design that was the norm in personal desktop computers prior to 2010 (the earliest forms of Aspen Plus date back to 1977). The fundamental framework, models, and numerical solution strategies that embody Aspen Plus were developed over four decades, taking advantage of academic and industrial research advances that allowed users to solve ever-larger and more complex problems. However, these algorithms were almost always built around a single computer processing unit (CPU) concept, where only one computer chip exists on the user's computer. This single-CPU mentality is embedded at the very core of Aspen Plus. From 1977 to about 2005, CPU speeds increased exponentially, going up by an order of magnitude every decade. However, after that, CPU speeds started to level off and have grown very slowly to the present day (2021). Instead, the CPUs have simply gotten smaller and cheaper, but not much faster, so to increase performance, computers simply started having more of them. My personal desktop has six physical CPU cores, for example. You probably have more. On Windows, you can find out how many cores you have on your computer by running System Information (I type "system" into Windows' Search or Start Menu) and looking at the number of cores listed in the description of the processor, about 10 items down from the top of the system list.[1] You could also have multiple processors, each with one or more cores.

The problem is that Aspen Plus is designed to only run on one core during simulation, since the underlying numerical methods were built that way.[2] On the plus side, you can run a very long, complex simulation of Aspen Plus all day and still have plenty of other cores available for other tasks, like listening to the music breaks, without interrupting the simulation. However, it (currently) is not easy to make that one simulation go faster. Instead, what we can do is run multiple different Aspen Plus simulations at the same time on the same computer, one for each physical core. Although some users may have licensing restrictions which prevent that, in the general case, you can run multiple Aspen Plus runs simultaneously in parallel if you have the cores to do so. Most commonly, one might want to do that to conduct case studies, sensitivity analyses, or optimization—all tasks in which the same model is run many times with different parameters each time. The act of running many cores simultaneously in parallel to solve one large problem is called parallel computing.

Therefore, knowing how to take advantage of the parallel computing cores you already have available to you is a great way to drastically speed up your productivity and tackle more challenging problems. Although Aspen Plus has case study, sensitivity, and optimization features already built-in (see Tutorials 3 and 6), these use legacy algorithms that do not take advantage of parallel computing. However, with enough know-how, you can use external programs that are parallelizable which call Aspen Plus automatically, thus giving you access to parallel computing power. This is going to be increasingly important as each new computer you purchase will likely have more and more cores available. If you have not already, you may also find yourself using Aspen Plus remotely by connecting to shared high-performance computers which have hundreds or thousands of cores available to you. For example, my own research group routinely solves large flowsheet problems on a 64-core machine running Windows. Prior to V12, we would have up to 64 copies of Aspen Plus running simultaneously in parallel, each one managed with our own in-house code. Now, with Multi-Case, we can automate that much more easily without needing 64 program instances.

---

[1] Do not confuse cores with logical processors. Cores are physical CPU chips and you can only run one process on one core at a time. Logical processors are tricks played with a feature called hyper-threading which lets the OS put two or more threads on one core, so if one process needs to pause for something like waiting for a hard disk read, the other thread can take over instantly.

[2] Well, actually, it will actually use two or three processors during a run sometimes, if they are available. But it is not scalable.

Tutorial

PART 1: ASPEN MULTI-CASE FOR LARGE PARALLEL COMPUTING JOBS

Aspen Multi-Case first appeared Version 12. It is a simple but useful tool, which will likely evolve to have additional functionality over time. Although launched from the Windows Start Menu (Aspen Multi-Case), it runs in a browser, and only works with Microsoft Edge and Google Chrome (attempts at other browsers will likely lead to an error such as "Can't Connect to the License Server"). Depending on your setup, you may be given the choice of browser when you launch Multi-Case from the start menu. Otherwise, if you have an unsupported browser as your system default (Firefox, Opera, Safari, etc.), that browser will launch and attempt to run Multi-Case but you most likely will get an error message. Just launch either Edge or Chrome separately, and then copy-paste the URL from the other browser into Edge or Chrome, and it should work.

 TOM'S TIP: Multi-Case licenses are not included by default in some license bundles, such as some academic ones, and may need to be purchased separately. If you do not have a license, you would see a "Couldn't Connect to the License Server" error when launching Multi-Case, and hitting Retry repeatedly would not help. If so, you can still use parallel computing approaches through custom code that you can make in Python. Skip to Part 2 to find out how.

The first time you run Multi-Case, you are presented with a login screen asking for your name. In most cases, you can just type in whatever you want (it is not connected to any kind of user account) and move on. Your browser will remember this on subsequent logins. Once logged in, you are taken to the main Projects screen which keeps a record of the projects you have worked on in the past (currently nothing). There are three kinds of projects:

1. *Case Study:* A case study is essentially a single simulation file on which you want to conduct a sensitivity analysis in parallel. Typically, you would provide a converged Aspen Plus flowsheet (or Aspen HYSYS for that matter), indicate the parameters for your analysis that you want to vary, run them in parallel with multiple cores, and then retrieve the data for some other purpose. The values of the parameters for the analysis are usually chosen as a mesh of evenly spaced points within ranges that you define, called a coarse-grained approach.

2. *Reduced Order Model:* This is like a Case Study except that you only define the ranges of the parameters that you want to vary. Multi-Case uses a Design of Experiments algorithm that automatically chooses the values of the point within that range in a strategic way that helps build reduced order models, or ROMs. Specifically, it chooses the values quasi-randomly using the Sobol algorithm to maintain a large diversity of points while avoiding large gaps in the parameter space. It is similar to Monte Carlo or Latin Hypercube approaches, and it is ok if you do not know what any of that means. At present, this feature is designed specifically for use only with Aspen Tech's AI Model Builder software which is outside the scope of this book, although an astute engineer could write a parser script that could extract the relevant data from the output files that result. For now, just be aware that this feature is here in case it suits your interests for further exploration.

3. *Multi-File Analysis:* This is similar to the Case Study tool except that you can work across multiple files simultaneously, even having both Aspen Plus and Aspen HYSYS files. This is useful if you have process sections isolated in different files, or if you have process variants with structural differences that are located in different files (for example, you have one file for an open loop system, and another file for a variant with recycle).

In Part 1 of this tutorial, we are going to create an Aspen Plus simulation file with a simulation of a methane partial oxidation reactor to produce syngas and then use Multi-Case to run a sensitivity analysis on it, collecting data. We will then use that data to create our own ROM in Microsoft Excel, which will give us a simple equation that we can use to predict the outputs.

In a partial oxidation of methane (POM) reactor, methane converts to syngas (a valuable product) as follows at high temperature:

$$CH_4 + \frac{1}{2}O_2 \rightarrow CO + 2H_2$$

However, complete oxidation of methane (undesired) can also take place:

$$CH_4 + 2O_2 \rightarrow CO_2 + 2H_2O$$

Also, at those high temperatures, the water gas shift reaction takes place:

$$H_2O + CO \rightleftharpoons H_2 + CO_2$$

As well as the steam reforming reaction;

$$H_2O + CH_4 \rightleftharpoons 3H_2 + CO$$

Ultimately, designers need to carefully decide upon the temperature and pressure of the reaction, as well as the balance of inlet chemicals, which affects the rates of all four equations. Typically, the designer seeks a certain H_2 and CO content, and wants to minimize wastes like H_2O and CO_2. What we are going to do is use Aspen Plus to create a model of this reaction system, and then create a simple equation (ROM) that can predict the product mole fractions as a function of temperature (between 600°C and 1000°C) and pressure (between 2 and 50 bar) for a particular feed composition. That way, others can work on their designs using this equation without having to use Aspen Plus repeatedly.

Start by creating a blank Aspen Plus file, with the following chemicals: CO, H2, CH4, H2O, CO2, O2. Use the PR-BM physical properties setup and be sure to click the binary parameters folder (PRKBV-1) to make sure they are all there. On the main flowsheet, create a single RGIBBS reactor model, which if you recall from Tutorial 7 assumes chemical equilibrium is achieved. The feed to the reactor should be 1000 kmol/hr, 400°C, 50 bar and consist of 20 mol% H_2O, 60% CH_4, and 20 mol% O_2. Set the reactor temperature to 800°C and pressure to 10 bar. Add an output stream and the model should run. The product stream should contain about 51 mol% H_2, essentially no O_2, and about 20 mol% CO.

 TOM'S TIP: If you want to, just download the completed Aspen Plus file to start with so you do not have to make it yourself. See the beginning of the Solutions section for the link.

Our next step is to run this simulation a thousand times or so in parallel, with different values of reactor temperature and pressure, and record what the reactor products are for each. In Aspen Multi-Case, create a new Case-Study Project, and give it whatever name you would like (see Figure B2.1). Choose the Aspen File that you just made as the Case File. Now everything you do with this Case Study will work with this one file. Then, click on the project you just made to open it. Click Add Analysis to create a new analysis (you can change its name after you create it by clicking the three vertical dots next to it and choosing Edit Analysis).

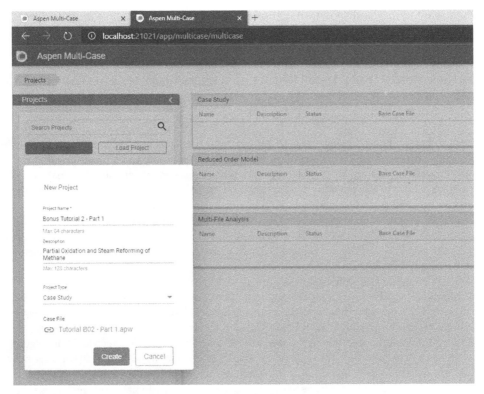

Figure B2.1   Starting a new project in Aspen Multi-Case.

An analysis is a collection of scenarios. A scenario is a large batch of runs that you will do using the chosen case study file using parallel computing. You can have many analyses inside of a Case Study, and many scenarios within an analysis. Each scenario can be run independently on command.

Click the Add Scenario button within the mostly empty Analysis box that appears on the right-hand side, and give it a name. Now what you want to do next is define the independent variables of your analysis. For this tutorial, let us choose to vary the temperature of the reactor from 600°C to 1000°C and the pressure of the reactor from 2 to 50 bar, with 35 evenly spaced samples taken in each direction (you can reduce this number later if you find your computer takes too long to run that many). You can add that in yourself by using the "+" button in the scenario in the Independent Variable section and adding each of those in (e.g. Reactor.Temp and Reactor.Pres where Reactor is your name for your RGibbs block in the Aspen Plus file). Choose a linear step style, meaning that the temperature and pressure points will be evenly spaced apart across the range. What you are doing is specifying that you want $35 \times 35 = 1225$ runs, one for each combination of the 35 temperature and 35 pressure options. Aspen Multi-Case will then take the Aspen Plus file, change the temperature and pressure of your reactor to the current combination, and run the file.

TOM'S TIP: You can also choose logarithmic spacing, which is useful in certain circumstances. Alternatively, you can choose Add Discrete Set from the three vertical dots dropdown menu for that scenario. This lets you simply type in what points you want to visit, and allows you to define those points with multiple independent variables so they do not have to align perfectly on a grid. However, each point must be typed in manually, and at present, it does not allow copy-paste from Excel or elsewhere. So although you can manually specify independent variable combinations that might result from Latin Hypercube or Monte Carlo sampling approaches, the manual interface in the present version is I think too tedious to be useful for this purpose.

Next, use the "+" button in the Dependent Variables section to add in as many dependent variables as you want. Dependent variables are the outcomes of the simulation that you want to track, and it will record the results of the dependent variables for each combination of the independent variables. You may notice that not all of the variables available in Aspen Plus are available for being included in the dependent variables, but hopefully, there is enough information there to be useful. In my case, I added the reactor calculated heat duty (QCALC), the product stream total molar flow rate (MolarFlow), and the mole fractions of the product stream for each chemical (MoleFractions.H2, etc.). Add at least the mole fraction of H_2 of the products here. The results should look something like Figure B2.2. Note that "base value" is the value currently stored in the simulation file.

Now, we are ready to run! This is optional, but you can go to the scenario menu (three vertical dots) and choose Launch Monitor, which brings up a little panel that updates during the run. When you are ready, hit the run button (triangle). Then watch your simulation run. I like to also look at the Windows Task Manager (right click the Windows task bar to launch it) and go to the Performance tab, or the CPU tab of the Resource Monitor (load from your Windows Start menu) to ensure that all cores are in use. My screen looks like Figure B2.3, showing that while Aspen is now in the midst of running 1225 simulations, all 12 logical cores are hitting 100% usage. This is a good way to confirm that the system is running in parallel. In my case, it took less than 3 minutes to complete. The Estimated Time to Completion shown in the Monitor was a little conservative for me, but if it is more than 4 or 5 minutes, feel free to stop the simulation (the square button), reduce the number of steps for independent variables on the scenario, and run again.

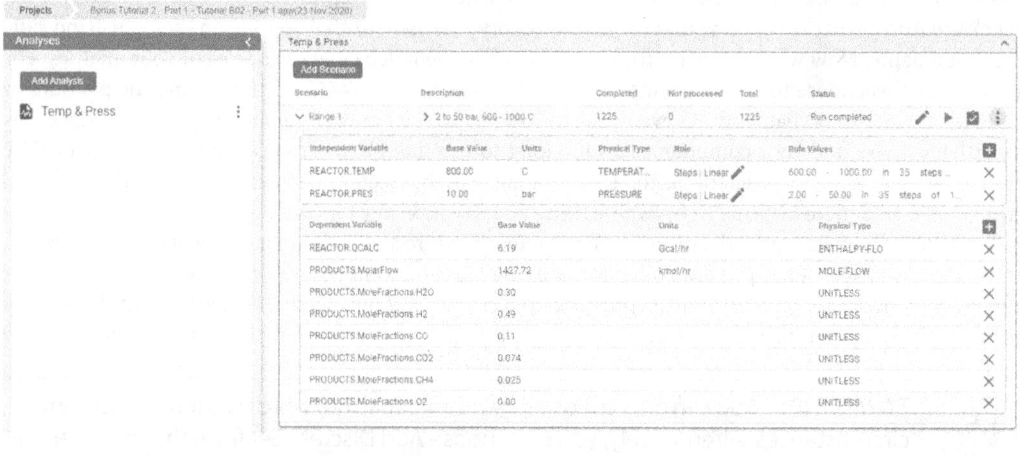

Figure B2.2 The scenario setup has been completed, showing the independent variables that will be varied, and the dependent variables that will be recorded.

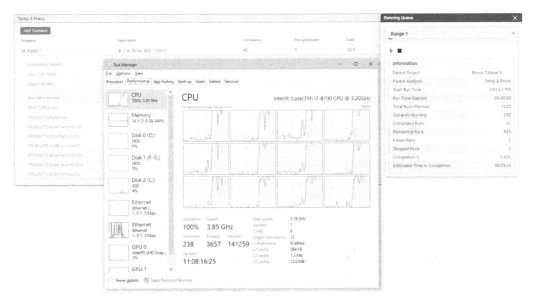

Figure B2.3   Multi-Case is now running. All of my 12 logical cores are currently at 100% usage.

   TOM'S TIP: By default, the software may not use all available cores. When you hit the run button on the Scenario (not in the Monitor), it will tell you how many cores it is going to use. If you want more, stop or cancel the run. Go to the settings menu in Multi-Case (click the Gear button on the top right of the window, next to the "?" button), and find the text field called Max Number of Parallel Runs. Change the value there. Obviously, you cannot go larger than the number found in Number of CPUs Available. If *that* number seems too small, you should contact your system administrator (and hopefully that person is not you). If you are on a large, multiuser system using a remote connection instead of your desktop, your system administrator has a setting that limits how many cores you can access. Also, if you are using a virtual machine, then you may not be able to use many cores simultaneously depending on how that is set up, even if they appear available.

Once the run has completed, click on the blue clipboard-with-checkmark icon in the Scenario information to display the results. You will be treated to a table of numbers. Each row is one simulation run, with the values of the independent variables used in that run shown in the first few columns, and the results of the dependent variables to the right of that. You will also see information about whether the run converged correctly or not, or if there were errors, etc. On the upper right, you will see a button labeled Table, and you can click this to switch to Chart view. Chart view makes a convenient plot of either 2D or 3D data; 2D is for one independent variable, and 3D is for two independent variables, with one dependent variable plotted. It is convenient for a quick view of the results and works well.

Q1)   Generate a 3D plot, with temperature and pressure as the independent variables, and heat duty as the dependent variables.

Q2)   Generate a parametric plot with CO mole fraction as the x-axis variable, heat duty as the y-axis variable, and reactor temperature as the parameter.

TOM'S TIP: Aspen Multi-Case runs using Windows services, which are programs that run in the background. Specifically, the services are called MultiCaseServer V12.0 and SimService V12.0. The browser-based interface Multi-Case is primarily just an information-gathering tool that submits job requests to the services; the web browser does not actually run the simulations. That means, however, if something goes wrong, restarting the browser is not likely to fix it. If you find that you are suddenly not able to do additional runs (especially after an interruption) you will have to restart the services. For example, if you see the "One Moment Please" message when trying to launch a run and it never goes away, or the monitor sits there showing that nothing is happening, then something may be held up in the service. To resolve this, you can restart those two services using the Services app that comes with Windows (type Services into the Search or Start Menu). Find MultiCaseServer V12.0 and SimService V12.0, right click on them, and click Restart. Then close your browser tab and reopen Multi-Case. Failing that, try the Dogbert Tech Support Solution.[3]

One of the things that happens for some users (and not others) is that you may have missing or obviously incorrect data. For example, Figure B2.4 shows an example 3D plot of H_2 product mole fraction as a function of reactor temperature and pressure. Although this should be a smooth plot, some data points are "missing," and others are just off for some reason. In truth, those outlying results shown in that plot are simply incorrect. There is nothing special about those points as to why they would be so different from their neighboring points and the overall trend. In fact, if you locate those suspect points in your results table, and then re-simulate them in Aspen Plus using the exact same temperature and pressure combination that failed in Multi-Case, you will not get the outliers and instead get the correct answers with no problem. You can also verify this by simply rerunning your Multi-Case over again and looking at the plots. You may still find that perhaps 1% of the runs are erroneous, but they will in fact be *different* runs this time that failed, meaning that this phenomenon is nondeterministic.

This is a known bug in V12.0 and has been addressed in V12.1.[4] However, I am leaving the buggy examples in this version because it is useful for learning how to deal with bad or missing data (which may be because of simulation problems as opposed to this particular bug), and because many users may be on old versions due to reluctance to upgrade. If you still experience this bug, I would say it is still worth it to deal with about 1% bad or missing data, so that is what we will use next.

To get the tabulated data in a useful form to use in other software, right click the data table (somewhere in the white space) and choose Export | Excel Export.[5] It will automatically put a copy in your browser's Downloads folder, so open that up in Excel. The first thing to do is validate your data, whether there were errors or not. First, I would start by identifying missing data. In our case, these are rows in which one of the

---

[3]Restart the computer.

[4]It was yours truly who found and reported it to AspenTech.

[5] One of the options is to export to CSV, which is a text format that is immediately understood when you open it in a text editor. Many programs, including Excel, will be able to open this format.

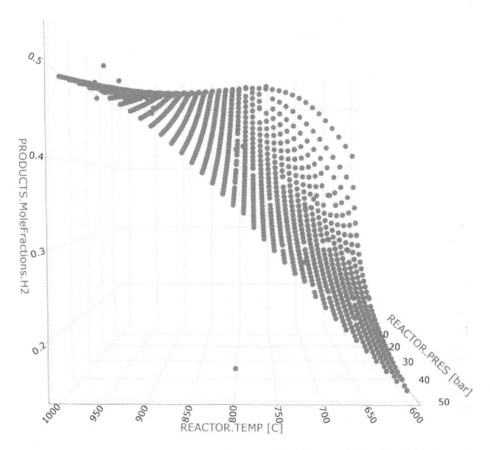

Figure B2.4 A 3D chart. You can grab and rotate to get a better understanding of the data.

dependent variable values is stored as "NaN," which means "Not a Number" and is a special bit pattern that is stored in a floating-point variable when a number is divided by zero. Delete the rows that have these in there. To find them, you can use the Find feature and search for "NaN." Or, do something more clever by using Excel's Data Validation tool (Excel | Data | Data Validation | Data Validation), where you require all of your mole fractions to be Decimals between 0 and 1, for example. If you click Data | Data Validation | Circle Invalid Data, it puts a red circle around the invalid data, so it is easy to find when scrolling down (see Figure B2.5). Other options include using IF statements to output warning text to an adjacent cell if data are not inside a certain range, or using Conditional Formatting in a similar fashion. Finding points that have data but just have wrong values is harder, especially when you have more than two independent variables such that you cannot plot it easily. One way to check for these is to find and flag points that differ substantially from their nearest neighbors (e.g., differ from the neighboring points by a much higher percentage than the two neighboring points differ from each other) using IF statements that spit out an error message if it is a suspect point. For our tutorial, as long as we get rid of the blanks or NaN's, we can actually keep the cells with numbers which are incorrectly computed because they are such a small percentage of the total that it will not have a too drastic effect on the reduced model we are trying to make. But if you know they are simply wrong, you could get rid of those too.

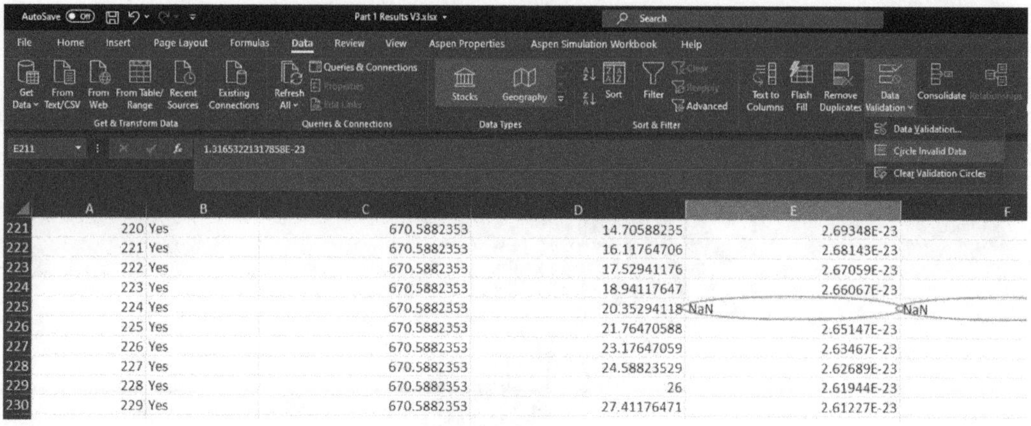

Figure B2.5 Using the Data Validation feature in Excel.

Now that we are left with only valid data, we are going to use it to build a reduced model that can be used to predict the reactor product composition as a function of reactor temperature and pressure, starting with the mole fraction of H_2 in the reactor product. There are many tools available for this task, and many forms that such a model could take. For example, software such as Matlab, Excel, Pylab, R, Aspen ProMV, and ALAMO all contain tools that create models out of a given set of data through various regression or machine learning algorithms. The models can be in various forms such as linear or nonlinear equations, artificial neural networks, tree ensembles, multivariate statistical models, principal components, and so on. For this tutorial, we will create a linear-in-the-parameters (LitP) model using Excel, since most readers will readily understand and have access to Excel.

In a LitP model, the user enters the independent variables and then can quickly and directly calculate the predicted dependent variables from it. To create such a model to predict the hydrogen product mole fraction (y_{H_2}) as a function of reactor temperature (T) and pressure (P), the engineer first proposes a model of the following form:

$$y_{H_2} = a_0 + a_1 f_1(T, P) + \cdots + a_N f_N(T, P)$$

where a_0 through a_N are unknown constant parameters and $f_i(T, P)$ are called **basis functions** of T and P. The engineer first selects the basis functions, which are things like $f_i(T, P) = P$, $f_i(T, P) = T^2$, $f_i(T, P) = P \log T$, $f_i(T, P) = \sin\left(\dfrac{2T}{\pi}\right)$, or any other function of the independent variables T and/or P that we can think of. The engineer then uses an algorithm of some kind to determine the best values of the unknown parameters a_i such that the resulting equation predicts the dependent variables as closely as possible to the known data that has been collected, either from experiment or from a rigorous model like Aspen Plus. It is called linear-in-the-parameters because the unknown parameters do not appear in the basis functions, resulting in a classic linear equation where the parameters are the unknowns. For this tutorial, we will start with the basis functions T, T^2, T^3, P, P^2, P^3, and TP, giving the following model equation:

$$y_{H_2} = a_0 + a_1 T + a_2 T^2 + a_3 T^3 + a_4 P + a_5 P^2 + a_6 P^3 + a_7 TP$$

Figure B2.6 Using the Regression feature in Excel.

This model equation is a guess. You can always try different basis functions and see how good it is. In Microsoft Excel, we will use the Regression feature to solve for the a_i parameters for us.

First, in the Excel file created by Aspen Multi-Case, add seven columns to the right of your data, one for each of the seven basis functions. For each column and row, compute the value of the basis function for that row using a formula. For example, for the first basis function (T) in row 2, the formula would simply be =C2, because the temperature column in my case is column C. For the basis function (P^2), my formula is =D2^2, because the pressure column in my case is column D. Then just copy those down to fill all rows.

Now, launch the Excel | Data | Data Analysis[6] | Regression tool.[7] For the Input Y Range (the dependent variable), choose the column that contains the H_2 mole fraction data from Multi-Case, and choose all the rows, including the header. For me this is I1:I1221 because column I contains my data for y_{H_2}, row 1 is the header, and I have 1220 rows of data (since I deleted a few bad ones). For the Input X Range, choose the seven rows and columns of the basis functions that you created in the previous paragraph, which for me is M1:S1221. Check the labels checkbox if you selected row 1 (the header information) in your range. This will use the names of the variables in the results for convenience. You can leave all the plotting options unchecked or check them, as you prefer. Mine looks like Figure B2.6.

The results are shown in Figure B2.7, and yours may differ slightly. The resulting parameters a_0 through a_7 are found in the Coefficients column of the regression output. The R^2 value, which is the most common metric for understanding goodness-of-fit, is about 0.975 in my example. The closer to 1, the better the model hits the experimental data, so this is ok. It could be better. However, it is important to understand that there is more to the story. For example, are there too many basis functions that might cause spurious results *in between* the data points? To check that, you can perhaps take more data points and see how well the model works for those new points (this is called a training set/testing set approach). Are any of the basis functions not really necessary? To check the latter, you can try deleting some of them and rerunning the regression and seeing if R^2 changes much. For example, when I remove P^3 as a basis function, the R^2 gets only trivially smaller, so I probably do not need it and thus it is better not to have it. Are there better basis functions that have not yet been tried? That takes more work, and finding the right combination is not a trivial task! There are some third-party packages which can help, like ALAMO,[8] which does the regression but also picks the best basis functions for you, which is quite nice.

♫ Music break[9]

---

[6]Do not use the Analyze Data tool (a lightning bolt symbol), that is different.

[7]If you cannot find it, you may have to enable the "Analysis ToolPak" in Excel's Options | Add-Ins section.

[8]Available from Carnegie Mellon University for free for students and some others at: https://minlp.com/alamo.

[9]Recommended listening: *Island in the Sun* by Weezer.

	A	B	C	D	E	F	G	H	I
1	SUMMARY OUTPUT								
2									
3	*Regression Statistics*								
4	Multiple R	0.987							
5	R Square	0.975							
6	Adjusted R	0.975							
7	Standard E	0.017							
8	Observatio	1220							
9									
10	ANOVA								
11		*df*	*SS*	*MS*	*F*	*Significance F*			
12	Regression	7	13.9971	1.9996	6666	0			
13	Residual	1212	0.3636	0.0003					
14	Total	1219	14.3607						
15									
16		*Coefficients*	*Standard Error*	*t Stat*	*P-value*	*Lower 95%*	*Upper 95%*	*Lower 95.0%*	*Upper 95.0%*
17	Intercept a_0	6.68E-02	1.88E-01	3.56E-01	7.22E-01	-3.01E-01	4.35E-01	-3.01E-01	4.35E-01
18	T a_1	-5.96E-04	7.19E-04	-8.29E-01	4.08E-01	-2.01E-03	8.15E-04	-2.01E-03	8.15E-04
19	T^2 a_2	3.07E-06	9.09E-07	3.38E+00	7.57E-04	1.29E-06	4.85E-06	1.29E-06	4.85E-06
20	T^3 a_3	-1.97E-09	3.78E-10	-5.22E+00	2.11E-07	-2.72E-09	-1.23E-09	-2.72E-09	-1.23E-09
21	P a_4	-1.66E-02	3.37E-04	-4.94E+01	8.29E-293	-1.73E-02	-1.60E-02	-1.73E-02	-1.60E-02
22	P^2 a_6	9.20E-05	6.30E-06	1.46E+01	1.53E-44	7.96E-05	1.04E-04	7.96E-05	1.04E-04
23	P^3 a_5	-2.32E-12	4.58E-13	-5.07E+00	4.51E-07	-3.22E-12	-1.43E-12	-3.22E-12	-1.43E-12
24	TP a_7	1.14E-05	2.93E-07	3.89E+01	1.06E-215	1.08E-05	1.20E-05	1.08E-05	1.20E-05

Annotations in figure:

This is the classic R^2 value. Numbers close to 1 means that model predictions match the experimental data closely. However, it says nothing about how the model behaves "in between" data points.

The values of the unknown parameters are in this column. They are labeled individually next to each one.

Figure B2.7   The results of the linear-in-the-parameters regression. The most important values are annotated.

PART 2: RUNNING ASPEN PLUS FROM PYTHON CODE USING PARALLEL COMPUTING

Multi-case is quite useful, but sometimes you need to automate your simulation by including simulation runs as a part of some larger algorithm. Although you can perform large sensitivity-analysis type studies as you can in Multi-Case, there are many other cases where you might want something more powerful:

- Connecting Aspen Plus to other software; e.g., the resulting stream conditions of Aspen Plus model might become the inputs to a Matlab model.
- Performing optimization, where you want to search in some intelligent fashion for the parameters that give you the best design (which we will do in Part 3)
- Handling sensitivity analyses with complex needs, such as binary decisions (such as the existence or non-existence of a unit operation), logic, or complex initial guess generation
- Performing a Monte-Carlo analysis, such as determining the effects of uncertain parameters on the outputs based on probabilities
- Advanced model identification strategies, especially those using iterative frameworks (more advanced than what we did in Part 1)

Tutorial 9 discussed how you can connect to Aspen Plus through the Excel interface, which you can then automate by writing code in VBA. In addition to Excel, it is possible to write a program that connects to Aspen Plus in any language as long as that language supports using the Windows COM interface ("Component Object Model"). The COM interface is a way that one program can connect to another program, even one you have not written yourself (like Aspen Plus). In my own group, we have used VBA, Matlab, C++, Fortran, and Python to do this. VBA and Matlab are popular platforms for engineers to use for coding because they are simple, handy, and available, but they are not full-featured languages and lack the ability to take full advantage of parallel computing with COM objects. Python, however, is rapidly growing in popularity among engineers because it strikes a good balance between the raw power of low-level languages like C++ and Fortran with user-friendliness and simplicity, like VBA and Matlab. Therefore, this tutorial uses Python as the chosen framework because it strikes the right balance for most engineers.

If you are not a Python expert, do not worry. If you are familiar with the basics of any programming language at all, you should be able to follow along, and readily adapt the example codes here for your own purposes. I have tried to make this as simple as possible. If you have no programming experience at all, you might be able to figure out how to make minor changes to the code and still take advantage of its usefulness. If not, well, that is what Multi-Case is for!

TOM'S TIP: If you do not have Python already installed, you can get the installer for free at https://www.python.org/downloads/windows/ and get the right version for your computer (most modern users will want 64-bit Python). When you run the install program, make sure you check the box for installing pip, which is a program that installs Python packages. I also recommend you check the box which adds Python to your Environment Variables. This means you can call Python from a Windows command line. Finally, you have to download and install pywin32, which is the package that lets you use COM objects (and thus call Aspen Plus). Once you have Python and pip installed, you can download and install pywin32 by typing `python -m pip install pywin32` at a Windows command prompt. If you are someone that uses Anaconda instead of pip (and obviously would have to know what that even means), you use this command instead: `conda install -c anaconda pywin32`.

First, make sure you have Python and the pywin32 package installed (see the Tom's Tip if you do not). We will use the same Aspen Plus file we made in Part 1, and if you would like you could use the provided one (see Preface for a link) instead (Tutorial B02.bkp). For Part 2 of this tutorial, the goal is to run a thousand or so simulations of this at various reactor temperatures and pressures, and get the resulting reactor heating duty and reactor outlet mole fraction of carbon monoxide for each. This will be just like Multi-Case (Part 1), except we are going to select our points using Latin Hypercube Sampling (LHS) instead.

The primary purpose of LHS sampling is to get as much sample diversity as possible within an unknown state space using a small number of samples. In Part 1, we did a Multi-Case example in which we sampled all combinations of reactor temperature from 600°C to 1000°C in 35 steps (about every 11°C) and pressure between 2 and 50 bar in 35 steps (about every 1.3 bar), resulting in $35 \times 35 = 1225$ samples. The result is a "rectangular" grid of evenly spaced points, similar to that shown in Figure B2.8 left, which is a "rectangular mesh" approach. The nice part about this is the regularity of the grid. The downside is that we have huge gaps in between points, and repeat samples of our values many times. For example, there are 35 cases where we sample 600°C, but no cases in which we sample anything in between 601°C and 610°C. Furthermore, as you increase the dimensionality of the problem (e.g., you have even more independent variables you are interested in), the size of the grid explodes to absurd numbers.

LHS, however, gives you a way of sampling much more of the interior points while still exploring the entire state space. In this approach, the two ranges are divided up in N increments, where N is the number of samples you want to take. So if you wanted to take $N = 1225$ samples, the temperature range of 600–1000°C would be broken up into 1225 pieces, or a width of about 0.32°C each. For the pressure range, the distribution is 0.04 bar apart. In LHS, we select N points, each of which has one of the N temperature points

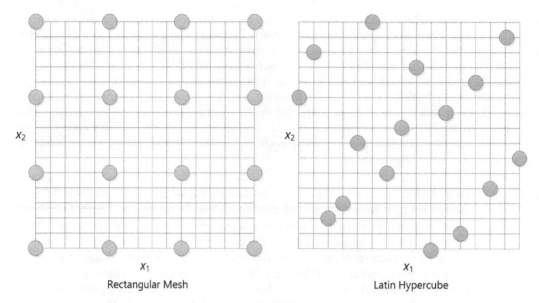

Figure B2.8 Two different ways of taking 16 samples of a state space, in this case, a two-dimensional space, with the two independent variables being x_1 and x_2. The left chart shows classic rectangular meshing. The right chart shows Latin Hypercube Sampling, where each evenly spaced row and each evenly spaced column has exactly one point in it.

and one of the N pressure points, chosen randomly, such that all of the N values of temperature are used exactly once, and all of the N values of pressure are used exactly once. The results look something like the chart on the right side of Figure B2.8. The goal of this is to allow you to explore a greater diversity of independent variables and not repeat anything twice, while keeping gaps to a minimum. It also scales well in large dimensional spaces, because you just choose N to be whatever you are willing to wait for. There are some disadvantages of course, but LHS is a very easy and effective way of taking samples.

Let's first look at the basic strategy. In parallel computing, the idea is that you have several processors in your computer and you want them to run simultaneously in parallel. So for example, if you have six processors, you might want to have six copies of Aspen Plus open, where each processor uses one to do work for you at the same time. To make this happen in Python, the easiest way to do this is to first have a Main program, where the purpose of the Main program is to create a long list of Jobs that need to be done, create Processes that do those jobs, and then when the Processes are all done, collect all of the resulting data and do something with it. For LHS with parallel computing, the overall strategy might be:

1. Start the main program:
   a. Create a list of all the samples we want to take, using LHS. Specifically, this means coming up with all the locations of all the points of interest. These are the Jobs.
   b. For each computer processor we will use, create a new Process.
   c. Give each Process a portion of the big LHS Jobs list. For example, if there are two Processes, each should have its own half of the list.
   d. Tell each Process to start, and then wait for them all to finish.

2. Within each Process:
    a. For each sampling point that was given to it, run an Aspen Plus simulation.
    b. Print the resulting simulation variables of interest to the screen.
    c. Stop when all the simulations have been run.
3. Then back in the main program, once all the Processes have finished:
    a. If desired, do further analysis with the resulting data from the Processes.
    b. End the program.

There are more complex ways to improve upon this, but it is an effective approach that a typical engineer could manage to code on their own. Now, look at some basic Python code that can help us do this, found in the file Part2.py. You can download this or type it yourself (I recommend typing so you get to understand each line yourself). This code is quite simple, and is a good point for learning. You can type it into any text editor. Everyone has Notepad, but I recommend some better free ones for engineers like PSPad, Notepad++, and EditPad Light. Professional coders will choose more advanced software, like Visual Studio, Emacs, or heaven forbid, vi.

Figures B2.9 and B2.10 contain the code (it is one file, just split into two images for the book). There are basically three sections. The first (lines 44–48 of Figure B2.9) is the import package section. The second (rest of Figure B2.9) is the function that runs a bunch of Aspen Plus simulations at given points and prints out the

```
         0       10       20       30       40       50       60       70       80       90      100      110      12
        ..|........|........|........|........|........|........|........|........|........|........|........|........|........|
43
44 import os                             # This connects to your operating system
45 import win32com.client as Windows     # This is a package that lets you call Windows programs
46 import multiprocessing as mp          # This is the multiprocessing package
47 import random                         # This lets us call the random number generator
48 import time                           # This has the timer function in it
49
50
51 # This runs the Aspen Sim for the given T and P, and returns the result. Does not do error checking.
52 def RunAspenPlusReactor(JobID, Points):
53
54     # Create the Aspen Plus Object
55     AspenPlus = Windows.Dispatch('Apwn.Document')
56
57     # Open the file
58     AspenPlus.InitFromArchive2(os.path.abspath('Tutorial B02.bkp'))
59
60     # Now, run a simulation for each point we were given
61     for point in range(len(Points)):
62
63         # Set the T and P variables for the current point as the T and P of the reactor in the simulation
64         AspenPlus.Tree.FindNode('\Data\Blocks\REACTOR\Input\TEMP').Value = Points[point][0] # Temperature in C
65         AspenPlus.Tree.FindNode('\Data\Blocks\REACTOR\Input\PRES').Value = Points[point][1] # Pressure in bar
66
67         # Run the simulation
68         AspenPlus.Engine.Run2()
69
70         # Get the results
71         Q = AspenPlus.Tree.FindNode('\Data\Blocks\REACTOR\Output\QCALC').Value        # The reactor heat duty in Gcal/hr
72         CO = AspenPlus.Tree.FindNode('\Data\Streams\PRODUCTS\Output\MOLEFRAC\MIXED\CO').Value # The mole fraction of CO
73
74         # Pint out the results
75         print(f"In JobID: {JobID} T: {Points[point][0]:0.2f} P: {Points[point][1]:0.2f} Q: {Q:0.4f} CO: {CO:0.4f}")
76
77     return (JobID)
78
```

Figure B2.9 The first half of the simple form of the LHS sampling code.

```
      0        10        20        30        40        50        60        70        80        90       100       110       120
      |....|....|....|....|....|....|....|....|....|....|....|....|....|....|....|....|....|....|....|....|....|....|....|....|....
 79  if __name__ == '__main__':
 80
 81      ###########################
 82      # USER CHANGE THIS PART #
 83      ###########################
 84      NumAspenCopiesOpen = 4
 85      PointsPerProcessor = 10
 86      ###########################
 87      ###########################
 88
 89      NPoints = NumAspenCopiesOpen * PointsPerProcessor
 90
 91      # Create some empty lists for us to fill in later
 92      PointsList = list()
 93      T = list()
 94      P = list()
 95
 96      # Starts the timer
 97      TimeStart = time.perf_counter()
 98
 99      # Create the LHS points
100      for i in range(NPoints):
101          T.append(600 + (400) * (i/(NPoints-1)))
102          P.append( 2 + (50) * (i/(NPoints-1)))
103      random.shuffle(P)
104
105      # Make job lists for each processor
106      for i in range(NumAspenCopiesOpen):
107          for c in range(PointsPerProcessor):
108              if (1 > c):
109                  PointsList.append(list())
110              PointsList[i].append([T[c + i * PointsPerProcessor], P[c + i * PointsPerProcessor]])
111
112      # Let's see how many processors we have (this is logical, not physical)
113      print("Number of processors available: ", mp.cpu_count())
114      print("Number of processes used: ", NumAspenCopiesOpen)
115
116      # Create the pool of processors. Each processor will have one copy of Aspen Plus to work with
117      pool = mp.Pool(NumAspenCopiesOpen) # This creates a pool of workers using all of the available ones.
118
119      # Start all the processors!
120      ResultData = [pool.apply_async(RunAspenPlusReactor, args=(JobID, PointsList[JobID])) for JobID in range(NumAspenCopiesOpen)]
121
122      # Closes the parallel worker pool
123      pool.close()
124
125      # This waits until all the above jobs are finished before continuing on
126      pool.join()
127
128      # Stop the timer. We are done!
129      TimeFinish = time.perf_counter()
130      print(f"Finished in {TimeFinish - TimeStart:0.2f} Seconds.")
```

Figure B2.10   The second half of the simple form of the LHS sampling code.

resulting stream conditions, and each processor in your parallel computing implementation will run its own copy of this function. The third section (Figure B2.10) is the main program that drives everything, and is where the program actually starts. Here is a breakdown of what happens when this code is run, in *chronological order:*

- *Lines 44–48: Import packages.* There are many Python packages which contain functions and code that you can use in your programs. The import commands bring these packages into your code so you can use them. Here, we need these packages for different uses later in the code. For example, the first package, os (operating system) has basic commands like changing directories. You can see that it was used in line 58 with the statement os.path.abspath('Tutorial B02.bkp') which basically gives me the full path name to the file Tutorial B02.bkp, which is the Aspen Plus simulation we are going to use.

- *Line 79: The main function.* The program actually starts at this point when you first run it. The previous code (lines 49–78) is the definition of a function, but it is not where the program actually starts executing. This `if` statement is here because of the way parallel processing works in Python. It basically ensures that the below code only runs in the main program, not in any of the Processes.
- *Lines 84–85: User settings.* The user can change these settings before running the program. The first is the number of parallel Processes that you want at a time (the number of copies of Aspen Plus you want running at once), and the second is the number of sampling points that you want each Process to run in series. Keep it small for now, perhaps 10. The total number of points that will be sampled (the N in LHS sampling) will be the multiple of these two numbers (line 89). You should pick a small value of `NumAspenCopiesOpen` at first (maybe two) and increase later to see what happens. You should not have this number be larger than the number of logical processors on your computer, because ideally you want each Process to have its own processor.
- *Lines 91–97: Housekeeping.* The `list()` command in Python creates an empty list. Here, we are creating three empty lists that will hold the values of the independent variables that we want to test later.
- *Lines 99–103: Create LHS points.* LHS points are very easy to create. First, you just make a list of evenly spaced numbers in each dimension on the range that you care about. In this case, making a list of N points between 600°C and 1000°C (line 101) and making N points between 2 and 52 bar (line 102). To make random pairs, you then need only randomize the order of each list of these numbers (or to save time, randomize each list except one). For example, if there were $N = 5$ temperatures and 5 pressures, you would start with a temperature list like 600°C, 700°C, 800°C, 900°C, and 1000°C, and a pressure list like 2, 14, 26, 38, and 52 bar. Then you randomize the pressure list to get: 14, 2, 52, 26, 38. Then your five LHS points would be (600, 14), (700, 2), (800, 52), (900, 26), and (1000, 38).
- *Lines 105–110: Assign jobs to each processor.* Each processor is assigned a portion of the big list. Ideally, you want each processor to have the same amount of work so they all start and finish at the same time. Although you cannot know how long it will take until you run it, it makes sense to give each processor the same number of jobs. The `PointsList` object is a list, containing one member for each Process. Each member of this list is another list: the list of points on which that Process will run simulations.
- *Lines 116–126: Start each process and wait for them to finish.* First, a pool of Processes is created (line 117) by telling it how many Processes you want open at the same time. Then, you use the `apply_async` command to start them (line 120). This line includes the information sent to the `apply_async` command which includes the name of the function that we want each Process to run (`RunAspenPlusReactor`), the data that we want to give to each Process that it will use (`PointsList`), and where we want to store the result (a new data object called `ResultData`). In lines 123–126, the `close()` and `join()` operations, which usually always appear together in this order, effectively wait for all of the Processes to finish before continuing, and close the pool when done.
- *Lines 129–130: Rest of the program.* This section happens after all Processes have finished. Here, it prints out the run time, but if we wanted to, we could take the information stored in `ResultData` and use it for something.

When this code is run and it gets to lines 116–126, it creates all of the Processes that run in parallel, and in each, the function `RunAspenPlusReactor` runs. That function works as follows:

- *Line 55: Create an Aspen Plus instance.* This opens a new copy of Aspen Plus on your computer, in the background, using the COM interface. There is no graphical user interface and you cannot use it through normal means. But you can see it in the Task Manager in windows. When the code runs, you should see one AspenPlus.exe process in the Task Manager for each Process that you create with this code.
- *Line 58: Load the file.* This takes the simulation file of interest and loads it into that background copy of Aspen Plus. We use the .bkp file instead of the .apw file because it is much smaller and consumes less memory.

You can use the .apw file though if you need to. Lines 55–58 can take 3–5 seconds and this is in many cases longer than the simulation itself. It is a major burden, and later in this tutorial, we will go to great lengths to do this as little as possible.

- *Lines 61–75: Run the simulation for each point given to the function.* This is a loop, and this loop happens once for each point that was given to the Process by the main program. Note that in Python, the indentation of the code is important (exactly four spaces per indent, no tabs!), and that tells you which code goes "inside" loops, if statements, and so on. The last line of code inside the loop is line 75 because it is indented—everything indented happens inside that loop. Line 77 is not indented and so it is not in the loop, and executes only after the loop has finished.

- *Lines 63–64: Set the independent variables.* These two lines set the values of the reactor temperature and pressure in this Process' copy of Aspen Plus, using the values given to it by the main program for the current point. Each time the loop iterates, a different point will be assigned here. You can see that the variable names should make intuitive sense once you look at them, but it can be hard to know what variable names to look for. You can find them by opening your simulation in the normal, graphical user interface for Aspen Plus, going to the Customize ribbon in the Simulations mode, and choosing Variable Explorer. You can navigate through the tree to find what you are looking for, and then on the right-hand side, the Call attribute tells you what Python code to use here. Figure B2.11 shows an example of the reactor temperature that is used in line 63.

- *Line 68: Run the simulation.* This runs the simulation! For complex simulations, this step can take a long time. For this simple simulation, it takes approximately 0.15 seconds on my desktop PC.

- *Line 71: Get the results.* The first two lines here collect the output data from the simulation, and then, in line 75, the results are written to the command terminal.

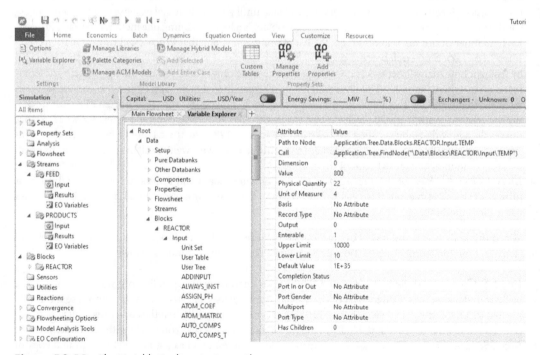

Figure B2.11   The Variable Explorer in Aspen Plus.

● *Line 77: End the process.* This is the end of the function. When all Processes have reached this point, then the main program can continue in line 129. Also, as each Process ends, Python closes whatever copy of Aspen Plus it opened in line 55. This is a nice feature of Python called "garbage collection." That is a fun way of saying that if the programmer forgets to delete up the objects it created, it deletes them for you.

TOM'S TIP: Once you start writing programs that launch other programs inside of loops, you have the potential to really screw things up for yourself in several ways. Here are some commands that may help you (don't type in the colon):

● `CTRL+Break` or `CTRL+c`: These keyboard combinations in the Command Prompt will stop the current program. However, they may leave programs and files open that you want to get rid of.
● `taskkill /IM "AspenPlus.exe" /f`: Python is good about shutting down the Aspen Plus copies that you opened after the program is run. However, if you have to break it with the previous keyboard combinations, that might not happen. You can close them manually in the Task Manager, but if that gets tedious, use this command to close all open copies of Aspen Plus. Warning! This will also close any other copies you have open, like the graphical user interface, as well.
● `del _*.*`: This dangerous but useful command will delete all files in this directory that start with the underscore "_". Aspen Plus creates lots of these temporary files when it is running and then deletes them when the program is closed. If it gets killed though, those files get left behind, and there is a lot of clutter. *Be very careful because if you forget to type the underscore, the command will delete all files in this directory, including your simulation and Python code!*

Now, it is time to run it. In the command prompt (launch one from the Start Menu), navigate to the directory in which contains your Python file (it should also contain the Aspen Plus simulation file as well). For example, if your files are in C:\Books\LAP24 you would type into your Windows Command Prompt:

```
cd C:\Books\LAP24
```

Then you can launch the program (mine is called Part2.py) by typing:

```
python Part2.py
```

After you run it, you might get something like:

```
Number of processors available:  12
Number of processes used:   4
In JobID: 0 T: 600.00 P: 10.97 Q: -12.8873 CO: 0.0372
In JobID: 0 T: 610.26 P: 22.51 Q: -13.6073 CO: 0.0284
In JobID: 0 T: 620.51 P: 48.15 Q: -14.2133 CO: 0.0211
In JobID: 0 T: 630.77 P: 19.95 Q: -12.6019 CO: 0.0392
```
(skipping some lines)
```
In JobID: 2 T: 876.92 P: 12.26 Q: 5.9135 CO: 0.2354
In JobID: 2 T: 887.18 P: 26.36 Q: 2.5706 CO: 0.2048
In JobID: 2 T: 897.44 P: 35.33 Q: 1.7355 CO: 0.1957
Finished in 16.65 Seconds.
```

So it worked! The first two lines told me that my computer has 12 logical cores but I am only using 4 of them for this example. The next set of lines of the output showed me which Processor was doing the work, what Temperature (T), Pressure (P) it used, and the resulting heat duty (Q) and product CO mole fraction (CO). At the end, it told me it was done and how long it took. If you wanted to use this data in Excel, for example, you could copy from the terminal and paste into Excel, and then use the Data | Text-to-Columns function in Excel to parse it. For large jobs, copy-pasting from the terminal is awkward, so instead, you can redirect the printed output to a text file instead of the screen, like this:

```
python Part2.py > PickAFileName.txt
```

You can then open the file in a text editor or even directly in Excel. If you are going to do this for anything serious, you can change the print statements in the program to give you a file format better suited for your needs, like comma delimited (CSV).

Ok, here's the rub. This above code may work for a small number of runs. But change the number of points per processor to something larger, like 100–300, and you may have a problem. The output may look fine at first, until you realize that there are a lot of points missing. What went wrong? Well, a lot of things actually, and it could be any of these:

- The simulation did not converge when you tried to run it. This could be because the initial guess used for the current run was not good. Note that it uses the results of the previous run as the initial guess for the next run, so the order of the points matters here. It could also be some other reason, such as any of the classic simulation convergence challenges, a poorly constructed model, or simply a bug within Aspen Plus.
- The simulation converged but some of the resulting data are missing, because of internal Aspen Plus errors or bugs.
- Aspen Plus crashed.
- There was some kind of error in the COM interface between Aspen Plus and your Python program.

The code above has no way of handling any of these errors. If any of the lines that communicate with Aspen Plus fails, it just shuts down the whole Process, although the other Processes will continue. So you might see in your output, for example, that it worked for a little while on one process but then stopped for some reason. On my desktop, one of the above issues tends to happen after every 10–30 runs, even on perfectly designed, simple simulations! Some test machines had more problems, fewer problems, or no problems at all. Empirical evidence suggests that the rate of these failures also depends on the particulars of your machine, the number of Processes you are using (the more you use, the more often it fails), the number of cores in your machine, and others. Therefore, because of the realities of dealing with OPC (other people's code), the difficulties of the COM interface, and the possibility of routine non-convergence issues, we need to make the code more robust.

To do this, we need some more robust code that checks for COM or other errors almost every time we interact with Aspen Plus. You can find this in the file Part2b.py, which is shown in Figures B2.12–B2.14. Part2b.py is basically the same thing as Part2.py, where the main code is basically the same, except that all of the Aspen Plus calls in the `RunAspenPlusReactor` function uses the `try` | `except` | `else` block to catch errors and handle them. For example, lines 72–76 have something similar to this:

```
try:
    AspenPlus.InitFromArchive2(os.path.abspath('Tutorial B02.bkp'))
except:
    print(f"Could not load the file #{JobID}.")
    return (None)
```

In the `try` section, put the COM calls that you want to try. In this case, the COM call tells Aspen Plus to load the simulation via `InitFromArchive2`. If an error happens, then the code inside the `except` block executes, which in this case, uses `print` to display an error message, and then completely exits the function using the `return` function such that no code beyond this point executes. If we did not have the `return` function there at the end of the except section, the program would continue after the `try` block on line 77, but in this case, we want it to just stop now because we cannot open the file. If the COM call was successful, then the `except` block is skipped completely, and the program continues in line 77. Other sections of the code use the `else` section of the `try` block, which is executed only when the `try` is successful, and then the program continues at the next line after the `try` block.

```python
59  def RunAspenPlusReactor(JobID, Points):
60
61      # Create the Aspen Plus Object
62      AspenPlus = Windows.Dispatch('Apwn.Document')
63
64      # These are variables that keep track of how many times certain errors happened
65      reinits = 0
66      redispatches = 0
67      varfails = 0
68      finalfails = 0
69
70      # Load the simulation from the file, and fail if it doesn't work the first time
71      # If it doesn't work the first time, usually this means the file name is wrong, or the license checkout failed.
72      try:
73          AspenPlus.InitFromArchive2(os.path.abspath('Tutorial B02.bkp'))
74      except:
75          print(f"Could not load the file #{JobID}.")
76          return (None)
77
78      # Create the empty list for storing results
79      Data = list()
80
81      # Now, run a simulation for each point we were given
82      for point in range(len(Points)):
83
84          # This keeps track of how many times we attempted to run the simulation for this point
85          attempt = 0
86
87          # Default values for the results are stored here, in case of failure
88          Q = 0
89          CO = 0
90
91          # Run The Simulation. We allow 3 attempts before giving up.
92          while (3 > attempt):
93
94              # If our first attempt failed, try re-initializaing the simulation and try again
95              # This could be just because the initial guess (the previous run) was poor
96              if (1 == attempt):
97                  print("Attempting to reinitialize ", JobID)
98                  try:
99                      # Reinitialize the simulation
100                     AspenPlus.Engine.Reinit()
101                     reinits = reinits + 1
102                 except:
103                     print("Could not reinitialize ", JobID)
104                     attempt = attempt + 1
105                 else:
106                     print("Reinitialize successful ", JobID)
107
```

Figure B2.12 Part one of the robust LHS sampling code.

```
107
108             # If our second attempt failed, try relaunching Aspen Plus entirely
109             if (2 == attempt):
110                 print("Attempting to redispatch Aspen Plus", JobID)
111                 try:
112                     AspenPlus = Windows.Dispatch('Apwn.Document')
113                     redispatches = redispatches + 1
114                 except:
115                     print("Could not dispatch on reopen ", JobID)
116                     break
117                 else:
118                     print("Dispatch successful ", JobID)
119
120                 # At this point, the dispatch worked, so try to open the file again.
121                 try:
122                     AspenPlus.InitFromArchive2(os.path.abspath('Tutorial B02.bkp'))
123                 except:
124                     print("Could not initialize on reopen ", JobID)
125                     break
126                 else:
127                     print("Initialize on reopen successful ", JobID)
128
129             # At this point, we have an open and working file.
130             # Set the T and P variables for the current point as the T and P of the reactor in the simulation
131             AspenPlus.Tree.FindNode('\Data\Blocks\REACTOR\Input\TEMP').Value = Points[point][0] # Temperature in Ce
132             AspenPlus.Tree.FindNode('\Data\Blocks\REACTOR\Input\PRES').Value = Points[point][1] # Pressure in bar
133
134             # Run the simulation
135             try:
136                 AspenPlus.Engine.Run2()
137             except:
138                 # If it did not run, print an error message
139                 print("Aspen Run Error in jobID ", JobID, " on attempt" , attempt)
140                 # Increment the attempt counter, so can try again in the next while loop iteration
141                 attempt = attempt + 1
142             else:
143                 # The simulation was successful. Let's record that it worked this time for later.
144                 worked = 1
145
146                 # Sometimes the simulation can be successful but some of the simulation varibles are missing
147                 # We need to check for that here.
148                 try:
149                     Q = AspenPlus.Tree.FindNode('\Data\Blocks\REACTOR\Output\QCALC').Value
150                 except:
151                     # Even though this simulation worked, there is some problem in Aspen Plus that prevents us
152                     # From seeing the variable. So We can set a default variable and indicate that it did not work.
153                     Q = 0
154                     worked = 0
155                     varfails = varfails + 1
```

Figure B2.13 Part two of the robust LHS sampling code.

Using this strategy, the provided code uses `try` blocks around almost Aspen Plus calls, and tries to handle the errors accordingly. The general strategy of this approach is to keep the current copy of Aspen Plus open as long as possible, and only reinitializing the simulation or reopening the file if necessary. The general algorithm is now as follows:

1. (Same as Part 1 before)
2. Within each Process:
    a. For each sampling point that was given to it:
        (1) Start a loop in which we attempt to run an Aspen Plus simulation up to three times, such that:
            (a) On the first attempt, try to run the simulation. This means it will use the results of the previous simulation as the initial guesses for the next simulation. If the simulation

```
156
157        # We need to check against each variable that we want to keep. A little tedious but important for n
158        try:
159            CO = AspenPlus.Tree.FindNode('\Data\Streams\PRODUCTS\Output\MOLEFRAC\MIXED\CO').Value
160        except:
161            CO = 0
162            worked = 0
163            varfails = varfails + 1
164
165        # If none of the variable pulls failed, then we can break from our while loop because the point was
166        if (1 == worked):
167            break
168        else:
169            # Otherwise we just try again
170            attempt = attempt + 1
171
172    # This checks to see if the previous while loop exited with a simulation run and all output variables obtai
173    # If not, we just keep track of the number of times this happened for convenience.
174    if (2 < attempt):
175        finalfails = finalfails + 1
176
177    # Store and print the data. Although I am not doing anything with the Data object, you could do something w
178    Data.append([Q, CO])
179    print(f" {Points[point][0]:0.4f}, {Points[point][1]:0.4f}, {Q:0.4f}, {CO:0.4f}; ")
180
181    # Close this out and spit out the data.
182    print("JobID: ", JobID, "Finished with this worker. Completed ", len(Data), " points, with ", reinits, " reinit
183    return (JobID, Data)
184
185 if __name__ == '__main__':
186
```

Figure B2.14 Part three of the robust LHS sampling code. The remaining code below this line (in the main section) is essentially the same.

executed without errors, *and* all COM interface commands worked, *and* all results data could be extracted from the simulation, then it worked, so stop the loop.

(b) If something went wrong after the first attempt, reinitialize the simulation, which is often very fast. Then, try to run the simulation a second time. If everything worked correctly with the simulation and all data extraction steps, stop the loop.

(c) If something went wrong after the second attempt, close the Aspen Plus program used by this Process. Open a new copy of Aspen Plus, and load a new copy of the file. This is often slow (3–5 seconds). Then, try to run the simulation a second time. If everything worked, stop the loop. If it still does not work, then the loop still stops (it gives up trying with this point), and instead some default data are recorded as a way to signal to the user that it did not work.

(2) Print the resulting simulation variables of interest to the screen.

3. (Same as step 3 in the original code)

Some key modifications in the code are:

- *Lines 65–68: Error statistics.* These variables keep track of the number of times we had to reinitialize and redispatch the code, as well as the number of times we had a successful simulation but had some error extracting data, and how many points we ended up on completely. These are here purely for informational purposes, so you can see what is going on behind the scenes, and are reported to the user in line 182.
- *Line 92: The attempt loop.* This is the start of the loop that will try to run up to three Aspen Plus simulations on a point before giving up. The `attempt` variable keeps track of how many times we attempted to run the simulation on this point.

- *Lines 94–106: Reinitialize.* This reinitializes before starting a second attempt. You can see how the `else` statement in the `try` block is used here; it reports to the user that the reinitialize was successful.
- *Lines 108–127: Redispatch.* This closes Aspen Plus and makes a new copy of it, and loads the simulation file. If either of these fails, the `break` command ends the attempt loop because there is no point in going further if we cannot even open the simulation.
- *Lines 148–170: Collect data.* This code happens after an Aspen Plus simulation is successful and is used to collect the simulation results we care about. The `try` statements here work in series because you need one attempt for each variable you are trying to collect. The `worked` variable lets the program know if all attempts were successful or not. If at least one failed, it breaks from the attempt loop because there is no point in going further if the simulation worked but there was a data collection error. However, we still try to get as much data as possible, because maybe we can get some of it.

Once you get the code running, you should see a much more robust run. You should be able to set up the program to run a thousand or so simulations and find that it works much more robustly, with few to no missing data points at all. You may find some things in the output that look like this:

```
    ...stuff...

615.5129, 45.4112, -14.3034, 0.0205;
615.6797, 9.0475, -11.7538, 0.0505;
Aspen Run Error in jobID  3   on attempt 0
Attempting to reinitialize  3
Could not reinitialize  3
Attempting to redispatch Aspen Plus 3
615.8465, 44.9108, -14.2809, 0.0207;
616.0133, 4.4187, -9.8889, 0.0746;
    ...stuff...

616.8474, 25.8949, -13.5608, 0.0285;
617.5146, 14.5521, -12.6147, 0.0397;
718.5988, 50.7907, -10.5889, 0.0594;
Dispatch successful  3
617.6814, 30.0651, -13.7342, 0.0265;
718.7656, 41.3244, -10.1030, 0.0661;
    ...stuff...
```

In this case, you can see that there was some problem that happened on Processor 3. It attempted to reinitialize, but that did not work, so it just redispatched the simulation. However, this takes some time, and the other processors will keep going as normal while that is happening. This is why there are results printed between the start and completion of the new Aspen Plus dispatch in processor 3.

You might also see a reinitialize that worked without needing to redispatch, like this:

```
619.0158, 16.3453, -12.7555, 0.0379;
720.1001, 16.0951, -7.2095, 0.1052;
Aspen Run Error in jobID  2   on attempt 0
Attempting to reinitialize  2
Reinitialize successful  2
619.1827, 42.8674, -14.1230, 0.0222;
 720.2669, 21.8916, -8.2447, 0.0914;
 619.3495, 26.7289, -13.5136, 0.0289;
```

In this case, the Aspen Plus simulation in Processor 2 failed, but, after a reinitialize, the simulation worked. This may often happen when there are no COM communication failures, but rather, the simulation itself does not converge because the results from the previous point made for a poor initial guess for this new point. You can see that the reinitialize was very fast, with no results appearing from the other Processors in between (although that could happen by luck).

At the end of the run, you might see something like this:

```
699.7498, 48.7473, -11.3309, 0.0509;
699.9166, 28.6472, -10.1358, 0.0671;
JobID: 1 Finished with this worker. Completed  300  points, with
5  reinitializes,  6  redispatches,  0  variable pull fails, and
2  complete failures.
Finished in 244.18 Seconds.
```

The end of every Process prints out a statement that showed how many problems it had. In this case, I had four Processes with 300 points each. Processor 1 was the last to finish. It attempted all 300 points, and needed to reinitialize five times and redispatch six times for it to work. There were no times when the simulation worked but for some reason we could not access the results, and two times we were not able to run the points at all. So what you would do in this instance is go back and rerun the two remaining points either in Aspen Plus or in a separate run using this code but with just the missing points (you can tell which points did not work because they have default values like 0 or infinity for them).

🎵 Music break[10]

PART 3: PARTICLE SWARM OPTIMIZATION IN ASPEN PLUS WITH PARALLEL COMPUTING

Although the optimization tools built into Aspen Plus are useful, they are not always robust and do not support parallel computing. For many years now, engineers and researchers have been designing their own algorithms to do optimization, including those using Aspen Plus automation via the COM interface. Particle Swarm Optimization (PSO) is one such algorithm that is extremely effective in solving optimal design problems with Aspen Plus in particular.

Although there are many ways to do optimization, using Aspen Plus calls via COM requires an optimizer that is classified as "derivative-free" or "black box," meaning, it does not require access to sensitivity values, derivatives, or the model equations. This limits our choices of optimizer algorithms and prevents the use of mathematical programming, for example. After 15 years of extensive use in my research group, we have found that PSO is usually the fastest and most consistent black box optimizer for Aspen Plus applications, compared to other popular options such as Genetic Algorithms, Taboo Search, Simulated Annealing, Differential Evolution, Ant Colony Optimization, Empire Colony Optimization, simplex searches, and countless others and variants. You are free to try these of course, but this tutorial will show you how to use PSO to optimize Aspen Plus while taking advantage of parallel computing. As best as I can tell, this is the first such publicly available PSO code suitable for calling Aspen Plus that uses robust error checking and parallel computing. Although there are many variants of PSO suitable for chemical process simulations specifically,[11] this tutorial shows the "plain vanilla" version, which is already quite good as it is.

PSO is a clever algorithm, originally developed by James Kennedy and Russel Eberhart in their classic 1995 paper.[12] Like all optimization algorithms, the goal is to search a multidimensional space for the location

---

[10]Recommended listening: *Long Way Home* by Gareth Emery.

[11]See Adams TA II, Seider WD. Practical optimization of complex chemical processes with tight constraints. *Comp Chem Eng*, 2008, 32:9:2099–2112 for more ideas.

[12]Kennedy J, Eberhart R. Particle swarm optimization. In *Proceedings of ICNN'95-International Conference on Neural Networks*, 1995, 4:1942–1948.

that has the smallest[13] objective function value. The difficulty with using a black box objective function (Aspen Plus) is that without having access to the model equations, you can only know the final objective function values at each point, and thus you have no extra information about what the state space looks like. You can only sample it blindly and hope you are not missing something, and as such, you can never truly know if the answer you found is the global best.

The PSO algorithm is a way of sampling this space blindly in an iterative but intelligent fashion. A "particle" in the PSO algorithm is simply a point in the state space where a sample is taken, and the objective function computed. There are N_p particles (typically N_p is between 20 and 40), each with its own position (where they are located in the state space) and velocity (the direction they are heading). In each iteration, the objective function is computed at the location of each particle. Then, when that is done, the particles move to new points. Their movement is determined by their current velocity, which includes the direction they are traveling and their "speed," or how far they will move after they compute an objective function.

The velocities change each iteration, based on a few factors. First, there is a small random perturbation in both direction and speed, just to keep things a little chaotic (entropy can be good when exploring the unknown). Second, the velocity changes to move toward the best point that the particle has found so far (the so-called "personal best"). So if the particle is far away from its personal best, it will eventually turn around and head back toward it. But at the same time, the velocity also changes to turn toward the "global best," or the best point than any particle has found thus far. There is some random weighting between the personal and global bests as well.

The net effect is that the particles tend to explore the state space broadly and coarsely at first. However, over time, a few promising locations become explored in more detail based on each personal best. Next, one region gets sorted out as the most promising and all particles migrate toward the global best. At that point, the global best is refined through small refinements as the rest of the state space is ignored. The algorithm terminates based on a user-supplied maximum number of iterations, or some early-exit criteria, such as all the particles are close together (which is what we will use), or the global best has not improved within a certain number of iterations.

For a video of this in action, check the link in the footnote,[14] with still captures shown in Figure B2.15. The video is short, but shows how the particles start out moving quickly and everywhere, and then eventually converge into a small region for small refinements. This was performed on a chemical process model (using Aspen Properties in this case) where unit cost was minimized, with two independent variables shown (two different unit temperatures).

The code for this example is found in Part3.py. The general strategy is as follows:

1. Create N_p particles and set their initial positions across the state space. I recommend using LHS to set the initial positions.
2. Set the initial velocities to be in random directions within the state space, with a magnitude randomly chosen between 0% and 25% of the range in each dimension.
3. For each iteration
   a. For each particle
      (1) Run the simulation using its current location as the independent variables. Compute the objective function based on the simulation results.

---

[13]We usually speak of optimization as wanting to minimize. If you want to maximize, you can just minimize the negative of the objective function.

[14]https://youtu.be/_bzRHqmpwvo.

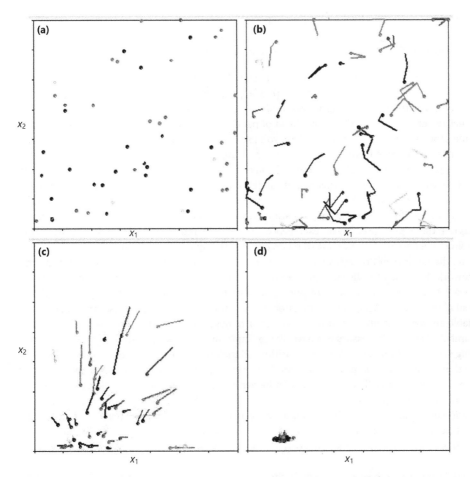

Figure B2.15 A graphical depiction of a PSO. (a) The particles start at different points in the state space (in this case just random). (b) After a few iterations. The lines depict the movement of the particles, with the kink in the line being the location of the point for that particle at the previous iteration, the circle point toward its next location, and the tail pointing toward two locations ago. The longer the line, the faster the particle is moving. (c) About halfway through, the particles have finished exploring the larger state space and are heading now toward the same area. (d) Toward the end, they have congregated around the global best and are moving slowly and making small improvements. The optimization could be stopped now since they are all close together. For the video, see https://youtu.be/_bzRHqmpwvo.

    b. For each particle

        (1) Update the personal and global bests in case any better points have been found since the last iteration.

        (2) Adjust the velocities according to randomly weighted influences of personal best and global best.

        (3) Adjust the particle locations by moving the particles according to their velocity.

    c. Check for early exit. In this example, if all points are within a small distance from the global best, then stop early. This is called "consensus."

4. When finished, the global best is reported as the optimal point.

Generally speaking, PSO parallelizes nicely because in step 3a, each particle can be run separately from the other and in any order. In theory, you could have up to N_p processors, and typically you would choose N_p to be an integer multiple of the number of processors you have. Step 3b is very fast, so fast that this does not need to be parallelized and is not worth the overhead of parallel computing. Therefore, our code will run everything on one processor, except step 3a, in which you use more of them.

We can reuse most of the code from Part 2b as well. The `RunAspenPlusReactor` function just runs a bunch of Aspen Plus simulations at different points. So, in each iteration, we can spawn up some Processes, each of which will take a portion of the particles, and run the Aspen Plus simulations to compute the objective function. For example, if we have $N_p = 32$ particles, and four Processes, each Process will be given eight points to simulate. This would then repeat each iteration. The downside of reusing our previous code without changing it is that the Aspen Plus copy for each Process needs to be redispatched at the start of each PSO iteration which slows it all down a bit, but it is so much easier that way so we will do it.

As a final note, the number of Processes you choose to use here makes a big difference when using this particular framework. Because of redispatching, I may want to use less Processes than I have processors because I may want to increase the number of points simulated per process to reduce dispatching times. Furthermore, a single standalone Aspen Plus simulation will also use two or three processors simultaneously if they are available anyway, meaning that it might be better to have the number of Processes be on half to one third the number of logical processors you have. For example, on my desktop with 12 logical processors, I found that using 4, 6, and 12 Processes for this algorithm, all took the same amount of time with the same results, and that they all used all 12 logical processors all the time anyway since each Aspen Plus simulation itself used one to three processors. The PSO took a lot longer when using one to three Processes, however, so there was definitely a speedup.

The optimization example we will use is the minimization of costs of operating the reactor (the same one as in the previous examples) by varying the reactor temperature (between 600°C and 1000°C) and pressure (between 2 and 50 bar). The objective function (cost) is computed from the sum of these elements:

- If the heating duty is positive (heating is needed), the cost of supplying this heat is $16/GJ.
- If the heating duty is negative (cooling is needed), the cost of cooling is $2/GJ.
- The value of the H_2 produced in this stream to my company is $0.10 per kg produced (value is a negative cost).
- The value of the CO produced in this stream to my company is $0.005 per kg.
- Because our reactor products will be needed at 51 bar further downstream, we will have to compress it. The lower the pressure of the reactor, the more we have to pay for compression. The compression costs are based on the reactor pressure P:

$$\text{Compression Costs} = \$50\sqrt{\frac{51\,\text{bar} - P}{5}}$$

The key parts of the code are shown in Figures B2.16 and B2.17. The `RunAspenPlusReactor` function has almost no changes from the previous example, except that at the end of the function, the objective

```
184
185        # This checks to see if the previous while loop exited with a simulation run and all output variables obtained.
186        # If not, we just keep track of the number of times this happened for convenience.
187        if (2 < attempt):
188            finalfails = finalfails + 1
189
190
191        # Compute the objective function using:
192        # +$16.00  per Gcal of heating duty (cost of natural gas)
193        # +$ 2.00  per GCal of cooling duty (cost of cooling system)
194        # -$ 0.1   per kg of H2 (value to us of the H2 we produce)
195        # -$ 0.005 per kg of CO produced (value to us)
196        # Need about $10 per hour per compressor stage for recompression to 51 bar after the reactor downstream
197        # Use 50 * sqrt((50-P)/5) as cost of low pressure
198        # Units on OBJ are $/hr
199
200        Obj = -0.100 * H2 - 0.005 * CO + 50 * math.sqrt( (51 - Points[point][1]) / 5 )
201        if (Q < 0): # If cooling duty
202            Obj = Obj + 2 * -Q
203        else: # heating duty
204            Obj = Obj + 16 * Q
205
206        # Store and print the data. Although I am not doing anything with the Data object, you could do something with it if you wanted.
207        print(f"Job ID: {JobID}, {Points[point][0]:0.4f}, {Points[point][1]:0.4f}, {Q:0.4f}, {CO:0.4f}, {Obj:0.4f}")
208
209        # Store the objective function result. In this case we want to have 0 heat duty so we minimize Q^2
210        Data.append(Obj)
211
212
213    # Now, going through the data and determine the best one out of this pool.
214
215    # Close this out and spit out the data.
216    print("JobID: ", JobID, "Finished with this worker. Completed ", len(Data), " points, with ", reinits, " reinitializes, ", redispatches, " redispatches, ",
varfails, " variable pull fails, and ", finalfails, " complete failures.")
217
218    return (JobID, Data)
219
220 if __name__ == '__main__':
221
222    #########################
223    # USER CHANGE THIS PART  #
224    # (see comments above)   #
225    #########################
226    MaxPSOIteration = 100
227    Consensus = 0.005
228    NumAspenCopiesOpen = 4
229    PointsPerProcessor = 8
230    #########################
231    #########################
232
233    N = 2 # This is the number of dimensions in the problem. In our case, 2 (T and P of the reactor)
234    Np = NumAspenCopiesOpen * PointsPerProcessor # These are the number of particles to use
235    minrange = [600, 2]    # This is the minimum permitted values of each of your decision variables
236    maxrange = [1000, 50]  # This is the maximum permitted values of each of your decision variables
237
238    # Initialize the location and velocity of each particle using Latin Hypercube Sampling.
239    x = [ [] for d in range(N) ] # Creates an empty list of locations for each particle
240    v = [ [] for d in range(N) ] # Creates an empty list of velocities for each particle
241
242    # For each particle, set its initial position and velocity.
243    for d in range(N):
244        for i in range(Np):
245            x[d].append(minrange[d] + (maxrange[d]-minrange[d]) * (i/(Np-1)))      # Creates a point on the grid
246            v[d].append(0.5 * (random.random() - 0.5) * (maxrange[d]-minrange[d]))  # Velocity is +/- 25% of the range, either positive or negative
247        if (0 < d):
248            random.shuffle(x[d])  # Randomizes each dimension (no need to do this on the first dimension)
249
250    # PSO Tuning Parameters. These values are classically chosen.
251    w1 = 2.8 # classic parameters
252    w2 = 1.3 # classic parameters
253    chi = -2 / (2-w1-w2-math.sqrt((w1+w2)**2 - 4*(w1+w2)))
254
255    # start the timer
256    TimeStart = time.perf_counter()
257
258    # Remember the personal and global bests of each particle. Start with the current point and Infinite
259    pbestx = x.copy();
260    pbestf = [float("inf")] * Np
261    gbestf = float("inf");  # This assumes the objective is to minimize the function.
262    gbestx = [float("NaN")] * N
263    prevx = list();
264
265    # Check to make sure that the user did not specify the wrong number of processors
266    if (0 > NumAspenCopiesOpen or mp.cpu_count() < NumAspenCopiesOpen):
267        NumAspenCopiesOpen = mp.cpu_count()
268    if (60 < NumAspenCopiesOpen):
269        NumAspenCopiesOpen = 60
270
271    # Let's see how many processors we have (this is logical, not physical)
272    print("Number of processors available: ", mp.cpu_count())
273    print("Number of processes used: ", NumAspenCopiesOpen)
```

Figure B2.16   Potions of the code highlighting how the objective function is computed and how the PSO is initializd.

```
274
275    # Start the PSO. We iterate until either the maximum has been reached or a consensus has been reached.
276    for i in range(MaxPSOIteration):
277
278        # Create the pool of workers for this iteration
279        pool = mp.Pool(NumAspenCopiesOpen)
280
281        # Remember the previous iteration
282        if (0 < i):
283            prevx.clear() # only need to clear it after the 1st iteration since it starts empty
284        prevx = x.copy()
285
286        # Create the job lists of the points that each processor will need to process
287        PointsList = []
288        for a in range(NumAspenCopiesOpen):
289            for c in range(PointsPerProcessor):
290                if (1 > c):
291                    PointsList.append(list())
292                PointsList[a].append( [ x[d][c + a * PointsPerProcessor] for d in range(N) ] )
293
294        # Evaluate the function for each particle
295        ResultData = [pool.apply_async(RunAspenPlusReactor, args=(JobID, PointsList[JobID])) for JobID in range(NumAspenCopiesOpen)]
296
297        # The ResultData does not contain our results in a way that we can access them.
298        # We will use this to access the results of each function
299        Results = [Result.get()[1] for Result in ResultData]
300
301        # Closes the parallel worker pool
302        pool.close()
303
304        # This waits until all the above jobs are finished before continuing on
305        pool.join()
306
307        # Check each point, and update personal and global bests accordingly
308        for a in range(NumAspenCopiesOpen):
309            for c in range(PointsPerProcessor):
310                k = c + a * PointsPerProcessor
311                if (gbestf > Results[a][c]): # update the global best
312                    gbestf = Results[a][c]
313                    for d in range(N):
314                        gbestx[d] = x[d][k]
315                if (pbestf[k] > Results[a][c]): # update the personal best
316                    pbestf[k] = Results[a][c]
317                    for d in range(N):
318                        pbestx[d][k] = x[d][k]
319
320        # update velocities of each particle and move them accordingly.
321        for k in range(Np):
322            for d in range(N):
323                # update the velocity (at current dimension)
324                v[d][k] = chi * ( v[d][k] + w1 * random.random() * (pbestx[d][k] - x[d][k]) + w2 * random.random() * (gbestx[d] - x[d][k]))
325                # move the particle (at current dimension)
326                x[d][k] = x[d][k] + v[d][k]
327                # enforce the bounds (using sticky bounds, leaving velocities intact)
328                if (minrange[d] > x[d][k]):
329                    x[d][k] = minrange[d]
330                if (maxrange[d] < x[d][k]):
331                    x[d][k] = maxrange[d]
332
333        # determine a normalized cloud size, to check for early exit.
334        # We can exit early when all points are within 0.5*Consensus % of the global best
335
336        stop = True
337        for k in range(Np):
338            for d in range(N):
339                if (gbestx[d] + 0.5*Consensus*(maxrange[d]-minrange[d]) < x[d][k] or gbestx[d] - 0.5*Consensus*(maxrange[d]-minrange[d]) > x[d][k]):
340                    stop = False
341                    break
342            if (False == stop):
343                break
344
345        if (stop): # all particles are within the box
346            print(f"Terminating early after {i} iterations since all particles have reached a consensus within {Consensus}")
347            break
348
349
350    #### END ITERATIONS
351
352    # Spit out the results
353    TimeFinish = time.perf_counter()
354    print(f"Finished with the parallel computing part in {TimeFinish - TimeStart:0.2f} Seconds.")
355
356    # Determine the consensus:
357    Cons = 0
358    for k in range(Np):
359        for d in range(N):
360            if (abs(gbestx[d] - x[d][k]) / (maxrange[d]-minrange[d]) > Cons):
361                Cons = abs(gbestx[d] - x[d][k]) / (maxrange[d]-minrange[d])
362    Cons = Cons * 2
363    print(f"Final consensus was: {Cons:0.4f}")
364    print(f"Global best was {gbestf:0.4f} at ")
365    for d in range(N):
366        print(f"x{d}: {gbestx[d]:0.4f}")
367
368    print(f"Program complete. {(i+1) * Np} Aspen Plus calls completed. Approximately {(TimeFinish - TimeStart)/((i+1)*Np):0.3f} Seconds per call." )
```

Figure B2.17   The end of the code, highlighting how the PSO algorithm is implemented.

function for each particle is computed and stored in the Data object, which is returned to the main program. The main program follows steps 1–4.

Some key parts of the code are:

- *Lines 1–190: Most of the Aspen Plus simulations management for each particle.* These lines are essentially identical to the previous code, see Figures B2.12 through B2.14. It is the meat of the function that calls Aspen Plus in a loop for each point it has been given, with robust error handling.
- *Lines 200–218: Compute the objective function.* This occurs after each point has been simulated. The objective function is computed according to the criteria listed previously, and stored in the Data object. Once this Process has finished with all the points it has been given, it returns the Data object to the main program. That way the main program knows the results of the simulations. Note that if there is a complete simulation failure, the code uses the default values for the heat duty, which is infinity. In other words, the objective function will be infinity for bad points, which means that the particles will ultimately ignore them. This is nice because if you have a few errors here and there in your run, the PSO will still work well, and if the failed point is a promising one, another point very nearby will likely be visited anyway by a different particle or in a different iteration.
- *Lines 226–229: User settings.* This is the beginning of the main function, and the code starts executing here. The NumAspenCopiesOpen and PointsPerProcessor settings are the same as before. The number of particles (N_p) will be the multiple of these two numbers (line 234). In my example, I am using four Processes (four Aspen Plus copies open), with eight particles/points simulated per iteration per processor, or 32 particles in total. MaxPSOIteration is the maximum number of PSO iterations (in step 3), and Consensus is a number which specifies how tight the particles need to be for early exit. The lower the number, the closer they have to be, and so the longer it may take but you may end up with slightly better results. A value of 0.005 for Consensus means that all particles should be within an N-dimensional box with dimensions equal to 0.005 of the range of that dimension. In other words, close together.
- *Lines 233–236: Problem settings.* The N variable is number of dimensions in the problem. For our example, it is two (the reactor temperature and pressure). Lines 235 and 236 are the bounds on the dimensions, given as a list of the minimums and maximums of each variable (you can see the 600–1000°C range and the 2–50 bar range here). For other N, grow or shrink this list accordingly.
- *Lines 238–248: Particle Initialization.* The starting locations are chosen by LHS. This is the same as in Part 2 but the program here is written generally for any N.
- *Lines 250–273: Housekeeping.* The parameters in the first three lines are the weighting factors for how the personal and global bests influence the particle's velocity. These are classically chosen based on heuristics. You can tweak them if you like but we find these values to be effective. The remaining code creates some empty spaces to store information (with some default values), and makes sure the user did not specify too many Processes.
- *Line 276: Start of PSO loop.* This is the start of step 3.
- *Lines 279–292: Parallel computing setup.* This creates the pool of Processes (workers) that will be used in line 279, and creates the list of jobs that each Process will handle in lines 287–929. Each process receives the same number (PointsPerProcess).
- *Lines 295–305: Run the simulations for this iteration for each particle.* This is step 3a. Each Process is given its jobs, the Processes do their work, and the results come back in the ResultsData object. The close and join operations make sure that all the Processes have finished and the pool of Processes is closed before continuing.
- *Lines 308–318:* These simply go through the results and update the personal bests of each particle and the global bests in case they got better.

- *Lines 320–331:* This updates each particle's velocity (line 324) and then moves it (line 326). Lines 328–331 ensure that it does not go outside of the bounds. If a particle leaves the bounds defined in lines 235–236, it is simply picked up and moved to the bound. The velocity update equation is one of the key drivers of PSO. For this variant, we use the classical equation:

$$v_{new,d,k} = x(v_{old,d,k} + w_1 r_1 (p_{best,d,k} - x_{d,k}) + w_2 r_2 (g_{best,d} - x_{d,k}))$$

where $v_{new,d,k}$ and $v_{old,d,k}$ are the velocities in the dth dimension of particle k after and before the Aspen Plus simulation, respectively; x, w_1, and w_2 are the weighting parameters in lines 251–253; $x_{d,k}$ is the dth dimension of the location of the current particle k, $p_{best,d,k}$ is the dth dimension of the location of the personal best of current particle k, and $g_{best,d}$ is the dth dimension of the location of the global best point found.

- *Lines 336–347:* This checks for early exit to see if the particles have established consensus according to the value in line 227. If even one particle is outside of a small box (i.e., is different from the rest), consensus has not been reached and so the PSO continues.
- *Lines 352–368:* After the PSO is finished, this code reports the results and some other stats.

Ok! If you can understand that, you should be able to modify this code to be used for your own purposes. When I ran this code with four processors, and eight points per processor (32 particles), I got a result like the following:

```
Number of processors available:  12
Number of processes used:  4
Job ID: 0, 600.0000, 22.1290, -13.9567, 777.4707, 106.2365
Job ID: 0, 612.9032, 26.7742, -13.7491, 827.4379, 95.4365
Job ID: 0, 625.8065, 28.3226, -13.3580, 949.9139, 88.1252
   ... lots of stuff...

Job ID: 3, 898.2660, 50.0000, -0.0053, 7210.4305, -142.6251
Job ID: 3, 898.3755, 50.0000, 0.0021, 7214.2229, -142.6750
Job ID: 3, 898.1386, 50.0000, -0.0139, 7206.0196, -142.5234
JobID:  3 Finished with this worker. Completed  8  points, with  0
reinitializes,  0  redispatches,  0  variable pull fails, and  0
complete failures.
Terminating early after 46 iterations since all particles have
reached a consensus within 0.005
Finished with the parallel computing part in 726.69 Seconds.
Final consensus was: 0.0050
Global best was -142.6867 at
x0: 898.3431
x1: 50.0000
Program complete. 1504 Aspen Plus calls completed. Approximately
0.483 Seconds per call.
```

The results are that we should use a reactor temperature of about 898°C with a reactor pressure of 50 bar, which would give us a minimum cost of –$143 per hour (i.e., we would be making $143 per hour in value)

Figure B2.18 shows an example of my computer processor use when running this code with four copies of Aspen Plus at once but with 12 logical processors. Almost all processors on my computer are in use most of the time, even though I have only four Processes running at once, since Aspen Plus will use up to three processors in each simulation if it is available. You can see that the usage pattern of each CPU is cyclic, corresponding to each iteration of the PSO loop. The peak moments of the activity cycle are when all the Aspen Plus simulations are running. The drop-offs in activity are when it is waiting for the other simulations to finish since they don't all stop at the same time, the velocities are updated, the particles are moved, and new

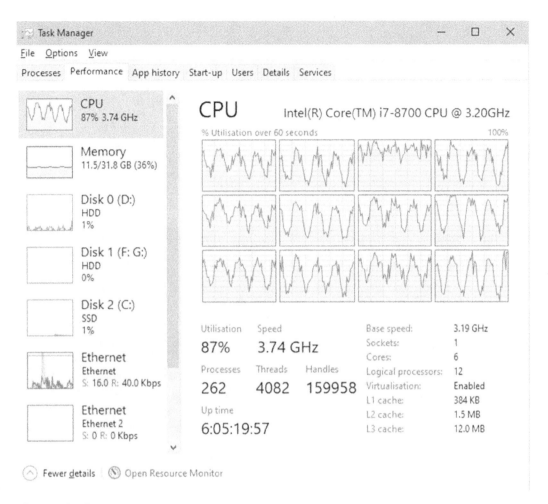

Figure B2.18

copies of Aspen Plus are launched at the start of the next iteration. You can see the regularity of how this works here. This code could be improved by a much more complicated program which does not close and reopen the Aspen Plus copies for each Process each PSO iteration, but this is much more complex and not suitable for a textbook. Overall, it works great, and takes about two to three times as long per simulation as it did in the Part 2 code when the entire job is known all up front, so that is about as much as could be improved here.

♪ Music break[15]

---

[15]Recommended listening: *I Alone* by Live.

Batch Operations in Aspen Plus

Objectives

- Learn basics of batch operations in Aspen Plus
- Learn how to use the new BatchOp feature for batch reactors and tanks
- Learn how to use BatchSep for batch distillation
- Integrate flowsheets with batch and continuous elements

Prerequisite Knowledge

This tutorial assumes you have a working knowledge of Aspen Plus, including Tutorials 1 through 4. Since batch operations are usually more complex than continuous ones, a working knowledge of the basic concepts of dynamic chemical processing and chemical process control will be required. It is also helpful to know how kinetics reactions are modeled in Aspen Plus (see Tutorial 7). Understanding how distillation models work (Tutorials 5 and 8) is also important for Part 3. The most important aspect is to understand how batch processes operate cyclically, usually according to some kind of recipe of steps and sequences. This tutorial uses a more realistic example to illustrate many of the different features of the batch process models, and so it is a little heavier than the others. Also, this tutorial may take around 3 hours, which is why it is a bonus.

Why This Is Useful for Problem Solving

Aspen Plus has been traditionally intended for modeling continuous processes. For dynamic processes (such as batch, semi-batch, and semicontinuous processes), other packages are generally more suitable, such as Aspen Plus Dynamics or Aspen Custom Modeler. These programs are quite powerful, but they are built upon a framework that feels very different from Aspen Plus and requires the user to overcome a significant learning curve in order to use it effectively. Over time, Aspen Plus has been expanded to include some common batch process operations such as batch reaction or batch distillation so that users can access these models without having to leave the relative comfort or simplicity of Aspen Plus. This allows Aspen Plus users to integrate continuous and batch process flowsheet elements without having to learn the much more complex Aspen Plus Dynamics framework. However, the compromise is that the batch capabilities within Aspen Plus (specifically `BatchOp` and `BatchSep`) are limited in scope and flexibility. For common batch operations which fit within the limitations of the batch models provided in Aspen Plus, these tools will work quite well. For others, you may be better off using Aspen Plus Dynamics, Aspen Custom Modeler, or some other product.

Process designers who are in the early stages of developing process concepts may not necessarily know yet whether batch, continuous, or semicontinuous processes are preferred.[1] For some kinds of processes, particularly for those involving biological reactors, pharmaceuticals, food, alcoholic beverages, or other processes in which small lots need to be kept strictly apart for safety or management reasons, batch processes are the only feasible choice. For others, the choice to use batch, semicontinuous, or continuous processing depends strongly on its intended scale, as shown in Figure B3.1. Generally, batch processing tends to be the most optimal at small scales, but because it does not scale-up well, continuous processing is preferable at large scales. Semicontinuous processing, which is a sort of hybrid of the two, is best for those in between. For these cases, engineers can use Aspen Pus simulations for batch and continuous variants in order to determine which strategy is optimal for their case.

Semicontinuous processes can come in many forms. In some cases, they are simply batch processes which have some continuous elements, such as a continuous feed or continuous product draw. In some cases, Aspen Plus batch modeling tools may be sufficient as they often can accommodate these features. More complex ones, such as semicontinuous distillation systems which have more rigorous definitions and a lack of common batch features such as column startup or shutdown, need more complex software like Aspen Plus Dynamics for simulation. Whatever the case, unlike continuous processes, batch processes and semicontinuous processes always have prominent cyclic elements by design, and depending on the complexity, the Aspen Plus batch tools may be suitable for your needs.

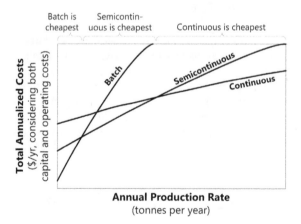

Figure B3.1 Typical curves for a process that could be produced through batch, semicontinuous, or continuous process variants. Usually, batch processes start out the cheapest at small scales but do not scale well and have the steepest slope. Continuous processes typically have high costs at small scales but increase in cost more slowly as plant capacity increases. Semicontinuous processes often lie somewhere in between.

---

[1]For an introduction to semicontinuous distillation, see Adams TA II and Pascall A, Semicontinuous Thermal Separation Systems, *Chem Eng Technol*, 2012, 35:1153–1170.

Tutorial

PART 1: SIMPLE TANKS WITH BATCHOP

In Tutorial 7, we covered the RBatch reactor model which is now depreciated in Aspen Plus V12. Although the models continue to exist within Aspen Plus so that previously made models will still function, users are now expected to use BatchOp instead. RBatch works well for what it is: a single batch reactor with buffer tank connections that enable continuous connections to its flowsheets. BatchOp is much more general and allows for more operations, arbitrary batch recipes, and more kinds of connections. Unfortunately, it can also be difficult to work with, can be a bit buggy at times, and is subject to the same growing pains as any new feature (after this tutorial was written, AspenTech fixed a bug in V12.1 that was discovered by yours truly during the preparation of this tutorial). However, once you understand it, it is quite useful, and this feature improves with each new version of the software. This tutorial will get you started using BatchOp just as a simple holdup tank used as a flash drum.

Figure B3.2 illustrates a simple batch process, consisting of two tanks and operating in the following simple cycle:

1. The first tank T1 receives a 2000 kg charge containing a liquid mixture of 25 mol% ethanol and 75% water from an upstream source over the course of 30 minutes.
2. At the 30-minute mark, the charge is stopped, and a heater in the tank will be turned on to provide 300 kW of heat. Over time, the liquid in the tank will begin to boil partially, with the vapor containing a higher percentage of ethanol than the liquid. The vapor remains in the tank. The heating continues until the liquid in the tank reaches 88°C, gradually purifying the liquid until it is at least 90 mol% water.
3. At that point, the liquid is drained from the tank and transferred into a second tank, which is the buffer tank that can connect to a downstream continuous process.
4. The heater is then stopped and the tank rests for 15 minutes while the remaining contents of the tank are collected. The remaining contents are the vapor phase which is only about 60 mol% water.
5. Repeat to the next cycle.

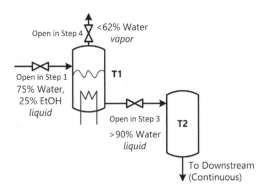

Figure B3.2   The process of Part 1. Tank T1 is a heated tank which is first charged with liquid ethanol and water, and then the heating starts. The liquid in the tank partially boils, but all exit ports are closed and so it stays within the tank as vapor. At a certain point, the liquid is drained from T1 and stored in the buffer tank T2, and then the vapor is drained from T1. T2 continuously feeds to a downstream continuous process. Pressure management is ignored for this example for simplicity.

Thus, this batch process contains a single-stage flash separation of ethanol and water.

Start a blank simulation. In the Properties section, add water and ethanol as chemicals, and choose NRTL–RK (the Non-Random Two-Liquid Method with the Redlich-Kwong equation of state) as the physical property method. Note that if you used a template with a default property method, like NRTL, consider deleting it entirely. Be sure to visit the Methods | Parameters | Binary Interactions tab and make sure there are parameters in the NRTL-1 section. You can check to make sure the properties make sense by generating a Txy diagram (Analysis | Binary), which should look like the one in Figure B3.3. It is important to check because some of the physical property methods do a poor job of modeling ethanol-water vapor-liquid equilibria (see Bonus Tutorial 4). Review Tutorials 1 and 2 if you need to know how to do these things.

Next, go to the Simulations section, so we can start creating the flowsheet. Up to this point in the book, all your process flow simulations have been the Continuous process type, which is the default type and historically the primary purpose of Aspen Plus. However, you can navigate to the Batch

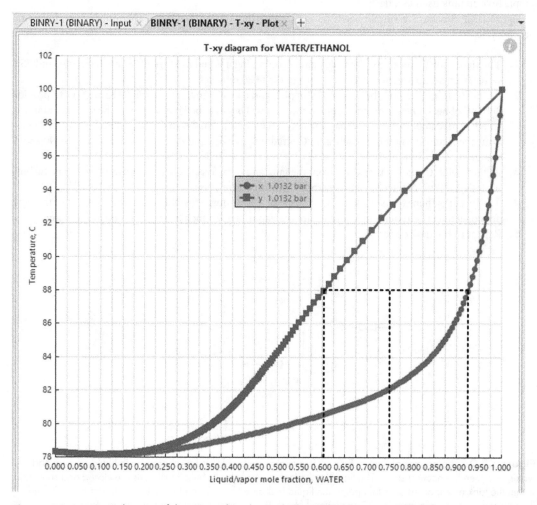

Figure B3.3 A Txy diagram of the water-ethanol system. The dashed lines indicate the kind of products we would expect if we took a 75% water mixture and flashed it at 88°C and 1 bar—about 61% water in the vapor product and about 92% water in the liquid product.

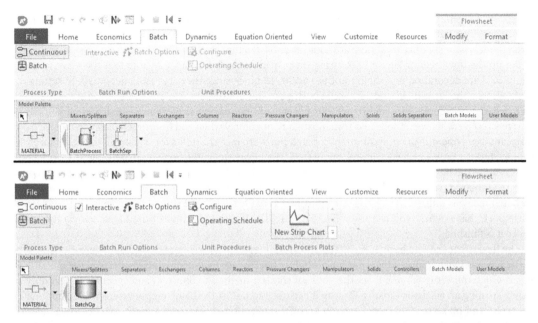

Figure B3.4 Different ribbon and Model Palette options when in Continuous mode (top) and Batch mode (bottom).

ribbon, where you will have the option to change the mode of the current flowsheet. When you are in continuous mode, the Continuous process type icon will be highlighted, as shown in Figure B3.4. If you look at the Batch Models tab of the Model Palette, you will see two models that you can add to your continuous-mode flowsheet. The second model is BatchSep which is a batch distillation model that includes the distillation column, pot, condenser, reflux, multiple distillate receivers, controllers, and batch recipes that you specify. We will cover this in Part 3. The confusing thing is that although it is a batch process, and it is found in the Batch Models tab of the Model Palette, it only works when your flowsheet is in Continuous mode. This is because this model was incorporated into Aspen Plus in earlier versions of the software when Batch mode did not exist.

The other model is BatchProcess, which is a Hierarchy block which starts in Batch mode by default. Hierarchy blocks (which you can create using the model found in Model Palette | User Models) are just flowsheets within flowsheets. They are a convenient way to group a bunch of models together into one block, and you can then connect streams from Hierarchy blocks to other Hierarchy blocks or ordinary models. This can really help with flowsheet organization for large flowsheets. You can copy-paste entire flowsheet sections, change the properties of an entire section, or just keep things neat. This is especially useful when you have some sections of your flowsheet operating in Batch mode and others in Continuous, because you can put them into two separate flowsheet hierarchies. You can "enter" a Hierarchy block by double-clicking on it, and then you see the flowsheet inside. One important restriction, however, is that you cannot have any Hierarchy blocks in Continuous mode that are inside one that is in Batch mode. So if you want to mix Continuous and Batch flowsheet elements in the same simulation, the main (outermost) flowsheet needs to be in Continuous mode.

You can make a BatchProcess hierarchy if you want, but for this tutorial, it is easier to switch the main flowsheet directly to Batch mode, as shown in Figure B3.4 on the bottom. If you look in the Model palette, you will see that the models that you can now put on the flowsheet are different. For example, the only Batch model available in Batch mode is BatchOp (surprisingly, BatchSep is not available in Batch mode).

`BatchOp` is a general tank-type model that can model reactions, including crystallization, and allows the user to specify a batch recipe. If you look around the Model Palette while the current flowsheet is set to Batch mode, you will notice that there are a lot fewer blocks that are available.

Set up the flowsheet of Figure B3.2, using `BatchOp` for both T1 and T2 (note that the valve icons in Figure B3.2 are decorative, you should not use a `VALVE` block since they are just conceptual). In setting up the flowsheet, the feeds to T1 and T2 should both go to the Continuous feed port. There is not much difference between the Continuous and Batch feed ports. Continuous ports are more general and let you either passively receive whatever is collected from upstream, or control when and how much feed you receive from upstream using operating recipe commands. Batch ports are nice shortcuts for when you know for sure you want to start the tank with a certain amount of feed from upstream at the beginning of the cycle. Continuous ports require the tank to start empty. The vapor product from T1 and liquid product from T2 should come from the required (red) Product ports; streams leaving Product ports contain whatever is in the tank *after* the batch recipe is finished. The draw stream from T1 that feeds into T2 should come from the Condensed Phase port; this is where you can draw liquid and solids (if they exist) out of the batch *during* the cycle, before it is finished.

The feed stream should be at 25°C, 1.05 bar, and contain 75 mol% water and the rest ethanol. Its flow rate can be any positive number—this is just a placeholder because we will specify this flow rate in the operations recipe. Go to the Setup tab of T1. Set the pot heat transfer method to Shortcut and specify the heat duty at 0, thus creating an adiabatic tank. In the batch cycle section, start the batch empty (there will be nothing in this tank), and make sure that the Reaction and Crystallization checkboxes are not checked—we will not have any reactions. Specify the pressure at 1.05 bar, and now you have essentially created a classic flash drum model in batch mode. Pick a batch discharge time of 15 minutes, which is the amount of time you say it will take for you to remove the contents of the vessel via the Products port. Pick a batch downtime of 0 minutes. This is the amount of time that the system will rest between cycles, and it does not affect the simulation, only the total cycle time that is reported after the run is completed. Finally, make sure that the Valid Phases is set to `Vapor-Liquid` (the default is just liquid). For T2, use all of the same settings as T1.

Start by creating a plot of the holdup of T1. Click on New Strip Chart in the Batch Process Plots portion of the Batch ribbon. Give it a name, and use the Find Variables button to search for the water and ethanol holdups in T1, and add them to the plot (see Figure B3.5). For added clarity, use the Axis Map button to put the two variables on the same axis. If you need to get to the plot again, you can find it either in the Flowsheet | Batch Process Plots section in the explorer at left, or as a new icon, you can click next to New Strip Chart in the Batch ribbon. At this point, you should be able to run the simulation. Run it, but stop the simulation

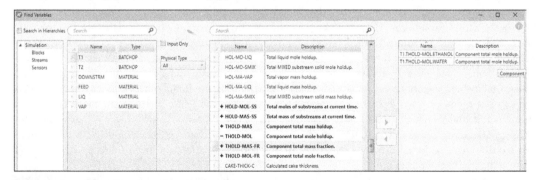

Figure B3.5 The Find Variables feature is a convenient way of choosing variables for Strip charts and certain Batch models.

(the blue square at the very top of the window) after a bit because it will just keep running forever. This is because your simulation has no batch recipe with stopping criteria. As it is now, your simulation will passively accept feed into T1 at whatever rate you specified. The tank has no maximum volume in this simulation so its holdup will increase in perpetuity. T2, on the other hand, starts empty and remains empty, because you have not yet told it to withdraw any feed from the condensed phase draw port of T1. You can verify this by looking in the Profile tab of the block, which shows a row by row breakdown of the tank contents at each timestep, along with other properties, depending on the tabs you choose.

Now that the basic flowsheet is working, we will set up the batch recipe. In the simulation explorer at left, find the Unit Procedures folder, which will either be in the main area if your main flowsheet is in batch mode, or else it will be inside the hierarchy block if you decided to use a `BatchProcess` hierarchy. Unit Procedures are recipes that you specify, and they refer to the flowsheet (in Batch mode) as a whole, rather than one block specifically. Create a new unit procedure. Inside it, you can specify when the unit procedure should start. In our case, we want it to start immediately, so leave it at the default of 0 hours after the start of the simulation (in more advanced cases, you can tell it to start at a different time, or after a process variable reaches some special number, which is useful especially if you want one unit procedure to start after some other unit procedure running in parallel has reached some point or has finished). A Unit Procedure is a listing of Operations, that occur in sequence, each one starting after the previous has finished. You also can choose to terminate the batch at the end of the recipe (leave this checked or else the simulation will not finish).

This simulation will have four Operations. Each Operation in a Unit Procedure contains a set of variable changes that will take place at the start of the operation. This is achieved by finding the variable you want to change and specifying the new value you want it to be. You also have the option to specify a ramp time. A ramp time of 0 means that the value changes instantaneously. A positive value means that the variable will change from its present setting to its new setting in a linear fashion over the specified time. Like other Aspen Plus tools that let you change variables programmatically (such as the Calculator, Design Spec, Optimization, and Sensitivity blocks), the variables that you can change are limited to the degrees of freedom of your flowsheet and are usually values you can type into the various forms directly. Finally, you can specify when the operation finishes, either by specifying a fixed time from the start of the batch or the operation, or by specifying a trigger condition by indicating a process variable and a target value—when that variable reaches the target value, the Operation ends.

Create each operation, one at a time, and run the simulation after each one to ensure that the operation is working correctly before you create the next one. The details are:

1. **Fill T1** at a rate of 4 tonne/hr (4000 kg/hr) for 30 minutes. To do this, create an operation which changes the value of `Mass-Flow.Mixed` of the feed stream going into T1. Use either the "Find Variables" button or the <Create New> option in the drop-down box in the Variables column of the Operation to find this easily. The stop criterion should be 30 minutes. Run a simulation to see that it worked (it should stop after 30 minutes of simulation time automatically), and you can check by looking at your new Strip chart you made and/or the various tabs in the Profiles folder in T1.

 TOM'S TIP: It is important to work in small steps and keep checking the results as you go. Check the basic stuff that you know for sure should be true, like vapor fractions, total holdup, and trajectories of basic variables like temperature. If you see anything out of the ordinary, stop and resolve it before you continue, otherwise you will just keep digging into a deeper hole.

TOM'S TIP: Although often confused, 1 "tonne" means a metric ton everywhere in the world, which is exactly 1000 kg or about 2205 lb. Do not confuse this with 1 ton, which in both North America and in Aspen Plus means a US short ton (2000 lb). In Great Britain, 1 ton means something else: it is a long ton (2240 lb) which in Aspen Plus is abbreviated L-ton.

2. **Stop the feed and heat the tank** at a rate of 300 kW until it reaches 88°C. The relevant variables you want in T1 are DUTY and TEMP-CALC. Do not use the variable TEMP, which is the temperature setpoint for a temperature controller (which you are not using). You want TEMP-CALC which is the temperature that Aspen Plus calculates at any given minute as it heats from 25°C to 88°C. This means that this value will approach from *below*, so set that accordingly. In other words, Aspen Plus will stop it as soon as the temperature is ≥88°C. Note that TEMP-CALC is a process variable, so you will not find it under operating changes. Do not forget to add an operating change that turns off the feed (sets its flow rate to 0). You can check to see if this worked by creating a plot of the tank temperature (TEMP-CALC) which should be at 25°C for 30 minutes, then ramp up to about 83°C quickly as it approaches the bubble point, and then slowly reaches 88°C by the end of the run as increasing amounts boil. At the end of this step, there should be both vapor and liquid in this tank, with compositions approximately equal to what we predicted in Figure B3.3. You can check that as well (answer Q1).

Q1)   What mole percentage of the matter inside the tank is in the vapor phase after 2 hours of batch operation?

Q2)   How much water is there in the vapor phase at the *end* of Step 2 (in kmol)? Hint: Check the Results folder of T1, not the Profiles folder.

TOM'S TIP: The Profiles folder is useful for getting a time trajectory of selected variables at nice intervals (currently at the default *communication interval* of every 0.01 hours). However, often more timesteps are computed behind the scenes, and are simply not reported to you in the Profiles tab for brevity. For this example, the last row reported in the Profiles tab is 2.37 hours, but Step 2 did not end until just after at 2.37003 hours (you can check the Control Panel output or T1's Results tab to find this). It is just a tiny difference in this case, but it illustrates that behind the scenes, the integrator may be taking very small steps, especially near discrete events like the start or end of an operating step.

3. **Turn off the heater and drain the liquid from T1** by changing the flow rate of the liquid phase draw stream of T1 to 4 tonne/hr (4000 kg/hr). The relevant variable is DRAWMASSFLOW.LIQ. The step should end when the T1 is nearly devoid of liquid (when the liquid mass holdup falls to 1 kg, whose relevant variable is HOL-MA-LIQ). Of course you can try to choose a smaller number as a stopping point but in practice, you would often leave a little in the tank to avoid cavitating the pump.

TOM'S TIP: At the time of writing, there is a bug in V12/V12.1 which AspenTech has acknowledged to the author and is working on a fix. If you delete an Operation from a Unit Procedure, the units on the variables in the Operating Changes section may change to some default values. For example, if you had specified a heating duty of 300 kW in Step 2, and then delete one of the other steps, the units on the heating duty would change to something else and you would not necessarily notice until you get unexplained behavior and go back and check. Hopefully, this will be corrected in your version by the time you see this in print, but if you are using older versions, be aware of this issue.

Q3) What is the mole fraction of water in the liquid product? Check the Results tab of T2.

Q4) What is the total mole fraction of water remaining in T1? Check the Results tab of T1.

So at this point, we have our separated products in T1 and T2. But now how do we get those products into the Products output ports? If you look at the stream results for the two product streams leaving T1 and T2, you will notice something strange. The product stream leaving T1 has an unrealistically high flow rate, and has exactly 75 mol% water. The product stream leaving T2 is empty. Neither of these makes much sense. So for this reason, it is common to add an Operation at the end of a Unit Procedure that literally does nothing, for a small amount of time, which is essentially a workaround that will load the appropriate stream conditions into those product streams. So, create a 4th operating step:

4. **Do nothing: stop the liquid collection.** Set the liquid product draw flow rate in T1 to 0, and set the duration of this operating step to something small (0.01 hours).

Now, if you check the stream results for the two product streams, they will make a lot more sense. For example, the mole fractions of the product streams of the two tanks should match your responses to Q3 and Q4. The flow rates also make sense. They represent the flow rate during the product draw, and they are based on the Batch Discharge Time that you specified for T1 and T2 (15 minutes each). So for example, since there is about 38.9 kmol in T1 at the end of the batch, with a 15 minute drain time, the flow rate during the drain is 155.6 kmol/hr. You can now connect this stream to other flowsheet elements downstream, even continuous ones. Remember though that this flow rate only occurs for only 15 minutes, and then outside of that, the flow rate is 0. So be careful when connecting to continuous elements downstream. Note also that the product stream from T1 contains both the liquid and vapor content of the tank at the end, meaning that the entire thing leaves through the port leaving a total vacuum inside the tank. This is not realistic of course, but still having access to the product stream can be useful for simulation and design purposes.

Finally, you should be aware that in Aspen Plus, Batch simulations use classic numerical integration methods for ordinary differential equations in order to compute the trajectories of the process variables over time. The default methods (such as Implicit Euler) usually work well. However, you may see some strange behavior, for example, as shown in Figure B3.6. The top of that figure shows the molar holdup of water and ethanol in T1 during the operation, and mostly makes sense, except for the little pip or bump in the circled area. That is nonphysical and strictly the result of numerical error in the method of integration. This sort of thing occurs often near discrete events, such as a sudden change of process variables that typically occur at the start of each new Operation. To combat this, you can try some strategies to reduce this error. One such strategy is to reduce the integrator step size, which you can do in Convergence | Batch Options | Sequential Modular | Integrator. The smaller the step size, the lower the error, but also, the longer the simulation will

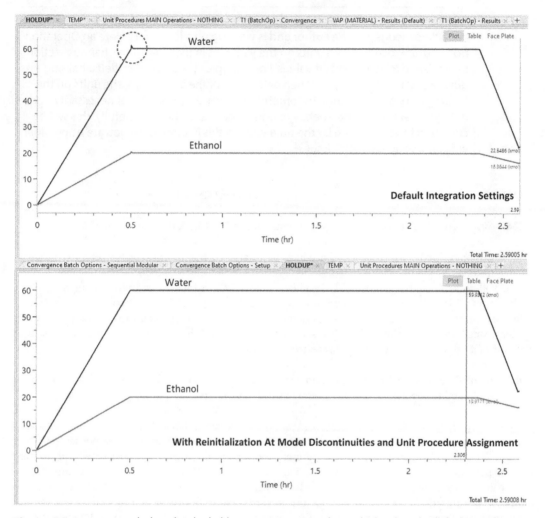

Figure B3.6   Water and ethanol molar holdup trajectories in T1 during the batch under default integration settings (top) and by forcing a reinitialization at every operational change (bottom). The reinitialization strategy takes essentially the same amount of run time and gives essentially the same final results as the default, but clearly avoids trajectory problems associated with discrete events.

run. If you are getting strange or erroneous results, this is one place to try. If your problems are occurring largely at model discontinuities, like step changes in your operating variables, you should try changing the reinitialization strategy. The default is to not reinitialize the simulation at each discontinuity, but on that same flow sheet if you change the reinitialization strategy to At model discontinuity or unit procedure assignment, then it will result in a much better plot like the bottom of Figure B3.6. Reinitialization helps get rid of these errors at discontinuities, but depending on the complexity of the flowsheet, can either significantly increase the run time, or even fail at that point (reinitialization is a hard task, which is why it is off by default).

So at this point, you should have a general sense of how Aspen Plus handles dynamic trajectories in the batch phase, with a simple holdup tank. The simulation does not repeat for multiple batches, so it is on you to determine how these simulation results are useful to you in designing or understanding the true batch process of interest. If you have complex recipes with several units working in parallel that require timing and interaction between them, or have parallel systems whose cycles do not neatly align together, then you would likely want to explore more detailed general dynamic simulations with Aspen Plus Dynamics or some other software.

♫ Music break[2]

PART 2: REACTIONS IN BATCHOP

One of the key purposes of `BatchOp` is to replace `RBatch`. For the rest of this tutorial, we will simulate the batch reaction of acetaldehyde and propylene glycol to produce water and 2,4-Dimethyl-1,3-Dioxolane (24DMD), and the subsequent batch distillation of the products.[3] 24DMD is a flavoring agent used in many foods, with a fruity, grassy flavor. The reversible chemical reaction is:

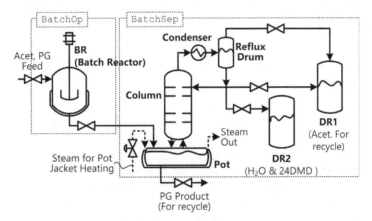

| Acetaldehyde | Propylene Glycol | Water | 2,4-Dimethyl-1,3-Dioxolane |

In Part 2, we will create a batch reactor model with `BatchOp` for the production, and in Part 3, we will model a batch distillation with `BatchSep` in order to recover it from the reaction broth. The process we will simulate is shown in Figure B3.7, with the corresponding batch recipe shown in Figure B3.8.

Figure B3.7   The process to produce 24DMD in a batch reactor and then recover it via batch distillation. The dashed boxes indicate that all of the units and streams inside the box are contained implicitly within either the `BatchOp` or `BatchSep` models, so there is no need to create separate blocks for the column, plot, and distillate receivers, for example.

---

[2]Recommended listening: *Montana* by Tycho.

[3]This tutorial is based on the paper Adams TA II, Seider WD. Semicontinuous distillation with chemical reaction in a middle vessel. *Ind Eng Chem Res*, 2006, 45:5548–5560.

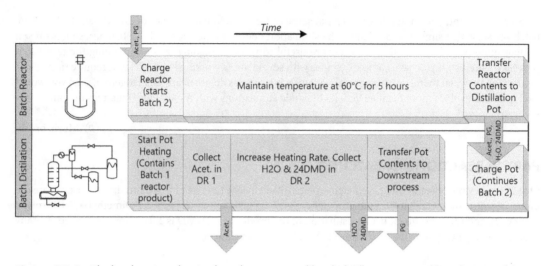

Figure B3.8   The batch recipe, showing how the reactor and batch distillation unit would work in parallel. In this figure, the reactor starting the beginning of batch 2 on the left with a charge, while the batch distillation unit starts its procedure by heating the pot, which contains the reactor product of batch 1.

In the real system, the batch reactor and the batch distillation system would be working in parallel, on two different batches. The batch distillation system would be drained of its products (which would be sent downstream), and then the reactor products which are sitting in the batch reactor (and hopefully ready by this point) would be transferred to the pot of the batch distillation system so that it can purify the next batch. For our simulation, however, we do not really need to model two separate batches running in parallel, it is enough for us to model one batch and then use the information to help design the real system which would be working in parallel.

Start a new blank simulation, with the four chemicals of interest. Note that propylene glycol is also called `PROPANEDIOL-1,2` and 24DMD has a chemical alias in the Aspen Properties databanks of `C5H10O2-N18`, but you have to first add the `APESV120 ACID GAS` databank to the list of selected databanks in the Enterprise Database tab of the Components | Specifications folder in Properties. Add nitrogen gas (N_2) as well (you will need this in Part 3). 24DMD is a specialty chemical and Aspen Plus does not have vapor-liquid equilibria parameters stored in its database, so we will have to make our own. Choose the `UNIQUAC` physical property method, and in the Properties | Methods | Parameters | Binary Interaction | UNIQ-1 folder, make sure it contains binary pair parameters for water with acetaldehyde and propylene glycol. There will be only two rows in this table, so you will have to enter additional data to model this complex system correctly. First, enter the custom binary parameters for 24DMD (component i) and water (component j)[4] shown in Table B3.1. Then, click the Estimate using UNIFAC checkbox on that same form, to allow Aspen to generate the rest of them. Then, go into the Components | Molecular Structure section and make sure the atomic structure data is there for each chemical. If it is not, for each chemical, click on the Structure and Functional Group tab and click the Calculate Bonds button. After performing a Properties run (hit the play button or `F5`), you should see this form populate with 10 rows. As a check, the Acetaldehyde + Propylene Glycol pair should have

---

[4]VLE parameters in Table B3.1 and experimental data are from Chopade SP, et al. Vapor-liquid-liquid equilibrium (VLE) and vapor pressure data for the systems 2-methyl-1,3-dioxolane (2MD) + water and 2,4-dimethyl-1,3-dioxolane (24DMD) + water. *J Chem Eng Data*, 2003, 48:44–47.

Table B3.1   **Property Parameters for Modeling Phase Equilibria of 24DMD (i) and Water via the UNIQUAC Method**

A_{ij}	A_{ji}	B_{ij}	B_{ji}	C_{ij}	C_{ji}
89.06	−4.36	−11380	1230	−4.97	−0.608
D_{ij}	D_{ji}	E_{ij}	E_{ji}	Temperature Units	Temperature Range
−0.083	0.013	0	0	C	Leave at default

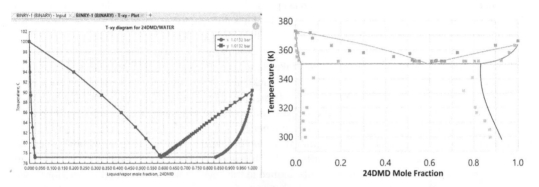

Figure B3.9   (Left) The results of a binary analysis for 24DMD + Water vapor-liquid equilibria. (Right) Those same results shown as lines, plus additional lines for liquid-liquid equilibria simulations, compared to experimental data, shown as the squares. Data are from Chopade et al. *J Chem Eng Data*, 2003, 48:44–47. Overall, the model does a good job of matching the experimental data.

−550.33 for B_{ij} (or B_{ji}) and the Acetaldehyde + 24DMD pair should have −133.859 for B_{ij} (or B_{ji}). If you set up a Txy diagram of 24DMD and water at atmospheric pressure, it should look like Figure B3.9. With that, you should be all set to simulate.

Because this simulation will have both batch elements (`BatchOp`) and continuous elements (`BatchSep`, which is unintuitively classified as a continuous model) on the same flowsheet, your main simulation flowsheet should remain in continuous mode. Instead, create a `BatchProcess` hierarchy block and enter it (by double-clicking), and inside that create a new `BatchOp` block. This block only needs one feed stream (to the Batch Charge port) and one product stream leaving it. The feed stream should be an equimolar mixture of acetaldehyde and propylene glycol at 20°C and 1.5 bar. The flow rate of this stream does not matter much because, in your batch reactor, you should specify that the batch charge is 100 kmol to start. Although the flow rate that you choose for this feed stream will affect the total simulation time for the reactor as shown in the reactor Results tab, it will not affect anything else. Aspen calculates the time it would take to charge this batch given the flow rate of this feed stream, but none of the simulations are affected.

In the Setup folder for the batch reactor, specify a constant pressure of 1.5 bar, and set the valid phases at `Liquid-Vapor` since although we want the reaction chemicals to stay in the liquid phase as much as possible where the reaction occurs, during the early stages of the cycle, the temperature is high enough such that some of the acetaldehyde is in the vapor phase.[5] Toward the end of the operation, enough of the acetaldehyde

---

[5]I recognize that there is the possibility for two liquid phases to form instead of just one. However, the reaction kinetics information we have is for well-mixed phases assuming one bulk liquid phase. Since the batch reactor assumes it is well-mixed (the real device would have a stirring impellor), assuming a single liquid phase here is ok.

is reacted away such that the mixture is below the bubble point at the reaction temperature of 60°C. Set the batch discharge time and the downtime to 0.1 hours. In the Model Detail section, check the Reaction box.

Next, choose a Specified Temperature of 60°C. To maintain the temperature, we are going to add a controller that runs a heating system since the reaction is endothermic. In the Controllers tab of the batch reactor, set the proportional gain to 10 and leave the integral time and derivative constants at their defaults of blank (infinity) and 0, respectively. What this will do is add a PID controller that will automatically adjust the heating rate to the batch reactor over time in an attempt to maintain the temperature at 60°C. If you are familiar with PID controllers, then you will understand that this is a proportional-only controller, and the units on the gain are J/kg/K, such that the gain is proportional to the mass of the batch charge in the tank. If you do not know what that means, you can just understand that at each timestep, this controller will try to either add or remove heat to the system to try to keep it at the setpoint of 60°C. The higher the proportional gain constant, the more aggressively it tries to respond—if it is not aggressive enough, it will take too long to reach its setpoint, but if it is too aggressive, it could really overdo it and cause the system to quickly cycle between being too hot and too cold in an undesirable oscillation.

Next, we need to define the reactions that will take place. Start by creating a new reaction set in the Reactions folder of the main simulation explorer (not the block), just like in Tutorial 7. Choose GENERAL as the reaction type. Although we could use other types of reactions, I recommend starting with this type as it is the easiest to use in my opinion. For example, it allows you to define reversible reactions either as two separate reaction laws governing forward and backward kinetics, or it lets you define a single forward reaction and then either specify an equilibrium constant expression or instruct Aspen Plus to use Gibbs free energy calculations to compute the equilibrium constant. Furthermore, it gives you the ability to both see and change the reaction rate and concentration units defined in the kinetic formulas; with other reaction model types, users often have to dig through the documentation to find it.

Anyway, within this new general reaction set, create two reactions, one for the forward and one for the backward reaction. This is because the forward and backward reaction rate equations have different activation energies. Enter the reaction data accordingly in the kinetics tab, noting that the reaction is equimolar in either direction. Use the following kinetic reactions[6]:

$$\text{Forward Rate} = 0.003 \frac{m^3}{kmol-s} \exp\left(\frac{-12{,}596.5 \frac{kJ}{kmol}}{RT}\right)[C_{acet}][C_{PG}]$$

$$\text{Backward Rate} = 0.018 \frac{m^3}{kmol-s} \exp\left(\frac{-24{,}461.3 \frac{kJ}{kmol}}{RT}\right)[C_{H2O}][C_{24DMD}]$$

where the driving force uses a molarity concentration basis ([Ci]) with units of kmol/m^3, the reaction is in the liquid phase, and the rate basis is reactant volumetric (not catalyst) based. An example is shown in Figure B3.10. Note that the concentration exponents are 1 for all components in the above equations.

---

[6]Kinetic data are derived from Dhale AD et al. Propylene glycol and ethylene glycol recovery from aqueous solution via reactive distillation. *Chem Eng Sci*, 2004, 59:2881–2890. Equations have been simplified for use in this tutorial and are for educational purposes only; consult original paper for other purposes.

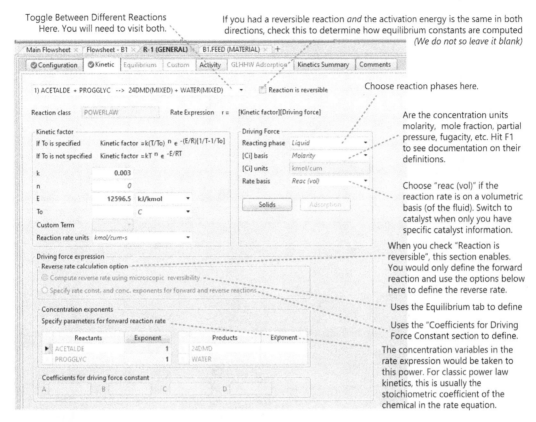

Figure B3.10 The kinetics tab of a General type reaction model.

Once this is in place, go back to your batch reactor and in the Kinetics tab, enable the reaction set you made. Then, we want the reaction to run for about 5 hours using the current settings. To do this, you essentially need to make a 5-hour "do-nothing" Operation in the Unit Procedures folder, in which the reaction proceeds and the controller attempts to maintain a constant temperature of 60°C. Go ahead and make this first operation and run the batch. One problem with this is that each Operation requires at least one Operating Change (a variable must have a value set). For a do-nothing operation, just pick a variable you already have set. For example, in the setup form, we set the temperature setpoint at 60°C, so we can just rewrite this same value (set the TEMP variable of the reactor to 60°C, which effectively does nothing since we already set it at 60°C to start with). Run the simulation and make sure it worked. Check the reactor product stream as well to ensure it contains the expected product.

If you plot the temperature (TEMP-CALC), temperature setpoint (TEMP), and heat duty (DUTY-CALC) of the reactor during the batch, you should get something shown in Figure B3.11 (top). You should see the reactor temperature starting at 20°C (the temperature of the charge) and then shoot up toward the 60°C setpoint. The controller starts by choosing a high heat rate (around 2 Gcal/hr) and then reduces it as it approaches the setpoint. It overshoots it a little bit, so the reactor actually pulls back the heat enough such that the duty becomes negative (i.e., it is cooling it, and in reality, you probably would not have a cooling system in addition to the heating one, you would just simply stop the heating entirely and let the endothermic reaction do the cooling; you cannot easily model this restriction in Aspen Plus but you can do it in Aspen Plus Dynamics).

Q5) What is the final mole fraction of water in the batch at the end of the 5-hour trajectory?

Q6) Has the reaction approached equilibrium, is it close, or should we extend the batch time beyond 5 hours? Consider Figure B3.11, bottom.

🎵 Music break[7]

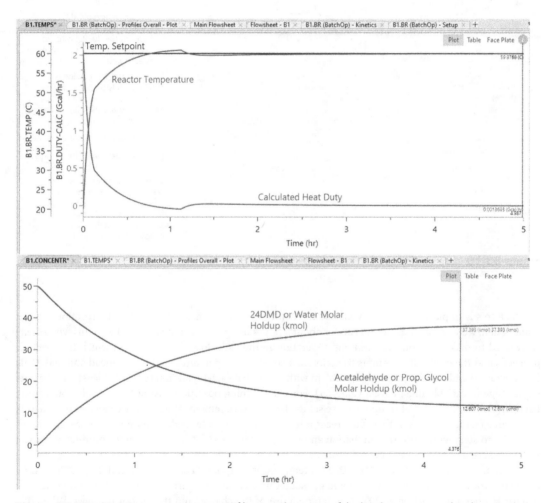

Figure B3.11   (Top) Reactor temperature profiles over the course of the batch, as compared to the temperature setpoint. The heat duty that is applied by the control system is also shown. (Bottom) The molar holdup of the reactants and products in the reactor, noting that because the feed is equimolar, the reactant and product trajectories overlap exactly.

---

[7]Recommended listening: *A Kiss to Build a Dream On* by Louis Armstrong.

PART 3: BATCH DISTILLATION WITH BATCHSEP

Now that you have your batch reactor product, create a batch distillation simulation using `BatchSep`. `BatchSep` operates in Continuous flowsheet mode, which is counterintuitive, so when you create your `BatchSep`, it has to be either in your main flowsheet outside of the `BatchProcess` hierarchy flowsheet, or you can create a new Hierarchy block in Continuous mode, place that in the main flowsheet, and then place the `BatchSep` block inside of that. If you cannot find the `BatchSep` model in the Model Palette then it means you are not in a flowsheet that is in Continuous mode.

The general strategy of the batch distillation (refer to Figure B3.9) is to start with an empty cold pot, charge the pot with batch reactor product, then turn up the heat until it partially boils and the product starts to appear in the distillate. The first distillate product will be rich in acetaldehyde. Eventually, once all this is collected, the next product will be a mixture of 24DMD and water, since they have an azeotrope between them and they cannot be separated through single-stage distillation like this. We will keep the acetaldehyde product in a separate distillate receiver (a tank) from the 24DMD + water product, and we will not attempt to separate 24DMD and water further since that would happen in a different unit downstream that we will not worry about. Once most of the 24DMD has been collected, the pot contents will be essentially just propylene glycol left, and so the heating can be stopped and the pot contents collected as the final product.

The challenge with any batch distillation is choosing a column design that works well in combination with an operating recipe that gives optimal results. There are many factors to consider, such as batch size, tray/packing details and counts, control parameters or settings, temperatures, pressures, and when to stop the collection of one product and start the next, potentially even with "slop" or off-spec cuts that are recycled to the next batch. For this tutorial, use the following settings for something that works well enough to learn from!

Start by connecting the batch reactor product to the pot charge inlet port, and connect to more material streams to the distillate streams outlet port, one for the acetaldehyde product and one for the 24DMD + water product. Connect a third stream to the Final Column Contents outlet port which will be the propylene glycol product. Then, navigate through the menus to specify the following:

1. *Column Setup | Configuration.* Use a Batch Distillation Column configuration (by the way, you can model just a pot with this, much like we did with `BatchOp`) with 22 stages (20 trays plus condenser and pot), vapor-liquid[8] valid phases, partial overhead (meaning that a vapor distillate product can be collected in addition to a liquid distillate product), two distillate receivers with the initial Condenser Receiver set to 1 for both liquid and vapor products (because early on we will get acetaldehyde in both the vapor and liquid distillate during operation), calculated pressure holdups from tray/packing hydraulics (because we may have variable pressures to deal with across product regimes), rigorous pot heat transfer calculations (because we want realistic heat profiles), and an empty initial condition (because we need to drain out the pot and clean it since 24DMD is a food product). Figure B3.12 shows a screen capture.

2. *Column Setup | Streams.* Set the pot charge to use your feed stream you already connected to the pot using a molar basis. Later, you will set the pot charge rate in an Operating step and then it will know which feed stream to use. Then connect your two distillate receivers to your two distillate product streams such that your first receiver connects to your aldehyde product stream and the second receiver connects to the 24DMD + water product stream.

---

[8]Again, it is possible to have two liquid phases form for this example, so in a more rigorous simulation, you would likely want Vapor-Liquid-Liquid instead of Vapor-Liquid. However, let's keep it simple for the purposes of this tutorial and assume that there is not enough time for phase separation to meaningfully occur such that we treat the liquid phases in bulk. Besides, we are not actually separating 24DMD and water from each other, they are being collected together.

Define which feed stream is the one used for charging, and which product streams connect to which distillate receivers

Condenser pressure specified here

Condenser temperature specified here

Define the details of the pot geometry

Specify reflux ratio, reflux drum geometry, and reflux controller parameters

Specify details of heating/cooling jacket and coils, and heating medium, like steam

Specify initial conditions of column (empty, charged, or reflux)

Details about trays, packing, etc.

Assume ideal (perfect) phase equilibrium, or specify sub-equilibrium approximations.

Shows the stage profiles at the end of the batch (or current timestep)

Time trajectories of selected variables

Create plots for variables not found in Profile or Time Results

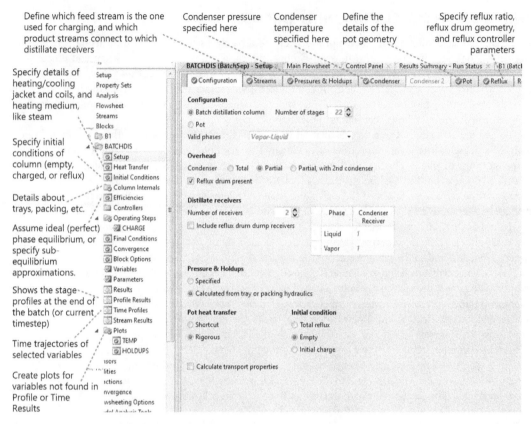

Figure B3.12   The BatchSep Setup form.

3. *Column Setup | Pressures & Holdups.* Set the condenser pressure to 1.5 bar (in reality, you would use a controller to maintain this pressure by manipulating perhaps the condenser duty or one of the product flow rates, but to model that you would have to move to the more complex Aspen Plus Dynamics software).

4. *Column Setup | Condenser.* Set the condenser temperature to 32°C. This was chosen because it is close to the boiling point of acetaldehyde at 1.5 bar (we need to keep it under pressure because acetaldehyde boils at room temperature and pressure). Again, you would normally use a controller to maintain this temperature by manipulating something like the reflux ratio or condenser duty, which is possible to model with the Controllers feature of BatchSep for some manipulated variables and not others. However, we will not use a controller for this example.

5. *Column Setup | Pot.* Set the tangent-to-tangent length of the pot (essentially, the height not including the round vessel heads) to 2.5 m and diameter to 2 m. This size was chosen to accommodate the batch charges we are using (100 kmol).

6. *Column Setup | Reflux.* Set the reflux ratio to 10 (this was chosen with trial and error, usually I recommend you start in the 3–5 range and then work your way up or down depending on how the results go) and check the box so that there will be a reflux drum present. The drum should be

vertical, 6 in. tall and 6 in. wide (usually the drum should be enough for roughly a 3-minute residence time). Scrolling down, you will see that there is a reflux level controller put in place by default. Leave the values at their defaults; the blank reflux level puts the level set point at about halfway up the drum and manipulates the reflux outlet to achieve that with a PI controller.

7. *Heat Transfer | Configuration.* Check the box indicating that you want a heating jacket around the pot that also covers the bottom (to increase heat transfer area). The box for a rigorous model should already be checked because you chose it on the main Setup form. Do not check the other boxes but you can see how you can add more details like heating coils, the heat capacity of the steel shell, losses to the environment, and extra heat exchangers.

8. *Heat Transfer | Jacket Heating.* Choose steam heating as your heating option at 300°C at a flow rate of 100 kg/hr. If not set, the limiting method should be set to Latent Heat (so it only transfers the latent heat from the steam, not specific heat due to subcooling). Some of these will be changed during operations and were chosen by trial and error as a good heating rate for the first stage of operation. Often you would have a controller that manipulates this rate to achieve some objective, like short cycle times while considering product purity constraints and column operability issues like flooding and weeping. For this example, we will just pick a single steam flow rate and work with that. Leave the other settings at defaults.

9. *Initial Conditions.* Here, specify the column to start empty, but since we do not want it to be an actual vacuum, and because there will be a nitrogen blowdown between batches in the real system, we want the column (as well as the pot, reflux drum, etc.) to start filled with nitrogen as the "pad gas" at 1.5 bar and 20°C. If you do not have nitrogen as an optional component, then go back and add it in the physical properties section of the simulation. Note that in general, it is easier to start with either the column in total reflux or with an initial charge, primarily because the numerical solvers sometimes have a hard time converging on an initial empty state. Even if the solver can successfully initialize the simulation, sometimes it has a hard time advancing to the earliest time steps when a batch charge is filled because many algorithms fail during sudden discreet changes to the system. For our case though, it should work.

10. *Column Internals.* Next, we have to specify what is inside the column (trays, packing, etc.). This works similar to `RadFrac` in which you can specify regions of the column and what is in that region. In our case, we only need one region, so add a new region (and double click to enter it). Inside it, you need only to define one section, and that section should start at stage 2 (the top tray since stage 1 is the condenser) and end at stage 21 (the bottom tray, since stage 22 is the pot). Specify trays with 1 ft. diameter and 18 in. tray spacing. You can see how with multiple sections you can define complex structures, such as dial-diameter columns or variable packing structures. You can click the View button if you want to see that the model will be considering default sizing parameters relating to the weir length, downcomer area, hole sizes, etc. Do not change this section, but know you have this level of detail if you need it.

11. *Operating Step 1: Charge Batch.* The Operating Step section of `BatchSep` is where you define the recipe for this unit, unlike `BatchOp` where you define Operations Inside of a general `Batch-Process` flowsheet hierarchy in Batch mode. However, the basic principle is the same, in that you define variables that change at the beginning of the operating step and conditions after which the step ends. Create a new Operating Step that charges the pot with your feed stream to receive 100 kmol in total. For example, specify that the Feed molar flow rate should be set to 1000 kmol/hr and set the stop criteria to be operation for 0.1 hours.

That was a lot, but at this point, you should have a column that is initialized and can charge the pot. Run it and make sure it works. You can go to the Results folder of the BatchSep block, check the pot holdup, and check the time profiles of holdup and temperature. I suggest creating a few plots of these time profiles in the Plots folder of the BatchSep block which can really help. I also recommend opening the control panel (F7) and checking the BatchSep messages. The key ones to watch out for are a notice that the initialization was successful and that it integrates in steps from 0 until the pot charge has been completed. You may notice that it may take small baby steps or get stuck at certain points in time and move very slowly. This is usually because during periods of rapid change, the numerical solvers need to keep timesteps small to keep numerical error from accumulating (a common strategy when using the Implicit Euler method, which is used by default).

 TOM'S TIP: You will likely find that, depending on the speed of your computer, the complete process simulation can take up to 30 minutes to complete! So, in order to speed it up, first loosen the relative integration tolerance from 0.0001 to 0.001, which can be found in the Convergence folder of the BatchSep block. What this does is allow a greater amount of numerical error (although it is still small), allowing the integrator to take bigger timesteps and/or require fewer convergence iterations per timestep, speeding up the simulation considerably. The tradeoff is that the results will have a larger error, but if you compare runs using the two tolerances, they are still quite close in our example. In practice, what you would do is develop your simulation and design using a looser tolerance (so simulation runs are faster), but then as you lock down your design, tighten the tolerance to get a final, more rigorous run. If the results change a lot, keep tightening the tolerance until the results stop changing as the tolerance gets tighter, meaning that you have tightened them enough!

Then, add the remaining operating steps, running the simulation after each one to make sure it works before adding the next.

12. *Operating Step 2: Heat and collect acetaldehyde.* At the end of the first step, the pot should be starting to boil and some vapor should be leaving the broth and heading upward. Now, increase the steam flow rate to 250 kg/hr (the variable location is Jacket Heating) and turn off the reactor feed (set its flow rate to 0). The step should end when the liquid mole fraction of acetaldehyde in the pot is 0.008 (0.8 mol%), approached from above of course. This means essentially all the acetaldehyde in the pot has been removed and collected in the first distillate receiver. Hopefully, our reflux ratio is high enough to prevent most of the water, 24DMD, and propylene glycol from coming with it. You can check the distillate receiver holdups to be sure, and note that the initial pad gas (nitrogen) that started in the pot should be with the acetaldehyde as well. That is fine, those can be easily separated with a condenser downstream.

13. *Operating Step 3: Collect 24DMD and water as the second product.* Now that the acetaldehyde has been collected, 24DMD and water are starting to appear in the distillate product. In this operating step, switch both the liquid and vapor distillate receivers to 2, change the reflux ratio to 5, the condenser temperature to 75°C, and the pot heating steam mass flow rate to 750 kg/hr. This step should end once the liquid mole fraction of propylene glycol in the reflux drum has risen (from below) to 0.01 (1 mol%), which means that most of the water on the trays has boiled upward and now only mostly propylene glycol remains on the trays. The condenser temperature is subcooled perhaps a little too far below the bubble point of 24DMD and water mixtures, but it is practical to be a little conservative. The lower reflux ratio of 5 is possible because it is easier to separate 24DMD and water from propylene glycol. The increased steam heating rate is needed to boil up all that 24DMD and water at a sufficiently high rate, but is low enough to prevent most of the propylene glycol from vaporizing. The total simulation time should be about 25.8 hours.

TOM'S TIP: Sometimes you need to stop the run because you can tell it is going to take too long, so feel free to hit the stop button (the blue square). However, I find that I almost always have to re-initialize the BatchSep block before running again. Simply making a change and rerunning without re-initializing almost always results in some strange numerical error.

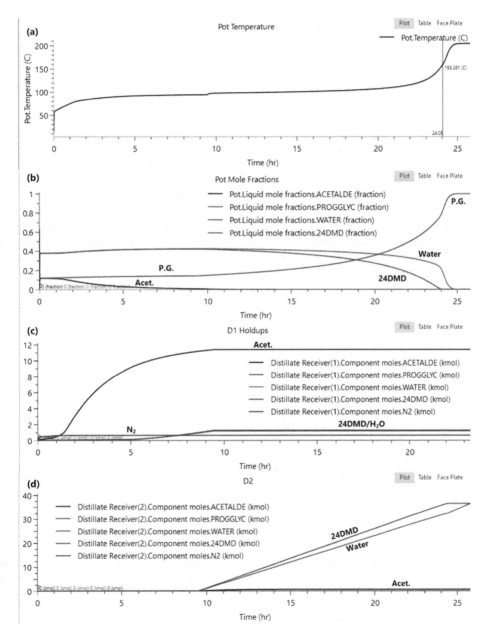

Figure B3.13 Trajectories of the completed run. (a) Pot temperature; (b) Pot mole fractions; (c) Distillate Receiver 1 holdups; (d) Distillate Receiver 2 holdups.

At this point, you should have essentially separated the three products successfully. But there is one problem: even though the pot liquid content is nearly 100% propylene glycol, the pot product stream is a lot less pure, since it contains a lot of water! That is because the liquid contained on the trays of the column and in the reflux drum has drained back into the pot after column shutdown at the end of the batch and ended up in this product stream. We can at least avoid letting the reflux drum contents (99% water) fall back into the pot. Since it contains mostly water, we want that to go into distillate receiver 2 instead. So, on the BatchSep block's Setup | Configuration form, check the box for "Include reflux dump receivers" and specify (in the new column that appears at right) that the reflux drum should empty into distillate receiver 2. Then rerun, and you should see a higher purity product in the final pot contents stream.[9]

Q7)   What are the final pot contents (the liquid molar holdups of each chemical)?

Q8)   How much cumulative heating energy did the batch distillation consume per cycle, in GJ?

If everything has worked correctly, you should have a system in which acetaldehyde has been recovered at about 78 mol% purity, 24DMD and water are recovered together at about 99% purity, and propylene glycol is recovered at about 98.5% purity, with no off-spec cuts or losses. Feel free to play with the system parameters to try to improve upon the results, particularly with regard to the acetaldehyde purity since it could benefit from larger reflux ratios and/or stage counts. The final plots should look similar to Figure B3.13. Congratulations, you have just completed a fairly complex batch distillation system featuring reaction, distillation, and lots of rigorous model considerations. There are more features to explore, and much to play with. However, if you need to go much farther beyond the complexity shown in this tutorial, such as using complex or unusual controllers, recipes, or coordinating between many different units, you may be better off using Aspen Plus Dynamics.

♫ Music break[10]

---

[9]In addition there is a feature in the Final Conditions folder of the block that lets you specify a reflux drum dump at the end of the batch into a distillation receiver. Unfortunately this feature is buggy in V12, although AspenTech has informed me that they are fixing it in a future version. Until then, enabling it may prevent column initialization when initializing from empty (for no obvious reason) even though the final conditions should not affect the initial conditions in this manner. You can try it but it may or may not work for you.

[10]Recommended listening: *Shut Up and Explode* by Boom Boom Satellites.

Choosing Property Packages

Objectives

- Learn strategies for choosing good property packages for your model
- Search the built-in database of physical property experimental data
- Test and validate your physical property model selections
- Handle normal liquids, normal gases, and normal gas/normal liquid pairs
- Validate complex situations such as tertiary phase equilibrium

Prerequisite Knowledge

This tutorial assumes you have completed Tutorials 1, 2, and 3 of this book. You should also have a basic understanding of vapor-liquid, liquid-liquid, and vapor-liquid-liquid equilibria, and concepts around model validation, such as R-squared and model fitting.

Why This Is Useful for Problem Solving

The models contained within Aspen Plus are quite well understood and are generally accepted by the process systems engineering community as being "correct" in terms of their derivation and implementation (when used in the proper circumstances). This is especially true for empirical and first-principles modeling for unit operations, physical properties for pure chemicals and chemical mixtures, and mass and energy balances. However, even though the equations themselves may be correct, in order for them to have any meaning or use, the parameters and assumptions on which they are based need to also be correct. Although Aspen Plus may run the simulations with no warnings or errors, you still have to decide if the numbers it provides you are valid and accurate. One wrong parameter and any model that uses it or uses information derived from it will be inaccurate.

For process modeling, the most fundamental models on which everything is based are the physical property models. These include things like pure component properties (heat capacity, surface tension, viscosity, vapor pressure, etc.), equations of state or other correlations between state properties (temperature, pressure, molar volume), and then the complex ways in which chemicals interact in mixtures (phase equilibria, solution chemistry, heats of mixing, etc.). In Aspen Plus, the built-in databanks contain this information for an extremely large number of chemicals. For the most part, the models and parameters for pure component properties (e.g., the equation for how the vapor pressure of a chemical changes with respect to temperature) are quite sound. Other types of property models can be either quite good or quite terrible, depending on the circumstances.

The proper way to validate any kind of model is to compare it against experimental data. That means rigorously comparing all aspects of the model: every single physical property model, empirical correlation, reaction and mass transfer kinetics, and unit operation. For most cases, especially with conceptual process design, that can be extremely hard to do. This is because in many cases the actual unit operations themselves may not have been built yet, or there are simply too many models embedded within a large flowsheet to test them all. The model developer has a limited amount of resources available for validation and so it has to decide which are the most important areas to validate. Although every situation is different, the modeling community has generally accepted this practice, and the risk that goes with it. However, there is one important type of model that is very often wrong and should be validated in almost all circumstances—phase equilibria.

Therefore, this tutorial is meant to help you determine the correct physical property models with regard to the phase equilibria of chemical mixtures. Because equilibrium calculations are a key component of nearly every unit operation model in the flowsheet involving fluids, choosing the wrong model can mean garbage results everywhere. If there is one thing you should spend time validating, it is this.

Tutorial

PART 1: PROPERTY METHOD TYPES

There are many types of physical property models within Aspen Plus. The Methods Assistant can help you filter the list to find some of the best ones for your specific case. You can find this in the Properties | Methods | Specifications form, with the button to the right of the method name dropdown. This takes you to a place in the help file where you can answer a few basic questions about your scenario, and after a few clicks it presents you with a list of suggested options for property packages. Usually, there are many, and further links are provided to find out more about them. This does not usually answer the question of what model to pick nor does it mean that these are necessarily the correct or even the best options for your circumstance. But this can be quite helpful for narrowing down the possibilities in your search.

It helps to understand the different types of methods. They can be classified roughly into three categories, and in each I have highlighted my favorites and some of the most common options. There are more than these to explore, but these are common starting points.

Equation of State Models

Equation of state models are sets of equations that relate state variables (i.e., temperature, pressure, and molar volume) to each other through a set of parameters. The parameters are determined through regression of experimental data combined with some fundamental thermodynamic theory. Most chemical engineers learn some of these early in their training and so are generally familiar with their use. What is remarkable (and even quite beautiful) about them is that you can derive analytical equations for enthalpy, entropy, and Gibbs free energy directly from the equation of state that relates temperature, pressure, and molar volume through the use of partial derivatives and thermodynamic theory. From those equations, you can get other properties like heat capacity, fugacity, and fugacity coefficients by doing a little more calculus. Although I will spare you the thermodynamics lesson,[1] it means that if you have very good parameters that map temperature, pressure, and molar volume together, you then also know entropy, enthalpy, heat capacity, and other properties in the liquid, gas, and supercritical states. It also means that if those parameters are not so good, then all of those properties will be incorrectly computed, and everything in your model can fall apart.

The most common and interesting models of this category are as follows:

- *Ideal Gas:* IDEAL is the classic ideal gas law. I would suggest that for most chemical processes, you should not use it because you have better models available to you at basically no extra effort. However, it is very useful for model validation or understanding basic principles. For example, if you are testing a new model, it would be useful to first use ideal gas because you could work the equations out yourself on paper and then determine if Aspen Plus is computing things in a way that you expect. It is useful in debugging sometimes because you can eliminate a physical property model as a variable.
- *Redlich-Kwong Variants:* This is a cubic equation of state, which means that it relates temperature, pressure, and molar volume in such a way that if temperature is known, the equation can be written such that pressure is a third-order polynomial function of molar volume. This structure makes it possible to represent liquid, vapor, vapor-liquid mixtures, and supercritical phases. There are many variants. The most common in Aspen are RK-SOAVE (Soave-Redlich-Kwong), RK-ASPEN (Redlich-Kwong-Aspen, which is basically RK-SOAVE but extended for better handling of polar molecules like water and alcohols), PSRK (Predictive Soave-Redlich-Kwong, which allows for UNIFAC prediction of unknown parameters—more on that later), RKS-BM (Soave-Redlich-Kwong with Boston-Mathias extensions) and others. You can read each help file entry to see if you think it fits your situation best. After 22 years of experience, my all-time favorite is PSRK. Although I have encountered situations where it does not work well, it very often was the best choice.
- *Peng-Robinson Variants:* The classic cubic Peng-Robinson equation of state (PENG-ROB) is available, but two variants are quite good: PRWS expands Peng-Robinson with Wong-Sandler mixing rules (which helps with binary phase equilibria) and has some UNIFAC predictive capability, and PR-BM has some additional parameters available through the Bost-Mathias modification which increases the potential for model accuracy. It is similar to RKS-BM. In my experience, I have found PR-BM to be the best choice in many circumstances and is one of the first methods I will usually explore.

Activity Coefficient Models

Although the theory behind equation of state models is quite elegant, in practice, it does not always work so well for complex phase equilibria, especially with multiple liquid phases. Activity coefficient models attempt to improve on accuracy by using separate models for the gas and liquid phases. In particular, the liquid phase model uses activity coefficients to compute liquid phase fugacity instead of an equation of state method (see Tutorial 2). The parameters for the liquid phase model are fit to the activity coefficients directly, and so they (in theory) could be more accurate than the equation of state method. On top of that, they no longer use equation of state information for the liquid phase, and so other empirical correlations are used instead for liquid phase properties, such as a polynomial equation to predict heat capacity as a function of temperature. Again, since each model is tailored to each specific property, each model should in theory have less error than an equation of state based approach. However, using separate models like this creates inconsistency between models because the elegant thermodynamic theory is not used. As a result, this will necessarily

---

[1]Unfortunately for them, my undergraduate students will not be spared this lesson.

introduce some error, although the hope is that the model consistency error it introduces is less than the error it avoids. Some examples are:

- *Classic Methods:* The WILSON and VANLAAR methods are some of the most famous activity coefficient methods. Plain vanilla WILSON and VANLAAR assume ideal gases. There are many variants (WILS-*something* or VANL-*something*) where what comes after the dash usually refers to an equation of state model for the vapor phase. However, I generally do not recommend any of these, as they have been supplanted by more modern methods.
- *Non-Random-Two-Liquid:* The NRTL method is a common activity coefficient method that is actually the default property method on many of the existing templates. However, I do not recommend this default NRTL since it uses the ideal gas law in the vapor phase (usually the first thing I do when I load up a template to start a new simulation is to go and delete the NRTL method and replace it before I do anything else). The variants are much better. These are intended for systems which can form two liquid phases, but often work well for vapor-liquid equilibria (VLE) generically. Personally, I find NRTL-RK often is the best activity coefficient model for my applications (the RK means the Redlich-Kwong equation of state). The electrolyte variants (ELECNRTL, ENRTL-RK, ENRTL-HOC, etc.) allow complex electrolyte chemistries and are explored more fully in Tutorial 12.
- *Universal Quasi-Chemical Models:* The UNIQUAC method is similar to NRTL, and uses the ideal gas law by default. Again, I recommend the UNIQ-RK or UNIQ-HOC variants. I have personally found UNIQ-RK to be very similar to NRTL-RK, and often indistinguishable.
- *UNIQUAC Functional Activity Coefficient Predictions:* The UNIFAC method is a way of estimating the binary interaction parameters for many of the above methods by using the shapes of the molecules themselves, which is extremely useful when binary pair information is missing (as it quite often is). UNIFAC can be incorporated into any of the UNIQUAC, NRTL, or PSRK methods and variants just by checking a box (see Figure B4.1). It also has its own property method definition called UNIFAC (which is UNIQ-RK with UNIFAC parameters instead of the ones in the database) and some variants. In most cases, I recommend that you start with UNIQ-RK and then only use UNIFAC to estimate the missing parameters, so you can take advantage of the regressed UNIQUAC parameters. But you can always try and see if UNIQFAC outperforms UNIQ-RK if you need to.

Specialty Models

While the above models are meant to be general for most situations, there are a lot of one-off specialty models that have been developed for certain subsets of chemicals or particular applications. There are more than I have had the opportunity to use, but if you work in a very particular or common type of process, there may be a model that has been designed specifically for it. The ones that I use most commonly are:

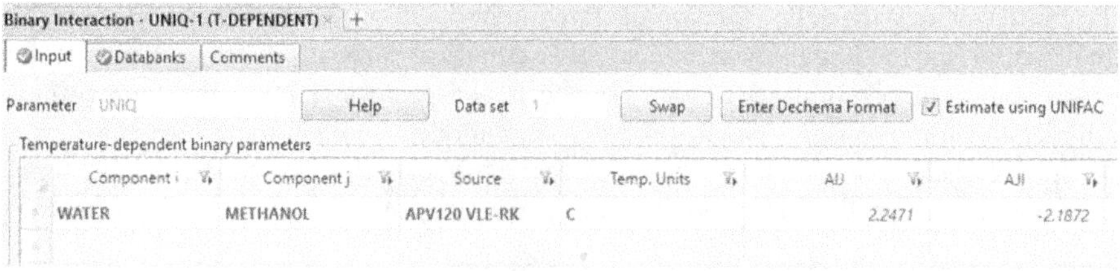

Figure B4.1   Check the box to estimate missing binary parameters using UNIFAC. When it works well, it is a lifesaver in the absence of property parameters!

- *Steam Tables:* Classic steam tables are equation correlations based on tabulated, experimentally determined physical properties of water at various phases. STEAMNBS should be your first choice for any model that uses pure water in the liquid, vapor, or supercritical phases. Although you usually will have other chemicals in your system, it is quite common that you would use STEAMNBS in individual unit operation models that use all water. For example, you may have a heat exchanger with a complex set of chemicals cooled by cooling water. You might choose PSRK for the hot side of the HEATX and STEAM-NBS for the cold side (you can individually specify your preference in the Block Options of any model). In fact, when adding cooling water or steam as utilities (see Tutorial 6), you will be asked if you want to add STEAMNBS to your simulation if you have not already so it can use it for all of its water utility calculations. STMNBS2 is the same thing, it just has a different solver algorithm which you can use in case you get root-finding errors. Legacy versions include RTOSTM and STEAM-TA and generally are only used in very particular circumstances such as free-water calculations (which are not covered in this book).
- *Amines:* AMINES is designed specifically for amines which are primarily used for gas sweetening applications (such as H_2S or CO_2 capture) specifically for monoethanolamine, diethanolamine, diglycolamine, and diisopropanolamine. Although this may seem very specific, it is of major interest right now due to its use in CO_2 capture systems. This is not necessarily the best method to use in all amine applications since ENRTL-RK can also be good. I cannot recommend one or the other in these cases; I use both and you should try them both if you are serious about gas sweetening.
- *Polymers:* The POLY*something* variants are intended for polymer and copolymer use. POLYSRK is an extension of PSRK for polymers. POLYSAFT model uses the statistical associating fluid theory (SAFT) equation of state for polymers, circa the 1990s. A more modern version from the 2000s, PC-SAFT, is intended for polymers, copolymers, and mixtures with normal liquids and gases. Although polymerization is not covered in the book, custom PC-SAFT models have found widespread use for the CO_2 capture solvent Selexol.

 TOM'S TIP: There's a lot to choose from. At the end of the day, I recommend starting with PSRK, PR-BM, and NRTL-RK as your first three candidates unless you have a special situation.

PART 2: VALIDATING NORMAL LIQUID SYSTEMS WITH EXPERIMENTAL DATA

Once you have selected a few candidate models using the guidance from Part 1, you will have to pick one to use based on experimental data. For this tutorial, we will focus on phase equilibrium data. You can find such data in scientific research publications in journals such as *Fluid Phase Equilibria* and the *Journal of Chemical & Engineering Data.* Common search terms should be something like "Water + Methanol Binary VLE" or "Water + Methanol + Ethanol Ternary Vapor-Liquid Equilibria," or "Methanol + Ethanol + Propanol + Butanol Quaternary VLLE." Those are often found in the titles or keywords of research papers which have this information. You can either search the journal websites directly or use indexing services like Engineering Village, Web of Science, or Google Scholar.[2] If your institution does not have a subscription to the journals, you may have to pay for the article.

However, before you stoop to reading the literature,[3] Aspen Plus comes with three tools, the NIST database, the DECHEMA database, and the DIPPR database, accessible from the Properties | Home ribbon (see Figure B4.2). The latter two are essentially links to external search engines where you can search for the data

---

[2]I recommend Engineering Village (which has the Inspec and Compendex databases) first and foremost. Many organizations and university libraries have a subscription. Major commercial research-specific search engines do a good job of getting high-quality results while avoiding the suspicious stuff. Google Scholar is getting better but I find it hits on too much junk, and different people can get different results from the same search based on their own settings and history.

[3]I worked at a company once with this sign over the library: "Two months in the lab saves an hour in the literature."

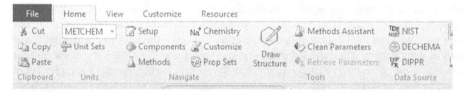

Figure B4.2 The NIST, DECHEMA, and DIPPR buttons (on the upper right) take you to databases full of valuable physical property data. We use NIST in this tutorial as DECHEMA and DIPPR require extra payment to use.

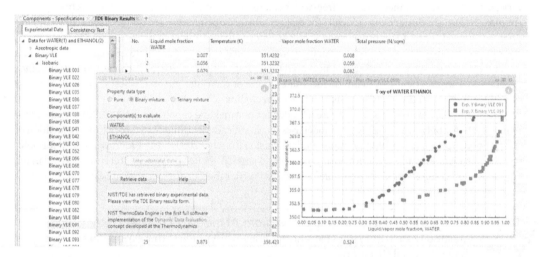

Figure B4.3 This is an example of using the NIST database to search for Water + Ethanol VLE. Enter your chemicals into the NIST ThermoData engine for a binary mixture and hit Retrieve Data. You will be presented with data collections shown on the Experimental Data tab at left. This example set is looking at Binary VLE 091, which is one set that contains VLE data at atmospheric pressure. This data appears in the right pane, behind the two popup windows. The T-xy plot button can be used to conveniently plot the data in the software, or it can be copy-pasted into Microsoft Excel or other software.

you want and then pay for access. However, NIST is provided with the data already available to you without additional charge, so let's use NIST!

Start up a new simulation and add water and ethanol to it. Let's use the NIST database to find binary physical property data for these two chemicals using the self-explanatory dialogs (see Figure B4.3). Click the NIST button in the home ribbon, choose the "binary mixture" radio button, enter the two chemicals, and hit retrieve data. After a wait, you should see a pop-under window with a left-hand pane that contains a very long list of experimental data, grouped into categories like azeotropic data, binary VLE, density, surface tension, etc. Those can be expanded to show individual sets of data collected in those categories. We are interested in Binary VLE, so go there, and specifically, we want isobaric (constant pressure) data. There are multiple pages of data sets in that example. Find an isobaric data set for 1 bar or 1 atm pressure (100,000 or 101,325 N/m^2; I chose data set 091 for this example). Clicking on a data set brings up the corresponding table of data on the right-hand side. Each data set contains experimental data from some publication, usually a peer-reviewed journal article, technical report, or conference proceeding, which is shown on the bottom of the rightmost pane. Each data set has different data because each experiment is different, so in practice you have to dig around for what you are looking for. You can visualize the data quickly with the T-xy (for isobaric) or P-xy (for isothermal) plot buttons in the home ribbon.

Q1)   According to Binary VLE Data Set 091, what will the mole fractions of water in the liquid and vapor phase be if a 50/50 mixture of water and ethanol is flashed at 1 atm and 356.123 K?

In some cases like this one, you might be overwhelmed with options. The key is selecting experimental data sets that represent the pressure and temperature ranges that are most relevant to your own application. It is a good idea to test out a few different experimental data selections, since these too could have different erroneous aspects to them. However, experimental thermophysical data in the peer-reviewed literature (even very old data) is usually trustworthy. Just because data is old, does not mean it is wrong.

TOM'S TIP: If you are having trouble selecting data sets, you can use the Consistency Test feature (the tab next to Experimental Data) and click the Run Consistency Tests button at the bottom of the right-hand pane. This performs a number of statistical assessments of the data that may or may not always be relevant, but it is a good guide. If you run this procedure, it produces an "overall data quality" score for each of the data sets. The higher (closer to 1), the better. Binary VLE 091 has a quality of 0.925, which is good, and one of the best actually, compared to the others. Keep in mind that this quality score is just a heuristic, useful for guidance.

Our next step is to compare this experimental data against model predictions using my three recommended starting-point models: PSRK, NRTL-RK, and PR-BM. There are a lot of ways to do this, and ultimately what we want to do is determine how well the models predict the data and select the best one. Sometimes a visual comparison is easiest, so let's use some outside tools to do that. Copy-paste the experimental data set into Excel or some other spreadsheet or graphing app. Then, set up your simulation to use PSRK as the property method, and use the Binary analysis tool to generate a T-xy plot (see Tutorial 2). Then, copy-paste those results into the spreadsheet as well, and plot them on top of each other. Typically, one easy way to communicate this is to show the experimental data points as points with no lines (since they are discrete quantities with experimental error and noise), and the model results as lines with no points (because they are continuous and have no noise, so are valid in the in-betweens). If the lines go through the points, then you have a very good model (well, for the range of pressures and temperatures that you have tested at least).

Repeat the Binary analysis for the two other property methods (PR-BM and NRTL-RK). Ideally, all you have to do is add them to the Methods section, load the binary parameters (if applicable), rerun the Binary Analysis (just change the property method in the Calculation Options tab of the Binary Analysis), and then copy-paste the results into your spreadsheet or plotting software to easily generate the new plot. I suppose it's easier said than done but if you set the first one up thoughtfully, it is trivial to do the rest.

My results are shown in Figure B4.4. You can see immediately that PSRK and NRTL-RK are both very good, but PR-BM is quite terrible. This is important to recognize because PR-BM is so widely used, and yet how often is it checked? Between the other two options, you can see visually that NRTL-RK fits the data more closely than PSRK because the left-hand side of the azeotrope fits it better. At this point, I would say that no further analysis is needed, NRTL-RK is the clear winner here.

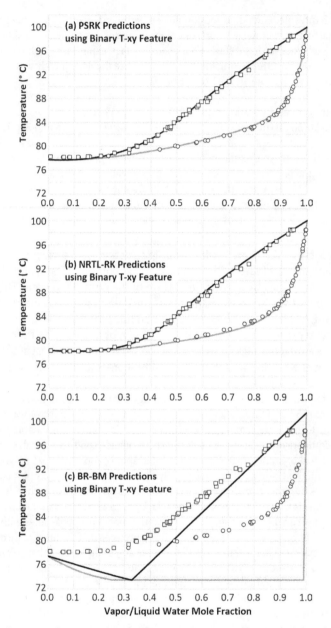

Figure B4.4   T-xy diagrams of water + ethanol VLE at 1 atm. *Points*: Experimental data from Otsuki H, Williams FC. Effect of pressure on vapor-liquid equilibria for the system ethyl alcohol–water. *Chem Eng Prog Symp Ser,* 1953, 49:55–67. *Lines*: Model predictions using the Binary T-xy feature using (a) PSRK; (b) NRTL-RK; (c) PR-BM with default parameters.

TOM'S TIP: What is going on with PR-BM? The first is that the azeotrope point is a few degrees too low, and so the VLE region to the left of it (i.e., having low water mole fraction) is quite off. To the right of the azeotrope, the large triangle shape actually indicates that the flash calculations simply did not converge, or converged at the trivial (azeotrope) solution. The Binary T-xy tool works by taking the range of liquid mole fractions you specify (by default it is 0–100 mol% of whichever is your primary liquid) and works through that range iteratively in small steps (default is every 2 mol%). For each of those points, it tries to solve a flash calculation problem in which it finds the temperature at which a mixture would flash into vapor and liquid phases, where the liquid phase has that desired composition. It doesn't always work. When you get T-xy plot results that look like the PR-BM example, you have to decide if the model is bad, or if the T-xy plot generator is just not working (or both!). You can test that by running a simulation of a flash drum explicitly on a flowsheet and seeing if you get the same kind of garbage results or if you get valid results.

♫ Music break[4]

PART 3: VALIDATING MORE COMPLEX SYSTEMS WITH EXPERIMENTAL DATA

Graphical Methods for Normal Gas + Normal Liquid Systems

A similar process can be used to select and validate physical property models in other circumstances. Let's use water + hydrogen VLE as an example. It is quite common in many kinds of chemical processes where water needs to be removed from a stream of light gases, such as hydrogen. Typically, simple condensation with a flash drum does the trick. Find some isothermal VLE data for water + hydrogen in the NIST database at about 38°C (311 K), which is a typical temperature to use for condensation since cooling water is readily available using cooling towers below that temperature. Make a P-xy plot of that data.

Q2) Consider the liquid mole fractions of H_2. What is the largest concentration (mol%) of H_2 in the liquid phase that was found experimentally at any pressure using this data?

From the plot, we can see that basically, H_2 has a very limited liquid presence, even at very high pressures. Let's use the Binary P-xy function to determine which of those three physical property methods are the best for this system. Note that if you use the P-xy function using its default values, you are likely to get a bunch of garbage, because it defaults to examining liquid mole fraction ranges from 0% to 100% of the primary chemical. Use the Binary P-xy plot generator, but alter the start points and end points of the analysis (i.e., the liquid water or H_2 mole fractions) so that they only go within a reasonable mole fraction range, based on the experimental data and your answer to Q2. Don't forget to select the temperature that matches the experimental data as well.

Now, you may notice that for some of the three models, you may get some errors, or cases even in which you have no results at all. This is most likely because the liquid mole fraction range that you have specified for your analysis contains points that, according to the model at least, are completely infeasible. If you are using the actual range of the experimental for your range on the Binary analysis and it is not working, then it likely means that the model is inaccurate at the edge of the range. You can try to bring your range in a little bit, and intentionally run it at points on the inside. Don't spend too much time messing with it though, it's not worth it.

---

[4]Recommended listening: *Bohemian Rhapsody* by Queen.

If the Binary Analysis function does not work quickly for one of your cases, you should instead try an alternate method. Make a simulation that consists of a single flash drum, with a feed containing a water and hydrogen mixture that is somewhere inside the phase envelope (in this case, the envelope is very wide so most mixtures will work, but a 50/50 mixture will be fine). Set the flash drum to the same temperature as the data you are comparing against. Then use a sensitivity block (see Tutorial 3) to run the analysis yourself by changing the pressure of the drum across a range and recording the vapor and liquid mole fractions of the output streams. Use that information in your plot. You can quickly rerun the simulation and sensitivity analysis for different property models by changing the property method in the Block Options tab for the flash drum.

Q3) Compare the experimental results to the model predictions for PSRK, NRTL-RK, and PR-BM. Which physical property method is the best choice for this system? Which is the worst? Can you make strong conclusions?

Graphical Methods for Ternary VLE (Three Chemicals)

The T-xy and P-xy plot approach is useful for a quick assessment of quality. However, there are more rigorous statistical approaches you can use, especially when you are looking at systems with ternary or quaternary data that you can't plot easily. Let's take a quick look at how you can similar things with ternary data. Add ethane, propane, and butane to your simulation, and update the three parameter models. Check the NIST database for ternary data. Ternary data are more rare, but they do have some here. Ternary VLE 004 has a collection of VLE at a few different temperatures and pressures. Let's take a simple example: data at 5 bar (500,000 Pa) and 260 K. There are six experimental data points spread over 12 rows (six as vapor data and six as liquid data, which go together). The liquid-vapor pairs are connected by tie lines, meaning that for any feed composition along that tie line, the vapor and liquid compositions after flashing will be equal to the endpoints of the tie line—that is, the experimental data. An example plot is shown in Figure B4.5.

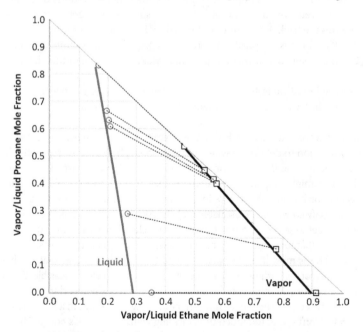

Figure B4.5 Ternary VLE at 260 K and 5 bar for ethane + propane + butane. *Points*: Experimental data from Clark AQ, Stead K. (Vapor + Liquid) Phase equilibria of binary, ternary, and quaternary mixtures of CH_4, C_2H_6, C_3H_8, C_4H_{10} and CO_2. *J Chem Thermodyn*, 1988, 20:413–427. *Dashed-Lines*: Tie lines between experimental data points. *Solid Lines*: Model predictions using flash drum simulations with the PR-BM method.

In order to test your property model, you can set up a flowsheet which would simulate a flash drum at 260 K and 5 bar, with different feed mixtures. You simply have to simulate feed mixtures along the inside of that envelope, and record the outlet vapor and liquid mole fractions (where the flash drum is always at 260 K and 5 bar). For example, if you simulate a feed that is 50 mol% ethane, 25 mol% propane, and 25 mol% butane, you will be inside the envelope, and close to the experimental tie line that is second from the bottom on the plot. You can do this with a sensitivity block as well, but you have to be more clever about how you

Table B4.1 **Experimental Conditions and Outcomes for a Quaternary System at 1 atm.**

		1	4	7	12	16
	Temperature (K)	334.35	336.95	339.85	343.55	346.05
Measured liquid mole fractions	Methyl acetate	0.141	0.137	0.200	0.056	0.053
	Methanol	0.678	0.497	0.188	0.289	0.250
	Isopropanol	0.044	0.076	0.243	0.354	0.221
	Isopropyl acetate	0.137	0.290	0.369	0.301	0.477
Measured vapor mole fractions	Methyl acetate	0.305	0.313	0.396	0.123	0.139
	Methanol	0.619	0.518	0.260	0.448	0.343
	Isopropanol	0.032	0.059	0.142	0.253	0.189
	Isopropyl acetate	0.044	0.110	0.202	0.176	0.330
Feed mole fractions*	Methyl acetate	0.223	0.225	0.298	0.090	0.096
	Methanol	0.646	0.508	0.224	0.369	0.296
	Isopropanol	0.038	0.068	0.193	0.304	0.205
	Isopropyl acetate	0.091	0.200	0.286	0.239	0.403
Simulated liquid mole fractions	Methyl acetate					
	Methanol					
	Isopropanol					
	Isopropyl acetate					
Simulated vapor mole fractions	Methyl acetate					
	Methanol					
	Isopropanol					
	Isopropyl acetate					

The experimental utcomes were reported in the paper: Xiao L, Wang QL, Wang HX, Qiu T. Isobaric Vapor–Liquid Equilibrium Data for the Binary System Methyl Acetate + Isopropyl Acetate and the Quaternary System Methyl Acetate + Methanol + Isopropanol + Isopropyl Acetate at 101.3 kPa. *Fluid Phase Equilibria*, 2013, 344:79–83.

*The feed compositions were made up (educational guesses) by yours truly for demonstration purposes but are likely not far from the real experimental starting mixture composition, which was not described in the paper, unfortunately.

set it up (maybe fix some molar flow rates for two of the chemicals in the feed, and vary the flow rate of the third chemical with the sensitivity block). My example plot of the experimental data, along with the results for the PR-BM method, is shown in Figure B4.5.

Q4) Create a plot similar to Figure B4.5 that shows how the NRTL-RK method compares to the experimental data. Is it better than PR-BM?

Statistical Methods for Any Number of Chemicals

The final approach is to use a statistical method. Essentially, what you want to do is replicate an experiment that tests phase equilibria in the simulator and see how that compares to the experimental data. This is hard to do with the data that is available in the NIST database because the feed compositions that correspond to each data point are not provided.

Table B4.1 has some example experimental data for a system of methyl acetate + methanol + isopropanol + isopropyl acetate at 1 atm. Each row of the table was a separate experiment. In each experiment, the feed mixture shown was heated to the temperature shown, the pressure was controlled at 1 atm, and the liquid and vapor phases that formed were collected and sampled.

Replicate these five experiments using Aspen Plus and the NRTL-RK method (or pick your favorite). Use a flash drum, and change the temperature and feed composition for each experiment accordingly. If you are using a digital version of this book, try copy-pasting from the table for convenience (either directly into Aspen Plus or into a spreadsheet as an intermediate). With this small data set, it might be easier to manually enter the data, but for larger sets, you can use the automation techniques through connections to Microsoft Excel (see Tutorial 9) or Python (see Bonus Tutorial 2). Then, fill in your simulation results in the empty table. You can then compare your simulation results to the experimental value using statistical analyses, since visualization with four dimensions is just too difficult. There are any number of statistical analyses that you can try, like the classic coefficient of determination (R^2).

Q5) Fill in Table B4.1 using flash drum simulations with NRTL-RK. Use some statistical method to decide if the fit is good or not. A simple coefficient of determination would suffice.

♫ Music break[5]

---

[5]Recommended listening: *Red Light Syndrome* by d.notive.

Solutions

Aspen Plus V12 simulation files for each of the tutorials can be found at the link below:

http://psecommunity.org/books/lap24

Note that the files are in the Aspen Plus Backup File format (.bkp). This format is meant to be forward-compatible such that future versions of Aspen Plus (which have not been released at the time of writing) may be used to open the file. If you are using a future version of Aspen Plus, you will likely be prompted with a notice that this file uses an older physical property model than currently available. You will likely be given a choice as to whether you want to use the original physical property models that I used (the "legacy" option) or whether you would like to use the most recent ("updated") properties. You can use either, but if you use the updated properties, then the solutions provided herein might be different. There is no guarantee of course that these files will work in later versions of the software but they most likely will.

All simulations are provided as-is with no warranty or guarantee of accuracy.

Tutorial 1

PART 1

Q1) 86.18 g/mol

Compounds found matching the specified criteria

Compound name	Alias	Databank	Alternate name	MW	BP <C>	CAS number	Compound class
HEXAMETHYLENEI	C6H16N2	APV120.PU	1,6-Diamino-n	116.205	199.85	124-09-4	OTHER-AMINES/
HEXANENITRILE	C6H11N	APV120.PU	N-HEXANENIT	97.1582	163.6	628-73-9	NITRILES
N-HEXANE	C6H14-1	APV120.PU	n-HEXANE	86.1754	68.73	110-54-3	N-ALKANES
N-HEXYLMERCAPT	C6H14S	APV120.PU	n-Hexanethiol	118.24	152.66	111-31-9	MERCAPTANS
PERFLUORO-N-HE	C6F14	APV120.PU	PERFLUORO-n	338.042	57.15	355-42-0	C,H,F-COMPOUN
1H-PERFLUORO-N	C6HF13	NISTV120.I	1H-PERFLUOR(	320.053	71.331	355-37-3	C,H,F-COMPOUN

Q2) 5

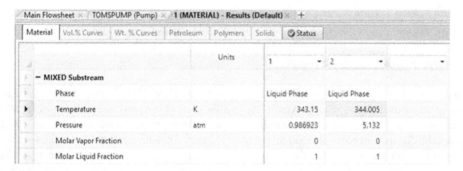

Q3 344 K

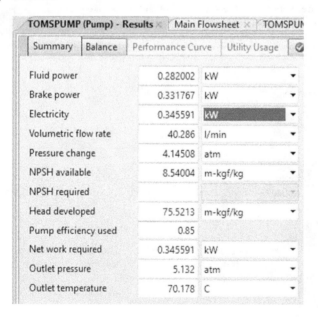

Q4) 0.346 kW

Q5) 5.20 kmol/hr

	Units	3	4
Molar Liquid Fraction		1	1
Molar Solid Fraction		0	0
Mass Vapor Fraction		0	0
Mass Liquid Fraction		1	1
Mass Solid Fraction		0	0
Molar Enthalpy	cal/mol	-41965.8	-52226.7
Mass Enthalpy	cal/gm	-486.971	-384.981
Molar Entropy	cal/mol-K	-140.248	-200.203
Mass Entropy	cal/gm-K	-1.62744	-1.47577
Molar Density	mol/cc	0.00557922	0.00335546
Mass Density	gm/cc	0.480802	0.455204
Enthalpy Flow	cal/sec	-60580.1	-98696.7
Average MW		86.1772	135.661
− Mole Flows	**kmol/hr**	**5.19681**	**6.80319**
N-HEX-01	kmol/hr	5.19681	0.803188
N-DEC-01	kmol/hr	4.56451e-07	6

Q6) 1.36 GJ/hr

Reboiler / Bottom stage performance		
Name	Value	Units
Temperature	496.065	K
▶ Heat duty	1.36156	GJ/hr
Bottoms rate	6.80319	kmol/hr
Boilup rate	29.2537	kmol/hr
Boilup ratio	4.3	

PART 2

Q7) −0.48 GJ/hr

Summary	Balance	Phase Equilibrium	Utility Usage	⊘ Status

Outlet temperature	298.15	K	▾
Outlet pressure	4.93462	atm	▾
Vapor fraction	0		
Heat duty	-0.478431	GJ/hr	▾
Net duty	-0.478431	GJ/hr	▾
1st liquid / Total liquid	1		
Pressure-drop correlation parameter			
Pressure drop	0	atm	▾

Q8) 0.9989

	Units	3 ▾
Mass Solid Fraction		0
Molar Enthalpy	cal/mol	-41986.9
Mass Enthalpy	cal/gm	-486.855
Molar Entropy	cal/mol-K	-140.326
Mass Entropy	cal/gm-K	-1.62714
Molar Density	mol/cc	0.00557575
Mass Density	gm/cc	0.480858
Enthalpy Flow	cal/sec	-60479.8
Average MW		86.241
+ Mole Flows	**kmol/hr**	**5.18561**
− Mole Fractions		
N-HEX-01		0.998863
N-DEC-01		0.00113748

Material | Heat | Load | Work | Vol.% Curves | Wt. % Curves | Petroleum | Polymers

Tutorial 2

PART 1

Q1) 0.3. It does not matter whether i and j are methanol or chloroform for this instance, since the C term is symmetric.

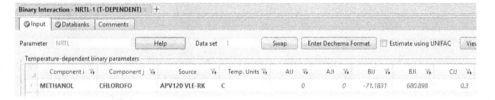

Component i	Component j	Source	Temp. Units	AIJ	AJI	BIJ	BJI	CIJ
METHANOL	CHLOROFO	APV120 VLE-RK	C	0	0	-71.7831	680.898	0.3

PART 2

Q2) −6904.5

PART 3

Q3) 153.5°C

PRES	MOLEFRAC METHANOL	TOTAL TEMP	TOTAL KVL METHANOL	TOTAL KVL CHLOROFO	LIQUID GAMMA METHANOL	LIQUID GAMMA CHLOROFO	VAPOR MOLEFRAC METHANOL	VAPOR MOLEFRAC CHLOROFO
bar		C						
5	0.891089	108.317	0.919406	1.65941	1.00586	2.13841	0.819272	0.180728
5	0.90099	108.564	0.924904	1.68337	1.00478	2.15772	0.833329	0.166671
5	0.910891	108.82	0.930751	1.70788	1.00383	2.17707	0.847813	0.152187
5	0.920792	109.084	0.936952	1.73294	1.00299	2.19647	0.862738	0.137262
5	0.930693	109.357	0.943511	1.75856	1.00227	2.21592	0.878119	0.121881
5	0.940594	109.639	0.950435	1.78478	1.00165	2.23541	0.893973	0.106027
5	0.950495	109.929	0.957729	1.8116	1.00113	2.25495	0.910317	0.0896832
5	0.960396	110.228	0.9654	1.83905	1.00072	2.27452	0.927167	0.0728334
5	0.970297	110.537	0.973455	1.86713	1.0004	2.29413	0.944541	0.0554593
5	0.980198	110.854	0.981902	1.89587	1.00018	2.31378	0.962458	0.037542
5	0.990099	111.181	0.990747	1.9253	1.00004	2.33346	0.980938	0.0190623
5	1	111.518	1	1.95542	1	2.35317	1	0
10	0	153.545	5.90391	1	4.13862	1	0	1
10	0.00990099	150.765	5.37037	0.956296	3.98485	1.00023	0.0531719	0.946828
10	0.0198019	148.421	4.92065	0.920795	3.83391	1.00093	0.0974385	0.902561
10	0.0297029	146.429	4.53715	0.89172	3.68741	1.00208	0.134767	0.865233
10	0.0396039	144.722	4.20674	0.867763	3.5464	1.00368	0.166604	0.833396
10	0.0495049	143.249	3.91949	0.847944	3.41147	1.00571	0.194034	0.805966
10	0.0594059	141.972	3.66772	0.831513	3.28292	1.00816	0.217884	0.782116
10	0.0693069	140.858	3.44546	0.817891	3.16084	1.01102	0.238794	0.761206
10	0.0792079	139.881	3.24799	0.806625	3.04518	1.01428	0.257266	0.742734

Q4) 1.056

PRES	MOLEFRAC METHANOL	TOTAL TEMP	TOTAL KVL METHANOL	TOTAL KVL CHLOROFO	LIQUID GAMMA METHANOL	LIQUID GAMMA CHLOROFO	VAPOR MOLEFRAC METHANOL	VAPOR MOLEFRAC CHLOROFO
bar		C						
5	0.623762	104.353	0.896699	1.17126	1.10021	1.6465	0.559328	0.440672
5	0.633663	104.419	0.893041	1.18501	1.09357	1.66337	0.565888	0.434112
5	0.643564	104.491	0.88977	1.19903	1.08725	1.68038	0.572624	0.427376
5	0.653465	104.569	0.886875	1.21332	1.08125	1.6975	0.579543	0.420457
5	0.663366	104.651	0.884351	1.2279	1.07555	1.71476	0.586649	0.413351
5	0.673267	104.74	0.88219	1.24276	1.07014	1.73213	0.59395	0.40605
5	0.683168	104.833	0.880385	1.25792	1.06501	1.74961	0.601452	0.398548
5	0.693069	104.933	0.878932	1.27338	1.06015	1.76721	0.609161	0.390839
5	0.70297	105.039	0.877824	1.28915	1.05554	1.78492	0.617084	0.382916
5	0.712871	105.15	0.877057	1.30524	1.05118	1.80274	0.625229	0.374771
5	0.722772	105.268	0.876628	1.32165	1.04706	1.82066	0.633603	0.366397
5	0.732673	105.392	0.876532	1.33839	1.04316	1.83869	0.642212	0.357788
5	0.742574	105.523	0.876767	1.35548	1.03948	1.85681	0.651065	0.348935
5	0.752475	105.659	0.877331	1.37291	1.03602	1.87502	0.66017	0.33983
5	0.762376	105.803	0.878221	1.39071	1.03276	1.89333	0.669535	0.330465
5	0.772277	105.953	0.879436	1.40887	1.02969	1.91173	0.679169	0.320831
5	0.782178	106.11	0.880975	1.42741	1.02682	1.93021	0.68908	0.31092

Q5) 0.485 (Roughly)

PRES	MOLEFRAC METHANOL	TOTAL TEMP	TOTAL KVL METHANOL	TOTAL KVL CHLOROFO	LIQUID GAMMA METHANOL	LIQUID GAMMA CHLOROFO	VAPOR MOLEFRAC METHANOL
bar		C					
5	0.39604	104.05	1.12233	0.919786	1.39015	1.30109	0.444486
5	0.405941	104.02	1.10483	0.928365	1.36963	1.31413	0.448496
5	0.415842	103.994	1.08829	0.93715	1.35011	1.32738	0.452556
5	0.425743	103.972	1.07265	0.946142	1.33154	1.34083	0.456671
5	0.435644	103.953	1.05786	0.955339	1.31387	1.35448	0.460848
5	0.445545	103.938	1.04387	0.964737	1.29704	1.36831	0.465098
5	0.455446	103.926	1.03066	0.974348	1.28103	1.38234	0.469416
5	0.465347	103.918	1.01818	0.984167	1.26577	1.39655	0.473813
5	0.475248	103.914	1.0064	0.994194	1.25125	1.41094	0.478295
5	0.485149	103.914	0.995294	1.00443	1.23741	1.42551	0.482868
5	0.49505	103.918	0.984823	1.01488	1.22422	1.44025	0.487536
5	0.504951	103.925	0.974962	1.02554	1.21166	1.45517	0.492305
5	0.514852	103.937	0.965687	1.03642	1.1997	1.47026	0.497181
5	0.524753	103.953	0.956975	1.04752	1.1883	1.48551	0.502168
5	0.534654	103.972	0.948795	1.05883	1.17743	1.50093	0.507277
5	0.544555	103.996	0.941142	1.07037	1.16709	1.5165	0.512504
5	0.554455	104.025	0.933992	1.08214	1.15723	1.53224	0.517857
5	0.564356	104.057	0.927326	1.09415	1.14784	1.54813	0.523343

PART 4

Q6) 0.990
Q7) 53.7°C

	Units	A-FEED	B-MEOH	C-AZEO
− MIXED Substream				
Phase		Liquid Phase	Liquid Phase	Liquid Phase
Temperature	C	30	64.1363	53.6584
Pressure	bar	1.01325	1.01325	1.01325
Molar Vapor Fraction		0	0	0
Molar Liquid Fraction		1	1	1
Molar Solid Fraction		0	0	0
Mass Vapor Fraction		0	0	0
Mass Liquid Fraction		1	1	1
Mass Solid Fraction		0	0	0
Molar Enthalpy	cal/mol	-49324.3	-55794.5	-39704.8
Mass Enthalpy	cal/gm	-846.877	-1698.91	-443.751
Molar Entropy	cal/mol-K	-53.5197	-54.4371	-48.2529
Mass Entropy	cal/gm-K	-0.918909	-1.65758	-0.539287
Molar Density	mol/cc	0.019454	0.0230605	0.0147338
Mass Density	gm/cc	1.13305	0.75734	1.31831
Enthalpy Flow	cal/sec	-2.74024e+06	-1.70943e+06	-989347
Average MW		58.2426	32.8414	89.4753
+ Mole Flows	kmol/hr	**200**	**110.297**	**89.7032**
− Mole Fractions				
METHANOL		0.7	0.990849	0.34238
CHLOROFO		0.3	0.00915144	0.65762

So far our flowsheet looks like this (when answering Q8 and Q9):

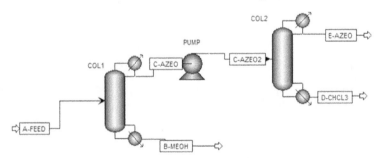

Q8) 0.94

Q9) 130.6°C

	Units	D-CHCL3	E-AZEO
− MIXED Substream			
Phase		Liquid Phase	Vapor Phase
Temperature	C	142.219	130.561
Pressure	bar	10	10
Molar Vapor Fraction		0	1
Molar Liquid Fraction		1	0
Molar Solid Fraction		0	0
Mass Vapor Fraction		0	1
Mass Liquid Fraction		1	0
Mass Solid Fraction		0	0
Molar Enthalpy	cal/mol	-30064.9	-36198.9
Mass Enthalpy	cal/gm	-262.885	-505.968
Molar Entropy	cal/mol-K	-39.3826	-28.2461
Mass Entropy	cal/gm-K	-0.344358	-0.394808
Molar Density	mol/cc	0.0106901	0.000337205
Mass Density	gm/cc	1.22258	0.0241249
Enthalpy Flow	cal/sec	-313705	-524279
Average MW		114.365	71.5438
+ Mole Flows	kmol/hr	**37.5633**	**52.1398**
− Mole Fractions			
METHANOL		0.0573854	0.5477
CHLOROFO		0.942615	0.4523

Q10) 0.94

Q11) 0.93

Material	Heat	Load	Work	Vol.% Curves	Wt. % Curves	Petroleum	Polymers	Solids

	Units	D-CHCL3	B-MEOH
− MIXED Substream			
Phase		Liquid Phase	Liquid Phase
Temperature	C	142.051	61.6337
Pressure	bar	10	1.01325
Molar Vapor Fraction		0	0
Molar Liquid Fraction		1	1
Molar Solid Fraction		0	0
Mass Vapor Fraction		0	0
Mass Liquid Fraction		1	1
Mass Solid Fraction		0	0
Molar Enthalpy	cal/mol	-30103	-54244.7
Mass Enthalpy	cal/gm	-263.493	-1411.49
Molar Entropy	cal/mol-K	-39.4004	-53.6835
Mass Entropy	cal/gm-K	-0.344874	-1.39689
Molar Density	mol/cc	0.0107019	0.0219902
Mass Density	gm/cc	1.22265	0.845098
Enthalpy Flow	cal/sec	-437018	-2.22607e+06
Average MW		114.246	38.4307
+ Mole Flows	kmol/hr	**52.2628**	**147.735**
− Mole Fractions			
▶ METHANOL		0.0587522	0.92685
CHLOROFO		0.941248	0.0731503

The final sheet should look like this:

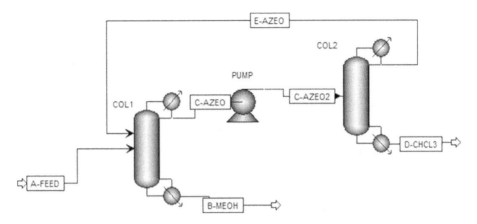

Q12) 115.1°C
Q13) 0.0943

Stream Point Analysis Results

Stream   D-CHCL3

Stream properties

PROPERTIES	UNITS	TOTAL	VAPOR	LIQUID
TEMP	C	142.051		
PRES	bar	10		
VFRAC		0		
MOLEFLMX	kmol/hr	52.2628		52.2628
MWMX		114.246		114.246
HMX ENTHALPY-FLO	cal/sec	-437020		-437020
HMX MOLE-ENTHALP	cal/mol	-30103		-30103
SMX	cal/mol-K	-39.4004		-39.4004
CPMX	cal/mol-K	29.8818		29.8818
RHOMX	mol/cc	0.0107019		0.0107019
RHOLSTD	mol/cc	0.0127967		0.0127967
MUMX	cP			0.233939
KMX	J/sec-m-K			0.0943384
SIGMAMX	dyne/cm			12.0148

Q14) 12.03 mol/L

Tutorial 3

PART 1

Q1)   403.9°C

	Units	3
Description		
From		HX
To		TURB1
Stream Class		CONVEN
Maximum Relative Error		
Cost Flow	$/hr	
− MIXED Substream		
Phase		Vapor Phase
Temperature	C	403.913
Pressure	bar	20.5
Molar Vapor Fraction		1

Q2)   318.9°C

	Units	3
Description		
From		HX
To		TURB1
Stream Class		CONVEN
Maximum Relative Error		
Cost Flow	$/hr	
− MIXED Substream		
Phase		Vapor Phase
Temperature	C	318.889
Pressure	bar	20.5
Molar Vapor Fraction		1

Q3)   14,495 kmol/hr

	Units	1 ▾	3 ▾
Phase		Liquid Phase	Vapor Phase
Temperature	C	95	360.002
Pressure	bar	1	20.5
Molar Vapor Fraction		0	1
Molar Liquid Fraction		1	0
Molar Solid Fraction		0	0
Mass Vapor Fraction		0	1
Mass Liquid Fraction		1	0
Mass Solid Fraction		0	0
Molar Enthalpy	cal/mol	-67048.7	-55173.3
Mass Enthalpy	cal/gm	-3721.77	-3062.58
Molar Entropy	cal/mol-K	-35.1676	-10.5186
Mass Entropy	cal/gm-K	-1.9521	-0.583872
Molar Density	mol/cc	0.0533951	0.000403544
Mass Density	gm/cc	0.961927	0.00726996
Enthalpy Flow	cal/sec	-2.69958e+08	-2.22144e+08
Average MW		18.0153	18.0153
+ Mole Flows	kmol/hr	14494.7	14494.7

Tabs: Material | Heat | Load | Work | Vol.% Curves | Wt. % Curves | Petroleum | Polymers | Solids

PART 2

Q4)   166.5 MW

Tabs: Summary | Balance | Phase Equilibrium | Utility Usage | ⊘ Status

Outlet temperature	95	C
Outlet pressure	1	bar
Vapor fraction	0	
Heat duty	-166.474	MW
Net duty	-166.474	MW
1st liquid / Total liquid	1	
Pressure-drop correlation parameter		
Pressure drop	0	bar

Q5) 33.5 MW

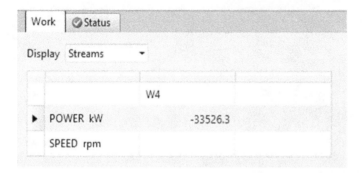

Q6) 33.5 MW

Row/Case	Status	VARY 1 TURB1 PARAM PRES BAR	TOTALW KW	TOTALW/1 000
24	OK	4.3	-33490.6	-33.4906
25	OK	4.4	-33497	-33.497
26	OK	4.5	-33502.7	-33.5027
27	OK	4.6	-33507.6	-33.5076
28	OK	4.7	-33511.9	-33.5119
29	OK	4.8	-33515.6	-33.5156
30	OK	4.9	-33518.7	-33.5187
31	OK	5	-33521.2	-33.5212
32	OK	5.1	-33523.1	-33.5231
33	OK	5.2	-33524.6	-33.5246
34	OK	5.3	-33525.6	-33.5256

Q7) 5.5 bar

Q8) 187.7 kW

Summary	Balance	Performance Curve	Utility Usage	Status
Fluid power		147.041	kW	▼
Brake power		187.685	kW	▼
Electricity		187.685475	kW	▼
Volumetric flow rate		4524.35	l/min	▼
Pressure change		19.5	bar	▼
NPSH available		1.64002	m-kgf/kg	▼
NPSH required				▼
Head developed		206.715	m-kgf/kg	▼
Pump efficiency used		0.783445		
Net work required		187.685	kW	▼
Outlet pressure		20.5	bar	▼
Outlet temperature		95.2614	C	▼

Q9) 17.0 MW

Compressor model	Isentropic Turbine	
Phase calculations	Two phase calculation	
Indicated horsepower	-16.9777	MW
Brake horsepower	-16.9777	MW
▶ Net work required	-16.9777	MW
Power loss	0	MW
Efficiency		0.72
Mechanical efficiency		1
Outlet pressure	1	bar
Outlet temperature	108.461	C
Isentropic outlet temperature	99.6324	C
Vapor fraction		1
Displacement		
Volumetric efficiency		

Q10) 33.5263 MW net power produced from Q7 result divided by 200MW = 16.8%.

Tutorial 4

PART 1

Q1) 69.5 kW
Q2) 120.2°C
Q3) 697.6 kg/hr
Q4) 59.8 kW

Summary	Balance	Exchanger Details	Pres Drop/Velocities	Zones	Utility Usage	⊗ Status

Heatx results

Calculation Model	Shortcut				
		Inlet		**Outlet**	
Hot stream:	BOH-IN			BOH-OUT	
Temperature	117.7	C ▼		51.7315	C ▼
Pressure	2	bar ▼		2	bar ▼
Vapor fraction	0			0	
1st liquid / Total liquid	1			1	
Cold stream	H2O-IN			H2O-OUT	
Temperature	25	C ▼		107.7	C ▼
Pressure	2	bar ▼		2	bar ▼
Vapor fraction	0			0	
1st liquid / Total liquid	1			1	
Heat duty	59.7519	kW ▼			

Q5) 107°C

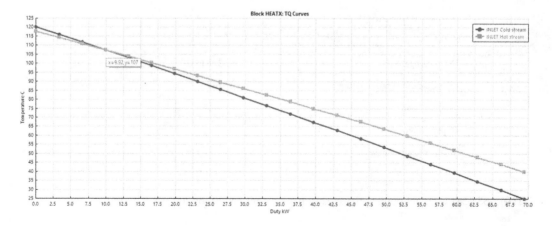

Block HEATX: TQ Curves

Q6) 697.6 kg/hr

Q7) 69.5 kW

Q8) 6.63 m^2

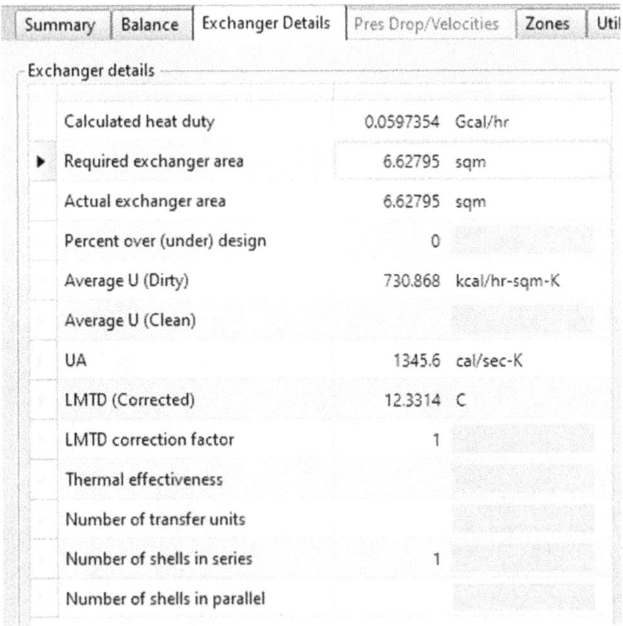

Summary	Balance	Exchanger Details	Pres Drop/Velocities	Zones	Util

Exchanger details

Calculated heat duty	0.0597354	Gcal/hr
▶ Required exchanger area	6.62795	sqm
Actual exchanger area	6.62795	sqm
Percent over (under) design	0	
Average U (Dirty)	730.868	kcal/hr-sqm-K
Average U (Clean)		
UA	1345.6	cal/sec-K
LMTD (Corrected)	12.3314	C
LMTD correction factor	1	
Thermal effectiveness		
Number of transfer units		
Number of shells in series	1	
Number of shells in parallel		

Q9) The heat exchanger is overdesigned by 22.2%.

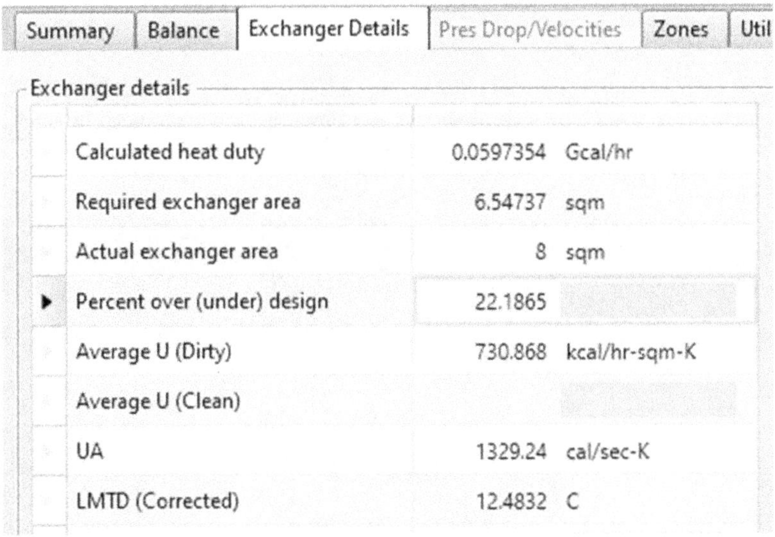

Summary	Balance	Exchanger Details	Pres Drop/Velocities	Zones	Util

Exchanger details

Calculated heat duty	0.0597354	Gcal/hr
Required exchanger area	6.54737	sqm
Actual exchanger area	8	sqm
▶ Percent over (under) design	22.1865	
Average U (Dirty)	730.868	kcal/hr-sqm-K
Average U (Clean)		
UA	1329.24	cal/sec-K
LMTD (Corrected)	12.4832	C

Q10) 43.3°C

Summary	Balance	Exchanger Details	Pres Drop/Velocities	Zones	Utility Usage	⊘ Status

Heatx results

Calculation Model	Shortcut			
		Inlet		**Outlet**
Hot stream:	BOH-IN		BOH-OUT	
Temperature	117.7	C ▼	43.3482	C ▼
Pressure	2	bar ▼	2	bar ▼
Vapor fraction	0		0	
1st liquid / Total liquid	1		1	
Cold stream	H2O-IN		H2O-OUT	
Temperature	25	C ▼	104.369	C ▼
Pressure	2	bar ▼	2	bar ▼
Vapor fraction	0		0	
1st liquid / Total liquid	1		1	
Heat duty	66.726	kW ▼		

Tutorial 5

PART 1

Q1) 0.592

Q2) −36.3

Component i	Component j	Source	Temp. Units	AIJ	AJI	BIJ	BJI	CIJ	CJI	DIJ	DJI	TLOWER
ACETONE	ISOBUTYL	APV120 VLE-RK	C	0	0	132.588	-210.63	0	0	0	0	25
ACETONE	ETHYLACE	APV120 VLE-RK	C	0	0	38.8767	-71.9375	0	0	0	0	39.3
ACETONE	N-HEPTAN	APV120 VLE-RK	C	0	0	42.921	-285.72	0	0	0	0	40
ISOBUTYL	N-HEPTAN	APV120 VLE-RK	C	0	0	33.5477	-160.959	0	0	0	0	45
ETHYLACE	N-HEPTAN	APV120 VLE-RK	C	0	0	12.3345	-119.796	0	0	0	0	50
ISOBUTYL	ETHYLACE	R-PCES	C	0.591581	-0.651678	-36.2797	1.23178	0	0	0	0	64.3

PART 2

Q3) 13

Q4) 13. Usually rounding up is preferred, but rounding to the nearest is fine if you are very close to an integer already.

Summary	Balance	Reflux Ratio Profile	✅ Status

▶ Minimum reflux ratio	1.31484	
Actual reflux ratio	45	
Minimum number of stages	12.5132	
Number of actual stages	12.916	
Feed stage	6.68376	
Number of actual stages above feed	5.68376	
Reboiler heating required	9.71569e+06	cal/sec
Condenser cooling required	9.62687e+06	cal/sec
Distillate temperature	57.8848	C
Bottom temperature	76.9801	C
Distillate to feed fraction	0.519188	
HETP		

Q5) 22

Summary	Balance	Reflux Ratio Profile

Theoretical stages	Reflux ratio
13	20.8153
14	8.04268
15	5.54625
16	3.6779
17	2.9832
18	2.66124
19	2.44504
20	2.27355
21	2.12164
▶ 22	1.97278
23	1.8174
24	1.69033
25	1.61756
26	1.57351
27	1.54312
28	1.5205
29	1.50273
30	1.48833

Q6) 11

Summary	Balance	Reflux Ratio Profile	⊘ Status

	Minimum reflux ratio	1.31484	
	Actual reflux ratio	1.97282	
	Minimum number of stages	12.5132	
	Number of actual stages	22	
▶	Feed stage	10.6812	
	Number of actual stages above feed	9.6812	
	Reboiler heating required	709561	cal/sec
	Condenser cooling required	620737	cal/sec
	Distillate temperature	57.8848	C
	Bottom temperature	76.9801	C
	Distillate to feed fraction	0.519188	
	HETP		

Q7) 0.465

	− Mole Fractions	
	ACETONE	0.481453
▶	ISOBUTYL	0.464668
	ETHYLACE	0.0168532
	N-HEPTAN	0.0370256

PART 3

Q8) 0.411

− Mole Fractions	
ACETONE	0.481104
ISOBUTYL	0.41138
ETHYLACE	0.0310183
N-HEPTAN	0.0764974

Q9) 0.401

− Mole Fractions	
ACETONE	0.480977
ISOBUTYL	0.40144
ETHYLACE	0.0371174
N-HEPTAN	0.0804664

Q10) 0.392

− Mole Fractions	
ACETONE	0.480728
ISOBUTYL	0.391849
ETHYLACE	0.0437131
N-HEPTAN	0.0837104

Tutorial 6

PART 1

Q1) 0.995

− Mole Fractions	
WATER	0
ACETONE	2.58514e-06
METHANOL	0.00458883
BUTANOL	0.995409

Q2) 0.995

− Mole Fractions	
WATER	0
ACETONE	0.994519
METHANOL	0.00548128
BUTANOL	2.06678e-08

Q3) $1.42/hr

Summary	Balance	Performance Curve	Utility Usage	Status

Utility ID	ELEC	
Utility duty	0.00922906	Gcal/hr
Utility usage	10.7334	kW
Utility cost	1.41681	$/hr
CO_2 emission rate	3.03905	kg/hr

Q4) $19.97/hr

Reboiler	HPS	
Duty	1.90772	Gcal/hr
Usage	4645.58	kg/hr
Cost	19.9681	$/hr
CO_2 emission rate	525.184	kg/hr

Q5) $11.57/hr

Condenser	BFWLP	
Duty	-1.46232	Gcal/hr
Usage	2793.23	kg/hr
Cost	-11.5714	$/hr
CO_2 emission rate	-402.567	kg/hr

Q6) 646.0 kg/hr

Condenser	CW			Reboiler	MPS	
Duty	-2.28269	Gcal/hr		Duty	2.34648	Gcal/hr
Usage	457811	kg/hr		Usage	4828.23	kg/hr
Cost	2.02612	$/hr		Cost	21.6133	$/hr
CO_2 emission rate				CO_2 emission rate	645.973	kg/hr

PART 2

Q7) 0.5

Q8) 0.514

Q9) about $62.3/hr

Q10) $47.7/hr. You can see from the below screen capture that this happens when the molar boilup ratio (BR) of column 2 is about 6.0, the BR of column 1 is about 2.9, and the molar reflux ratio (RR) of column 3 is about 8.1. These are very different from the starting conditions and all of the objectives are met, but with much lower cost! Note that you may have slightly different numbers, since different initial guesses can lead to different optimizer results (your initial guess would be the variable values from your most recent run, so depending on what you were doing before you hit run, you could have different guesses than others).

Main Flowsheet | C-2 | C-3 | **O-1 - Results** | +

Summary | Manipulated Variables | Constraints | Tear Variables | Iterations | ⊘ Status

Iteration	Constraint / * Tear stream / #Tear-variable with CMAX	CMAX	OBJECT-IVE FUNC TION	KUHN-TUCKER ERROR	LAGRANG-IAN FUNCTION	CONSTRA-INT 1 C-1 TOL= .1000-03	CONSTRA-INT 2 C-2 TOL= .1000-03	CONSTRA-INT 3 C-3 TOL= .1000-03	VARY 1 DC2 COL-SPEC MOLE-BR	VARY 2 DC1 COL-SPEC MOLE-BR	VARY 3 DC3 COL-SPEC MOLE-RR
0			62.4944			-1.62366e-...	0.00647691	0.014519	6.06109	5.1	2
1	*9	0.318241	61.0878	0.192666	61.154	0.00111606	-0.0001795...	0.0148522	6.12763	4.85362	2.19656
2	*9	13.3411	60.0099	0.115453	59.8655	0.00219353	-0.0006400...	0.014949	6.12387	4.71458	2.37052
3	*9	35.4937	57.5812	0.313661	57.8868	0.00367073	-0.0022566	0.0150255	6.17568	4.33408	2.86677
4	C-2	22.9424	53.5639	1.33263	53.3901	0.00558139	-0.0133768	0.014929	6.26614	3.7081	3.94428
5	C-2	134.647	49.9405	0.965147	49.6187	-0.00702233	-0.00297075	0.0117751	6.0068	3.23065	5.86955
6	C-1	71.063	47.3864	0.653018	47.0706	0.00940308	0.00801804	0.0024816	5.92348	3.08177	7.37705
7	*9	480.904	48.1217	0.509719	48.0387	-0.00873917	0.00814997	-0.0007802...	5.84245	3.05895	7.5773
8	C-1	87.281	48.632	0.190758	48.2784	-0.00211149	0.0106232	-0.00049601	5.93605	3.11893	7.25963
9	C-1	21.1212	48.5099	0.0317964	48.5335	6.23084e-05	0.0089321	0.000491903	5.98888	3.08453	7.37296
10	*9	1.28804	47.8344	0.0996795	47.9249	0.00113604	0.000487879	0.00140811	6.05167	2.95712	7.8693
11	*9	17.8807	47.7037	0.034957	47.7373	0.000239623	0.00015666	0.000258946	6.03735	2.93952	8.02951
12	*9	1.23895	47.6714	0.00796756	47.6761	6.26314e-05	3.34654e-06	1.95844e-05	6.03461	2.93508	8.06184
13	*9	0.329997	47.6718	0.00083231	47.6754	6.21653e-05	2.2609e-05	7.73544e-06	6.03461	2.93508	8.06184

Tutorial 7

PART 1

Q1) 939 seconds

Summary	Balance	Distributions	Polymer Attributes	⊘ Status

Stop criterion number	1	
Operation time	939.014	sec ▼
Heat load per cycle	-5.38816e+06	cal ▼
Average heat duty per cycle	-5393.48	cal/sec ▼
Minimum temperature	30	C ▼
Maximum temperature	33.5616	C ▼
Maximum volume deviation		
Maximum volume deviation time		▼

Q2) 17.8 kg

	Overall	Composition	Feed	Properties	Component Attr.	User \

Composition profiles

View Reactor accumulated mass ▾ Substream M

	Time	ALLYLALC	N-PROPYL	ACETONE	▲
	sec ▾	kg ▾	kg ▾	kg ▾	
▸	150	3.955	16.09	8.755	
▸	160	3.75358	16.4928	8.55358	▮
▸	170	3.56693	16.8661	8.36693	
▸	180	3.39356	17.2129	8.19356	
▸	190	3.23208	17.5358	8.03208	
▶	200	3.08138	17.8372	7.88138	
▸	210	2.94042	18.1192	7.74042	
▸	220	2.80835	18.3833	7.60835	
▸	230	2.68438	18.6312	7.48438	
▸	240	2.56781	18.8644	7.36781	
	250	2.45804	19.0839	7.25804	

Q3) Exothermic. The negative heating duty means cooling is required to maintain constant temperature.

PART 2

Q4) Aspen Plus reports $k = 2.71 \times 10^7$ which is in m³/kmol-sec. In m³/kmol-min this is about 1.63×10^9. Although the result can change considerably from run to run. To significant figures it is 27,000,000.

	Summary	Manipulated Variables	Fitted Data	Iteration History	✔ Status

	Vary no.	Initial value	Estimated value	Standard deviation	95% confidence interval	
					Lower limit	Upper limit
▶	1	2.5e+07	2.71336e+07	5523.97	2.71228e+07	2.71444e+07

Q5) 863 seconds

Summary	Balance	Distributions	Polymer Attributes	⊘ Status
Stop criterion number		1		
Operation time		862.977	sec	▼
Heat load per cycle		-5.38843e+06	cal	▼
Average heat duty per cycle		-5838.1	cal/sec	▼
Minimum temperature		30	C	▼
Maximum temperature		33.9362	C	▼
Maximum volume deviation				
Maximum volume deviation time				▼

Q6) 18.4 kg

Overall	Composition	Feed	Properties	Component Attr.	User Var

Composition profiles

View Reactor accumulated mass ▼   Substream MIX

Time	ALLYLALC	N-PROPYL	ACETONE
sec ▼	kg ▼	kg ▼	kg ▼
150	3.66751	16.665	8.46751
160	3.47172	17.0566	8.27172
170	3.29075	17.4185	8.09075
180	3.12307	17.7539	7.92307
190	2.96728	18.0654	7.76728
200	2.8222	18.3556	7.6222
210	2.68678	18.6264	7.48678
220	2.56019	18.8796	7.36019
230	2.44164	19.1167	7.24164
240	2.33041	19.3392	7.13041
250	2.22586	19.5483	7.02586

PART 3

Q7) 3.77 m. A design spec is needed.

PART 4

Q8) 0.00588 kmol/kg. You can get this by taking the moles in the outlet (80 kmol/hr) and dividing it by the mass flow rate (13615 kg/hr).

Q9) 457.2 K (184.1°C)

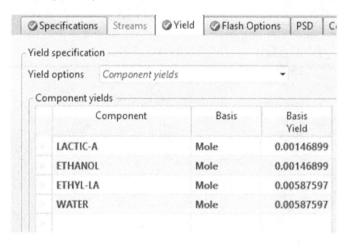

PART 5

Q10) 457.24 K (184.09°C). It should be exactly the same as Q9.

Q11) 8 kmol/hr.

PART 6

Q12) 33.35% conversion. You can calculate this by looking at the output stream component flow rates. Since we used exactly 100 kmol/hr of lactic acid in this example and 66.6452 kmol/hr are in the output, the difference (33.3548 kmol/hr) is what was reacted. Divide this number by the inlet (100 kmol/hr) and you get the percent conversion.

Q13) 33.07% conversion

Q14) 33.35% conversion (should be the same as the REQUIL case).

Material	Heat	Load	Vol.% Curves	Wt. % Curves	Petroleum	Polymers	Solids

	Units	EQUILOUT ▾	GIBBSOUT ▾
Mass Solid Fraction		0	0
Molar Enthalpy	cal/mol	-95435.8	-95435.8
Mass Enthalpy	cal/gm	-1401.94	-1401.94
Molar Entropy	cal/mol-K	-55.1319	-55.132
Mass Entropy	cal/gm-K	-0.809883	-0.809885
Molar Density	mol/cc	2.64204e-05	2.64173e-05
Mass Density	gm/cc	0.00179854	0.00179833
Enthalpy Flow	cal/sec	-5.30199e+06	-5.30199e+06
Average MW		68.0739	68.0739
− Mole Flows	**kmol/hr**	**200**	**200**
LACTIC-A	kmol/hr	66.3609	66.6452
ETHANOL	kmol/hr	66.3609	66.6452
ETHYL-LA	kmol/hr	33.6391	33.3548
WATER	kmol/hr	33.6391	33.3548
+ Mole Fractions			
+ Mass Flows	**kg/hr**	**13614.8**	**13614.8**
+ Mass Fractions			

Tutorial 8

PART 1

Q1) 2 column sections

Q2) in the range of 0.68 to 0.73 m (2.25 to 2.4 ft.)

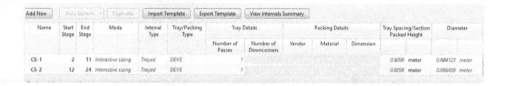

Name	Start Stage	End Stage	Mode	Internal Type	Tray/Packing Type	Tray Details			Packing Details			Tray Spacing/Section Packed Height	Diameter
						Number of Passes	Number of Downcomers		Vendor	Material	Dimension		
CS-1	2	11	Interactive sizing	Trayed	SIEVE	1						0.6096 meter	0.684123 meter
CS-2	12	24	Interactive sizing	Trayed	SIEVE	1						0.6096 meter	0.666459 meter

Q3) 0.68 to 0.73 m (2.25 to 2.4 ft.). Both methods essentially predict the same thing.

Name	Start Stage	End Stage	Mode	Internal Type	Tray/Packing Type	Tray Details			Packing Details			Tray Spacing/Section Packed Height	Diameter
						Number of Passes	Number of Downcomers		Vendor	Material	Dimension		
CS-1	2	11	Interactive sizing	Trayed	SIEVE	1						0.6096 meter	0.684449 meter
CS-2	12	24	Interactive sizing	Trayed	SIEVE	1						0.6096 meter	0.666461 meter

Q4) 0.77 to 0.8 m (2.5 to 2.6 ft)

Name	Start Stage	End Stage	Mode	Internal Type	Tray/Packing Type	Tray Details			Packing Details			Tray Spacing/Section Packed Height	Diameter
						Number of Passes	Number of Downcomers		Vendor	Material	Dimension		
CS-1	2	11	Interactive sizing	Trayed	SIEVE	1						1.5 ft	0.77229 meter
CS-2	12	24	Interactive sizing	Trayed	SIEVE	1						1.5 ft	0.732903 meter

PART 2

Q5) 0.049 bar

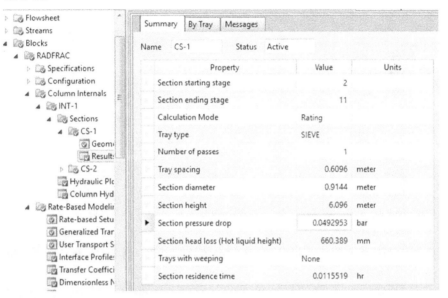

Tutorial 9

PART 1

Q1) 1417 kmol/hr

	Units	SYNGAS
Molar Solid Fraction		0
Mass Vapor Fraction		1
Mass Liquid Fraction		0
Mass Solid Fraction		0
Molar Enthalpy	cal/mol	-22904.3
Mass Enthalpy	cal/gm	-1769.47
Molar Entropy	cal/mol-K	4.06826
Mass Entropy	cal/gm-K	0.314292
Molar Density	mol/cc	0.000322369
Mass Density	gm/cc	0.0041728
Enthalpy Flow	cal/sec	-9.01515e+06
Average MW		12.9442
+ Mole Flows	**kmol/hr**	**1416.96**
+ Mole Fractions		
+ Mass Flows	**kg/hr**	**18341.4**
+ Mass Fractions		
Volume Flow	l/min	73257.7

Q2) 224 kmol/hr

Q3) 565 kmol/hr

	Units	STEAM ▾	SYNGAS-D ▾
Mass Liquid Fraction		0	0
Mass Solid Fraction		0	0
Molar Enthalpy	cal/mol	-56385	-8461.65
Mass Enthalpy	cal/gm	-3129.84	-935.828
Molar Entropy	cal/mol-K	-13.5652	0.582977
Mass Entropy	cal/gm-K	-0.752984	0.0644751
Molar Density	mol/cc	0.000855984	0.000994771
Mass Density	gm/cc	0.0154208	0.00899461
Enthalpy Flow	cal/sec	-3.50509e+06	-1.8821e+06
Average MW		18.0153	9.04189
— Mole Flows	**kmol/hr**	**223.789**	**800.736**
METHANE	kmol/hr	0	11.5263
WATER	kmol/hr	223.789	35.315
CO	kmol/hr	0	188.474
H2	kmol/hr	0	565.421

My calculator block looks as follows:

PART 2

Q4) 0.991

	Units	PERMEATE ▾
Molar Entropy	cal/mol-K	-4.93148
Mass Entropy	cal/gm-K	-2.18962
Molar Density	mol/cc	0.00099961
Mass Density	gm/cc	0.00225133
Enthalpy Flow	cal/sec	43830.3
Average MW		2.25221
− Mole Flows	**kmol/hr**	**526.96**
METHANE	kmol/hr	0.00476075
WATER	kmol/hr	0.310131
CO	kmol/hr	4.59738
H2	kmol/hr	522.048
− Mole Fractions		
METHANE		9.03436e-06
WATER		0.000588527
CO		0.00872435
H2		0.990678

My calculator block looks as follows:

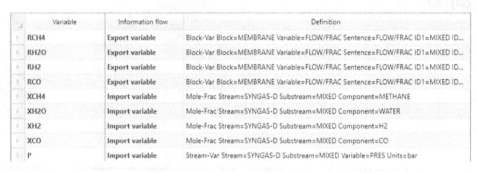

Variable	Information flow	Definition
RCH4	Export variable	Block-Var Block=MEMBRANE Variable=FLOW/FRAC Sentence=FLOW/FRAC ID1=MIXED ID...
RH2O	Export variable	Block-Var Block=MEMBRANE Variable=FLOW/FRAC Sentence=FLOW/FRAC ID1=MIXED ID...
RH2	Export variable	Block-Var Block=MEMBRANE Variable=FLOW/FRAC Sentence=FLOW/FRAC ID1=MIXED ID...
RCO	Export variable	Block-Var Block=MEMBRANE Variable=FLOW/FRAC Sentence=FLOW/FRAC ID1=MIXED ID...
XCH4	Import variable	Mole-Frac Stream=SYNGAS-D Substream=MIXED Component=METHANE
XH2O	Import variable	Mole-Frac Stream=SYNGAS-D Substream=MIXED Component=WATER
XH2	Import variable	Mole-Frac Stream=SYNGAS-D Substream=MIXED Component=H2
XCO	Import variable	Mole-Frac Stream=SYNGAS-D Substream=MIXED Component=CO
P	Import variable	Stream-Var Stream=SYNGAS-D Substream=MIXED Variable=PRES Units=bar

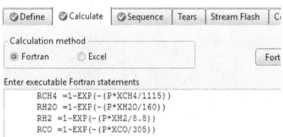

⊘ Define   ⊘ Calculate   ⊘ Sequence   Tears   Stream Flash   C

Calculation method

◉ Fortran   ○ Excel   Fort

Enter executable Fortran statements

```
        RCH4 =1-EXP(-(P*XCH4/1115))
        RH2O =1-EXP(-(P*XH2O/160))
        RH2 =1-EXP(-(P*XH2/8.8))
        RCO =1-EXP(-(P*XCO/305))
```

Q5) 0.984

Q6) 166.7 kmol/hr

Tutorial 10

PART 1

Q1) $75,100

Q2) $516,000

Q3) $158,200

Q4) $116,100

Q5) $220,300

Item	User T...	Item Description	Model	St...	I...	Equipment Cost (CAD)	Direct Cost (CAD)
1		Reflux Pump	CP CENTRIF			40200	75100
2		Distillation Column	TW TRAYED			250800	516000
3		Condenser	HE PRE ENGR			46800	158200
4		Reflux Drum	VT CYLINDER			16800	116100
5		Reboiler	RB KETTLE			99500	220300

Q6) 37 workers

```
         TOTAL MANPOWER SCHEDULE

              0    5   10   15
            40+....+....+....+  40
             :            :
             :            :
             :      *     :
             :      *     :
            35+      *    +  35
             :      *     :
             :      *     :
             :     **     :
             :     **     :
            30+     **    +  30
             :     **     :
             :     **     :
             :     **     :
    M        :    ***     :
    E       25+   ***     +  25
    N        :    ***     :
             :   ****     :
    P        :   ****     :
    E        :  ******    :
    R       20+  ******   +  20
             :  ******    :
    D        : *******    :
    A        :********    :
    Y        :********    :
            15+********    +  15
             :********    :
             :********    :
             :********    :
             :********    :
            10+********    +  10
             :********    :
             :********    :
             :********    :
             :********    :
             5+********    +   5
             :********    :
             :********    :
             :********    :
             :********    :
             0+....+....+....+   0
              0    5   10   15
```

PART 2

Q7) $1,089,300

Template: <Default> ▾ Save Save as new	

Summary	Utilities	Unit operation	Equipment	C

	▼
Total Capital Cost [USD]	5,320,310
Total Operating Cost [USD/Year]	1,739,840
Total Raw Materials Cost [USD/Year]	0
Total Product Sales [USD/Year]	0
Total Utilities Cost [USD/Year]	698,365
Desired Rate of Return [Percent/"Year]	20
P.O. Period [Year]	0
Equipment Cost [USD]	442,600
Total Installed Cost [USD]	1,089,300

Q8) 55 trays

Q9) 1.07 m (3.5 ft). Notice that it rounds to standard 0.5 ft increments.

Summary	Utilities	Unit operation	Equipment	Centri

	▼
User tag number	RADFRAC-tower
Remarks 1	Equipment mapped from 'RADFRAC'.
Quoted cost per item [USD]	
Currency unit for matl cost	
Number of identical items	
Installation option	
Application	
Base material Bottom	
Diameter Bottom section [meter]	1.0668
Bottom tangent to tangent height [meter]	37.1856
Design gauge pressure Bottom [barg]	5.52046
Design temperature Bottom [C]	196.785
Operating temperature Bottom [C]	169.007
Number of trays Bottom section	55
Bottom Tray type	SIEVE
Bottom Tray material	
Bottom Tray spacing [meter]	0.6096

Q10) 1.52 m (5 ft)

Summary	Utilities	Unit operation	Equipment	Centri

	▼
User tag number	RADFRAC-tower
Remarks 1	Equipment mapped from 'RADFRAC'.
Quoted cost per item [USD]	
Currency unit for matl cost	
Number of identical items	
Installation option	
Application	
Base material Bottom	
Diameter Bottom section [meter]	1.524
Bottom tangent to tangent height [meter]	37.1856
Design gauge pressure Bottom [barg]	5.52046
Design temperature Bottom [C]	196.785
Operating temperature Bottom [C]	169.007
Number of trays Bottom section	55
Bottom Tray type	SIEVE
Bottom Tray material	
Bottom Tray spacing [meter]	0.6096
Bottom Packing material Section1	

Tutorial 11

PART 1

Q1) 1,728,000 kJ/hr
Q2) 10,800 kJ/C-hr

Name	Inlet T [C]	Outlet T [C]	MCp [kJ/C-h]	Enthalpy [kJ/h]	Segm.	HTC [kJ/h-m2-C]	Flowrate [kg/h]	Effective Cp [kJ/kg-C]	DT Cont. [C]
Hot 1	320.0	220.0	1.728e+004	1.728e+006		720.00		---	Global
Hot 2	260.0	120.0	1.080e+004	1.512e+006		720.00		---	Global
Hot 3	140.0	139.0	1.260e+006	1.260e+006		720.00		---	Global
Cold 1	120.0	220.0	1.152e+004	1.152e+006		720.00		---	Global
Cold 2	70.0	240.0	5760	9.792e+005		720.00		---	Global
Cold 3	170.0	220.0	2.952e+004	1.476e+006		720.00		---	Global
Cold 4	235.0	236.0	1.260e+006	1.260e+006		720.00		---	Global
New									

Q3) A_Design4 has the lowest Total Cost Index of 3.33×10^{-3} \$/s.

Q4) There are 10 heat exchangers, six of which are process to process heat exchangers.

Q5) 1.26×10^6 kJ/hr.

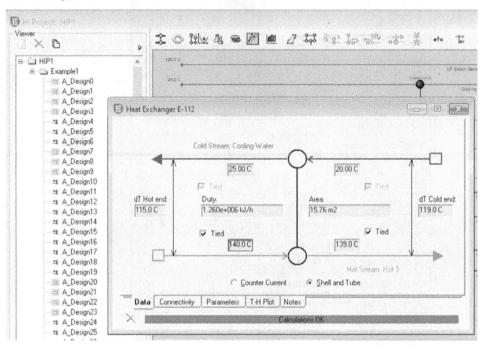

Q6) 3.24×10^6 kJ/hr.

Heat Exchanger		Load [kJ/h]	Cost Index [Cost]	Area [m2]	Shells	LMTD [C]	Overall U [kJ/h-m2-C]	FFactor
E-109		6.840e+00	8.533e+004	187.4	6	12.76	360.0	0.7945
E-111		1.239e+00	1.613e+004	10.71	2	37.30	360.0	0.8616
E-113		1.476e+00	2.592e+004	42.04	1	50.54	696.8	0.9969
E-114		2.902e+00	1.443e+004	8.486	1	95.74	360.0	0.9923
E-106		2.942e+00	2.176e+004	24.20	2	39.94	360.0	0.8453
E-108		1.260e+00	4.522e+004	95.36	2	37.09	360.0	0.9896
E-110		1.778e+00	1.498e+004	9.835	1	52.95	360.0	0.9485
E-115		4.099e+00	2.961e+004	45.87	2	30.43	360.0	0.8155
E-107		1.512e+00	1.390e+004	6.092	2	35.66	696.8	0.9991
E-112		1.260e+00	1.726e+004	15.76	1	117.0	683.5	0.9999

Q7) 1.49 Gcal/hr or 415,000 cal/sec

Tutorial 12

PART 1

Q1)   0.014 to 0.015 depending on convergence tolerances.

	Units	LIQ	VAP	FEED
Mass Liquid Fraction		1	0	0.300209
Mass Solid Fraction		0	0	0
Molar Enthalpy	kcal/mol	-68.5697	-94.2556	-81.2534
Mass Enthalpy	kcal/kg	-3728.11	-2144.62	-2620
Molar Entropy	cal/mol-K	-38.1474	-7.14397	-22.8379
Mass Entropy	cal/gm-K	-2.07406	-0.162548	-0.736402
Molar Density	kmol/cum	52.6945	1.96533	3.83343
Mass Density	kg/cum	969.19	86.3758	118.886
Enthalpy Flow	Gcal/hr	-0.0347096	-0.0465431	-0.0812527
Average MW		18.3926	43.9498	31.0128
+ Mole Flows	**kmol/hr**	**0.506195**	**0.493797**	**0.999991**
− Mole Fractions				
H2O		0.985478	0.0023068	0.499987
CO2		0.0144881	0.997693	0.499996
H3O+		1.69951e-05	0	8.60284e-06
HCO3-		1.69951e-05	0	8.60283e-06
CO3--		2.42042e-12	0	1.2252e-12

Tabs shown: Main Flowsheet | B1 (Flash2) - Block Options | **LIQ (MATERIAL) - Results (Default)** | FEED (MATERIAL) | +

Sub-tabs: Material | Vol.% Curves | Wt. % Curves | Petroleum | Polymers | Solids | Status

Q2) 0.5. The flash drum model incorrectly predicted only a single liquid phase with no vapour phase.

	Units	LIQ	VAP	FEED
Molar Vapor Fraction		0		0
Molar Liquid Fraction		1		1
Molar Solid Fraction		0		0
Mass Vapor Fraction		0		0
Mass Liquid Fraction		1		1
Mass Solid Fraction		0		0
Molar Enthalpy	kcal/mol	-85.152		-85.152
Mass Enthalpy	kcal/kg	-2745.73		-2745.73
Molar Entropy	cal/mol-K	-34.0621		-34.0621
Mass Entropy	cal/gm-K	-1.09833		-1.09833
Molar Density	kmol/cum	33.172		33.172
Mass Density	kg/cum	1028.75		1028.75
Enthalpy Flow	Gcal/hr	-0.085152		-0.085152
Average MW		31.0125		31.0125
+ Mole Flows	**kmol/hr**	**1**	**0**	**1**
− Mole Fractions				
CO2		0.5	0	0.5
H2O		0.5	0	0.5

Main Flowsheet × LIQ (MATERIAL) - Results (Default) × +

Material | Vol.% Curves | Wt. % Curves | Petroleum | Polymers | Solids | ⊘ Status

Q3) The ENRTL-RK model is more accurate.

PART 3

Q4) Roughly 831 to 832 tonne/hr. Results may vary.

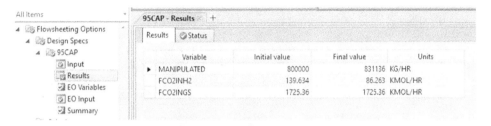

Q5) Should be the same or slightly lower than Q4. Results may vary. If you had something else, make sure you updated your stage efficiencies and reaction definitions for the new stage count.

Q6) about 21 mol% water and 75 mol% CO_2.

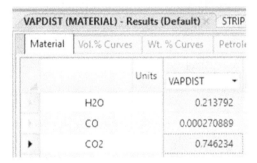

PART 4

Q7) 842 to 843 tonne/hr. Results may vary, but should be higher than before.

Q8) 820 to 860 kg/hr of water (small!) and just 1.3 to 1.5 kg/hr of MDEA (tiny!). Ranges are wider because of convergence tolerances—these are small amounts compared to the large solvent flow rates in the loop. Thus, the water loss rate through the product streams is only about 0.2% per hour of the total water flow in the loop, which is fine because it is cheap. The MDEA loss rate is only 0.0004% per hour, which is great because it is expensive!

MAKEUP (MATERIAL) - Results (Default)	95CAP	STRIPPER (RadFrac)	STRIPPER (RadFrac

Material	Vol.% Curves	Wt. % Curves	Petroleum	Polymers	Solids	⏀ Status

	Units	MAKEUP ▾
— Mass Flows	kg/hr	858.367
H2O	kg/hr	856.823
CO	kg/hr	0
CO2	kg/hr	0
CH4	kg/hr	0
H2	kg/hr	0
N2	kg/hr	0
MDEA	kg/hr	1.51308

Bonus Tutorial 1

PART 1

Q1) 31.8 MW

STEAM (MATERIAL)	OXIDIZER (RGibbs) - Results	Main Flowsheet

Summary	Balance	Phase Composition	Pure Solids	Atom Matrix	K

Outlet temperature	559	C ▾
Outlet pressure	30	bar ▾
Heat duty	-31.8433	MW ▾
Net heat duty	-31.8433	MW ▾
Vapor fraction	1	
Number of fluid phases	1	
Maximum number of pure solids	3	

Q2) 947°C

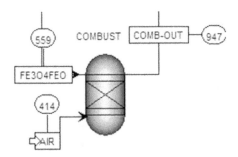

Q3) 0.819

	Units	FEOFE3O4 ▾
Mass Entropy	cal/gm-K	-0.0547412
Molar Density	mol/cc	0.0303199
Mass Density	kg/cum	5008.8
Enthalpy Flow	Gcal/hr	-352.367
Average MW		165.198
+ Mole Flows	**kmol/hr**	**2309.49**
+ Mole Fractions		
+ Mass Flows	**kg/hr**	**381525**
– Mass Fractions		
H2O		0
H2		0
CO		0
CO2		0
CH4		0
N2		0
O2		0
FE2O3		0
FEO		0.180673
FE3O4		0.819327

Q4) 0.17 bar

FE203-IN (MATERIAL) × / Main Flowsheet × **CYCLONE1 (Cyclone) - Results** ×		

Summary	Balance	Efficiency	⊘ Status

Results

Type of cyclone	Lapple-GP	
Number of cyclones	1	
Efficiency	1	
Pressure drop:	0.166018	bar ▾
Diameter of cylinder	0.5	meter ▾
Length of cylinder	1	meter ▾
Length of cone section	1	meter ▾
Diameter of overflow	0.25	meter ▾
Length of vortex finder	0.3125	meter ▾
Width of gas inlet	0.125	meter ▾
Height of gas inlet	0.25	meter ▾
Diameter of underflow	0.125	meter ▾
Axial inlet gas velocity	28.2571	m/sec ▾
Axial outlet gas velocity	17.9891	m/sec ▾
Circumferential velocity outer radius	22.2277	m/sec ▾
Circumferential velocity vortex radius	12.2389	m/sec ▾

Q5) This has an efficiency of 1, meaning there are no solids in the gas phase. So the assumption is perfectly accurate in this situation.

Q6) 99.97% (So if we had used a perfect separation assumption in this case by using an SSplit block instead, it would be a little bit inaccurate.)

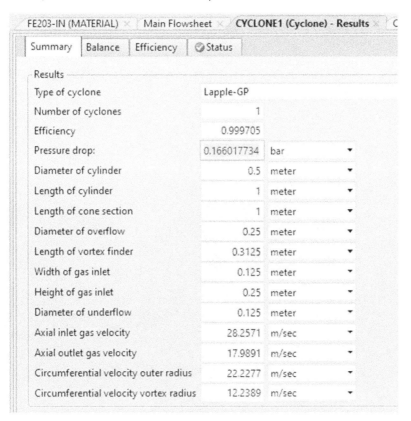

Q7) About 0.019 to 0.02 mm

Bonus Tutorial 2

PART 1

Q1) You may or may not have missing data, and if so, may have it at different locations.

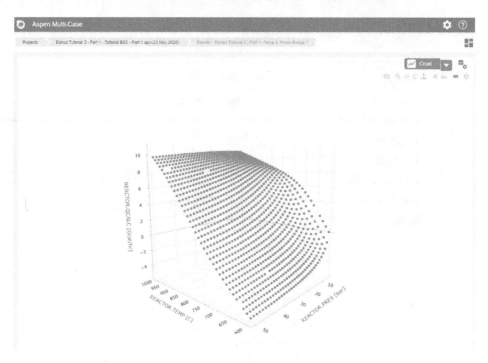

Q2) Outliers are evident in this plot, which is the result of either communication errors between Aspen Plus and Multi-Case, or numerical errors in the simulation in Aspen Plus. These errors are not consistent from run to run. Your outliers may be different, or may not exist. This is a known bug in V12.0 which appears to be fixed in V12.1. I have left the examples with the bugs in on purpose because many users routinely use older versions of the software. Outliers aside, this plot shows that the mole fraction of CO in the reactor outlet is strongly correlated with reactor heat duty. Reactor temperature has a moderate impact when the CO mole fractions is high, but when it is low it has almost no impact.

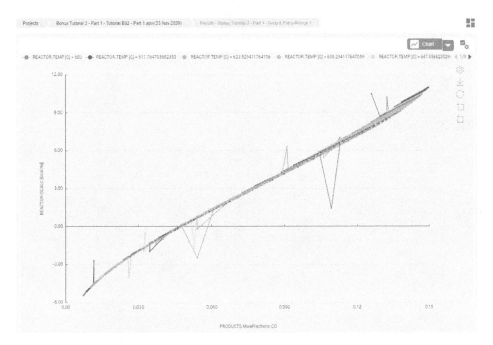

Bonus Tutorial 3

PART 1

Q1)   0.364 (36.4 mol%)

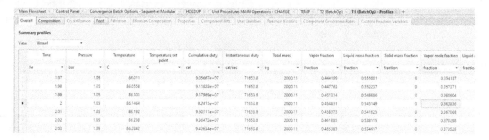

Q2)   37.67 kmol

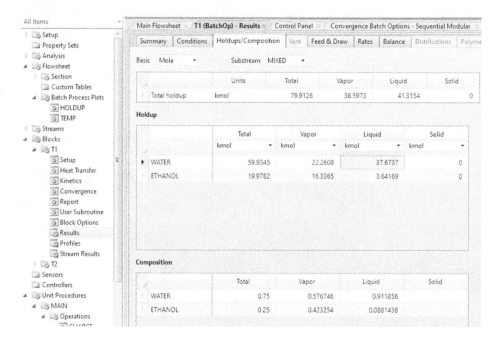

Q3) 0.912

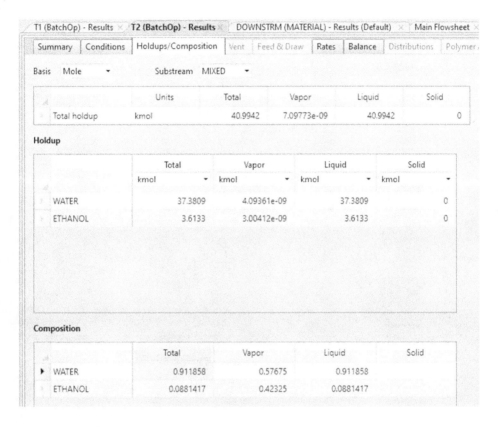

Q4) 0.580

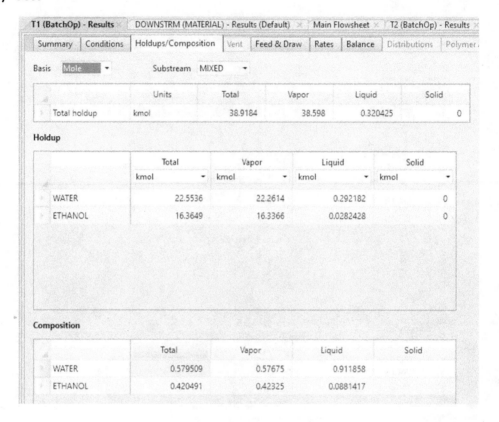

	Units	Total	Vapor	Liquid	Solid
Total holdup	kmol	38.9184	38.598	0.320425	0

Holdup

	Total	Vapor	Liquid	Solid
	kmol	kmol	kmol	kmol
WATER	22.5536	22.2614	0.292182	0
ETHANOL	16.3649	16.3366	0.0282428	0

Composition

	Total	Vapor	Liquid	Solid
WATER	0.579509	0.57675	0.911858	
ETHANOL	0.420491	0.42325	0.0881417	

Q5) 0.379

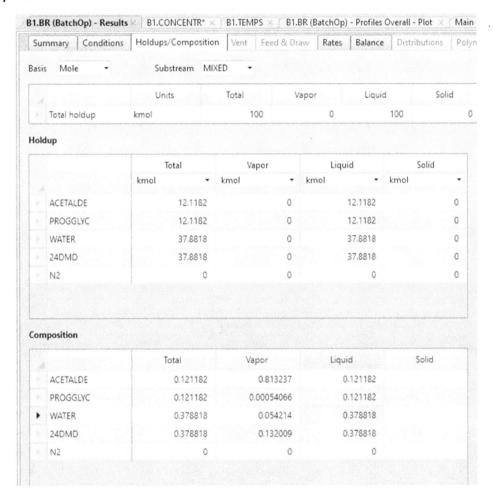

Q6) After 5 hours, the reaction is continuing quite slowly, so this is probably very close to equilibrium. It would probably not be worth running the reaction longer if it was the bottleneck of the larger process.

Q7) 10.92 kmol of propylene glycol liquid left in the plot, and some trace water. More propylene glycol is on the liquid trays and in the vapor phase throughout the column. You may have different values slightly if you used different convergence settings or stopping criteria.

Q8) 20.1 GJ. Note that this is the cumulative duty, not the instantaneous heat duty at the end of the cycle. That number would be about 207 kW and is found in the Jacket tab of the Results folder. This number is less useful, it just what the heat duty happens to be at the moment the last operating step ends.

BATCHDIS (BatchSep) - Profile Results	BATCHDIS (BatchSep) - Results	RCTPROD (MA

Summary	Pot	Distillate Receivers	Side Draw Receivers	Condenser	Condenser 2	F

Batch number	1	
Time from start of batch	25.7738	hr
Time from start of first batch	25.7738	hr
Cumulative pot duty	20.1052	GJ

Operating steps

Step	Step end time	Units	Value at step end	Units	TI
CHARGE	0.100009	hr			
— ACET	9.44891	hr			
POT L-MOLEFRAC ACET...			0.00799999	fraction	
— 24DMDH2O	25.7738	hr			

Bonus Tutorial 4

PART 2

Q1) Liquid phase is 79 mol% water, vapor phase is 47.4 mol% water.

Components - Specifications	TDE Binary Results	+

Experimental Data	Consistency Test

	No.	Liquid mole fraction WATER	Temperature (K)	Vapor mole fraction WATER
Binary VLE 078				
Binary VLE 079	15	0.68	354.923	0.423
Binary VLE 080	16	0.703	355.023	0.427
Binary VLE 082	17	0.759	355.923	0.454
Binary VLE 084	18	0.785	356.323	0.464
Binary VLE 091	19	0.79	356.123	0.474
Binary VLE 092	20	0.791	356.323	0.471
Binary VLE 093	21	0.797	356.423	0.476
Binary VLE 094	22	0.828	357.123	0.488

Q2) Binary VLE 010 has what we need. The highest H_2 concentration is 0.18 mol%, which occurs at the highest pressure in the range.

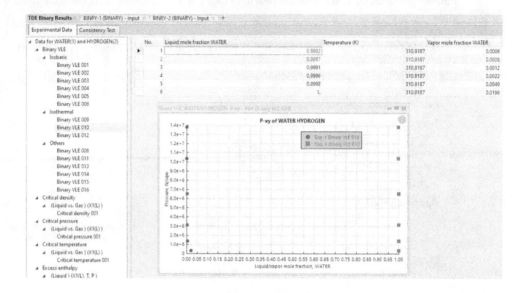

Q3) All three model the vapor phase excellently. NRTL-RK and PR-BM predict exactly zero hydrogen in the liquid phase and so they are both equally the worst. PSRK fits the liquid phase data closely. For low pressures (3 bar and lower) all three methods are good enough for most applications, unless H_2 purity is of great concern. Note that other experimental data (not shown) at close to atmospheric pressure shows that H_2 has liquid phase mole fractions on the order of 10 parts per million. Experimental data in Figure SB4.03 are taken from Gillespie PC, Wilson GM, Vapor-Liquid Equilibrium Data on Water-Substitute Gas Components: N_2-H_2O, H_2-H_2O, CO-H_2O, H_2-CO-H_2O, and H_2S-H_2O. *Gas Processors Association,* Research Report No. 41 (1980).

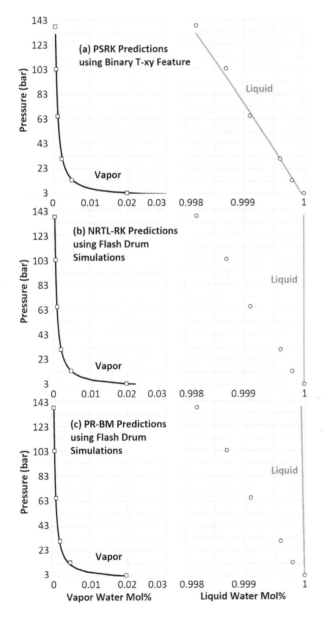

Q4) NRTL-RK is just a little better than PR-BM (by the way, PSRK is mostly indistinguishable from PR-BM). However, all three are pretty good to be honest, ternary models can be hard to perfect because the parameters are based on binary interactions only.

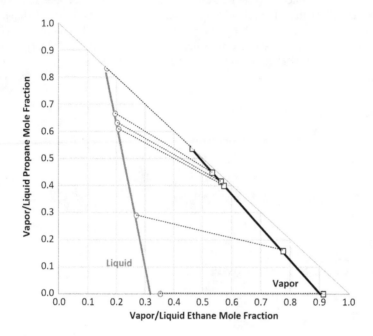

Q5) My completed table is below. The fit is quite good, and the R^2 is a very good 0.976.

		1	4	7	12	16
	Temperature (K)	**334.35**	**336.95**	**339.85**	**343.55**	**346.05**
Measured Liquid Mole Fractions	Methyl Acetate	0.1410	0.1370	0.2000	0.0560	0.0533
	Methanol	0.6780	0.4970	0.1880	0.2890	0.2495
	Isopropanol	0.0440	0.0760	0.2430	0.3540	0.2206
	Isopropyl Acetate	0.1370	0.2900	0.3690	0.3010	0.4766
Measured Vapor Mole Fractions	Methyl Acetate	0.3050	0.3130	0.3960	0.1230	0.1385
	Methanol	0.6190	0.5180	0.2600	0.4480	0.3430
	Isopropanol	0.0320	0.0590	0.1420	0.2530	0.1887
	Isopropyl Acetate	0.0440	0.1100	0.2020	0.1760	0.3298
Feed Mole Fractions	Methyl Acetate	0.2230	0.2250	0.2980	0.0895	0.0959
	Methanol	0.6485	0.5075	0.2240	0.3685	0.2963
	Isopropanol	0.0380	0.0675	0.1925	0.3035	0.2047
	Isopropyl Acetate	0.0905	0.2000	0.2855	0.2385	0.4032
Measured Liquid Mole Fractions	Methyl Acetate	0.1535	0.1709	0.2485	0.0651	0.0690
	Methanol	0.6734	0.5006	0.2095	0.3440	0.2647
	Isopropanol	0.0475	0.0772	0.2083	0.3249	0.2097
	Isopropyl Acetate	0.1255	0.2513	0.3337	0.2661	0.4566
Measured Vapor Mole Fractions	Methyl Acetate	0.3134	0.3289	0.4297	0.1614	0.1577
	Methanol	0.6161	0.5207	0.2627	0.4408	0.3688
	Isopropanol	0.0256	0.0489	0.1505	0.2406	0.1930
	Isopropyl Acetate	0.0449	0.1014	0.1571	0.1572	0.2805

Command Index

Below is a selected list of commands used in this book. This is not even close to an exhaustive index of commands available in Aspen Plus or related products. Consult the programs' user guides for more information.

Unit Operation Models

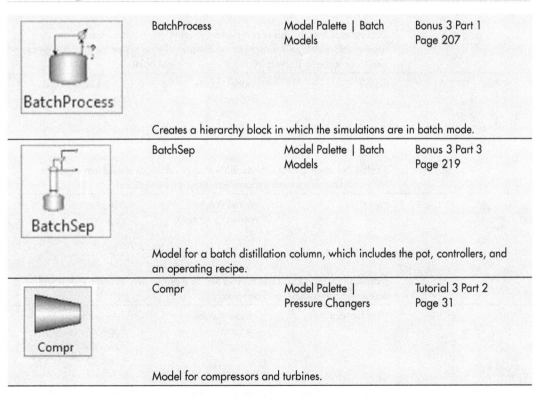

	BatchProcess	Model Palette | Batch Models	Bonus 3 Part 1 Page 207

Creates a hierarchy block in which the simulations are in batch mode.

	BatchSep	Model Palette | Batch Models	Bonus 3 Part 3 Page 219

Model for a batch distillation column, which includes the pot, controllers, and an operating recipe.

	Compr	Model Palette | Pressure Changers	Tutorial 3 Part 2 Page 31

Model for compressors and turbines.

	Cyclone	Model Palette | Solids Separators	Bonus 1 Part 2 Page 164

A rigorous model for a cyclone.

	DSTWU	Model Palette | Columns	Tutorial 5 Part 2 Page 55

A shortcut distillation model using the Winn-Underwood-Gilliland method.

	Dupl	Model Palette | Manipulators	Tutorial 5 Part 3 Page 58

Duplicates a stream. This is completely nonphysical, but it is useful to create superstructures on your flowsheet, which basically represent multiple design or simulation options starting from the same original point.

	Flash2	Model Palette | Separators	Introduction Page xv

Models the splitting of a stream containing a solid and a fluid into separate solid and fluid streams (a cyclone with simple assumptions).

	FSplit	Model Palette | Mixers/Splitters	Introduction Page xvi

Models the splitting of a stream into two or more smaller streams with equal pressure, temperature, and composition.

	Heat (stream)	Model Palette | Streams	Tutorial 3 Part 1 Page 27 Tutorial 4 Part 1 Page 42

Models a heat flow. Note: A Heat flow can be negative, which mean heat flows in the opposite direction that the arrow is pointing.

	Heat Mixer	Model Palette | Mixers/Splitters | Mixer	Tutorial 3 Part 2 Page 32

Models the combination of multiple heat streams. Temperature information of the inlet streams is usually lost. Functionally, this is the simple summation of inlet heat duties and can be positive or negative.

	Heater	Model Palette | Exchangers	Tutorial 4 Part 1 Page 40

A "half" or "one-sided" heat exchanger model.

	HeatX	Model Palette | Exchangers	Tutorial 4 Part 2 Page 44

A heat exchanger model that considers both hot and cold sides. Checks for temperature crossover.

	Hierarchy Block	Model Palette | User Models	Bonus 3 Part 1 Page 207

A hierarchy block is its own flow sheet that can connect to other units or hierarchy blocks. Thus you can have sub-flowsheets within flowsheets, helping to organize large models.

	Material (stream)	Model Palette | Streams	Tutorial 1 Part 2 Page 5

Models a material flow.

	Mixer	Model Palette | Mixers/Splitters	Tutorial 3 Part 2 Page 31

Models the combination of multiple material streams into one output stream. Functionally, this is the simple addition of mass and enthalpy of the inputs, followed by a flash calculation.

Pump	Pump	Model Palette | Pressure Changers	Tutorial 1 Part 2 Page 5

Models a pump, with varying degrees of rigor.

RadFrac	RadFrac	Model Palette | Columns	Tutorial 1 Part 2 Page 9 Tutorial 5 Part 3 Page 58

This is a stage-by-stage model of a distillation column with many features. Includes both equilibrium-based and rate-based models.

RBatch	RBatch	Model Palette | Reactors	Tutorial 7 Part 1 Page 75

Kinetic-based model for a batch reactor system with connecting buffer tanks.

REquil	REquil	Model Palette | Reactors	Tutorial 7 Part 6 Page 83

Generic reactor (no particular type) with user-specified reactions and associated equilibrium or equilibrium-approach conditions.

RGibbs	RGibbs	Model Palette | Reactors	Tutorial 7 Part 6 Page 84

Generic reactor (no particular type) without specified reactions using minimize Gibbs free energy approach to finding chemical equilibria.

RPlug	RPlug	Model Palette | Reactors	Tutorial 7 Part 3 Page 80

Kinetic-based model for a plug flow reactor.

	RStoic	Model Palette | Reactors	Tutorial 7 Part 5 Page 83

Generic reactor (no particular type) with user-specified reactions and associated specified conversions.

	RYield	Model Palette | Reactors	Tutorial 7 Part 4 Page 81

Generic reactor (no particular type) with user-specified output conditions.

	Sep	Model Palette | Separators	Tutorial 9 Part 2 Page 107

This is a generic separator where you literally specify the outputs of the separation and it performs a flash calculation on the outputs.

	SSplit	Model Palette | Mixers/Splitters	Bonus 1 Part 1 Page 164

Models the splitting of a stream containing a solid and a fluid into separate solid and fluid streams (a cyclone with simple assumptions).

	Work (stream)	Model Palette | Streams	Tutorial 3 Part 2 Page 31

Models the flow of mechanical or electrical work.

	Work Mixer	Model Palette | Mixers/Splitters | Mixer	Tutorial 3 Part 2 Page 31

Models the combination of multiple work streams. Functionally, it is the simple addition of the power of each input work stream.

Model Analysis Tools

Data Fit	Data Fit	Home | Analysis	Tutorial 7 Part 2 Page 80
	A collection of tools for model parameter fitting, such as in identifying the parameters of a kinetic reaction rate equation from experimental data.		
Design Specs	Design Spec	Model Analysis Tools	Tutorial 3 Part 1 Page 26
	Specify an output that you want, and what input or parameter you want to change so that the desired output is achieved. Aspen Plus will run a guess-and-check solver loop until either the desired output is found, or, it gives up.		
	Evaluate Project	Aspen Capital Cost Estimator	Tutorial 10 Part 1 Page 121
	Evaluates all equipment models in Aspen Capital Cost Estimator.		
Optimization	Optimization	Model Analysis Tools	Tutorial 6 Part 2 Page 67
	This is a sophisticated guess-and-check algorithm that attempts to find the combination of certain inputs or parameters that lead to the "best" outputs as you define it.		
Sensitivity	Sensitivity Analysis	Home	Tutorial 3 Part 2 Page 31
	Runs the simulation repeatedly in a loop, changing values of parameters each time according to your specification. Records key model results in a table.		
TQ Curves	TQ Curves	Home | Plot	Tutorial 4 Part 2 Page 46
	Generates "TQ Curves" (temperature vs. duty), also known as composite curves, which show the temperature of a heat exchanger as a function of where it is inside the reactor. Because the model does not consider the actual length or dimensions of the reactor, it uses the heat duty transferred so far instead.		
Process Utilities	Utilities	Economics	Tutorial 6 Part 1 Page 63
	Defines utilities that you will use in your process.		

Physical Properties

Binary	Binary Analysis	Properties | Home	Tutorial 2 Part 3 Page 15
	Generate a T-xy or P-xy vapor-liquid equilibria diagram using model predictions.		

	Calculate Bonds	Properties | Components | Molecular Structure	Tutorial 5 Part 1 Page 54
Calculate Bonds			

Retrieves the molecular structure information about a chemical from the Aspen Properties database and converts it to the tabular form needed in order to estimate UNIQUAC or NRTL binary interaction parameters using UNIFAC.

	Components	Properties | Home	Tutorial 1 Part 1 Page 3
Components			

This lets you choose which components (such as chemicals) you have in your model.

	Draw Bonds	Properties | Components | Molecular Structure	Tutorial 5 Part 1 Page 55
Draw/Import/Edit			

This lets you define a molecular structure visually, either starting from scratch or using an existing structure.

	Electrolyte Wizard	Properties | Components | Specifications	Tutorial 12 Part 1 Page 140
Elec Wizard			

The electrolyte wizard is a guided process to help add electrolyte reaction models into a physical property package.

	Methods	Properties | Home	Tutorial 1 Part 1 Page 5
Methods			

Go to the form where you select the default property methods for your simulation.

	NIST	Properties | Home	Bonus 4 Part 2 Page 229
TDE NIST			

Launch the National Institute of Science and Technology database of physical property experimental data

	Retrieve Parameters	Properties | Home	Tutorial 2 Part 2 Page 14
Retrieve Parameters			

Aspen Properties uses a large number of physical property parameters that are hidden from the user by default, such as molecular weight, critical pressure, vapor pressure correlations, and more. Use this button to get access.

	Setup	Properties | Home	Tutorial 1 Part 1 Page 2
Setup			

The setup form is where you enter the most basic information about the simulation.

User Commands

Axis Map	Axis Map	Plot Format Ribbon	Bonus 3 Part 1 Page 208
	Adjust which variables appear on which y-axis on a plot. You can have a different y-axis for each variable or you can combine variables together into one y-axis.		
Batch	Batch Mode	Batch	Bonus 3 Part 1 Page 207
	Switches the current flowsheet into batch mode.		
Exchange Icon CTRL+K	Change Icon	(When Block or Stream Selected | Right-Click)	Tutorial 1 Part 2 Page 9
	Changes the icon of the current flowsheet block. This *does not* change the model itself; it is purely for show.		
METCHEM ▾ Unit Sets	Change Units Set	Setup | Specifications	Tutorial 1 Part 1 Page 2
	Change the default units set using the drop-down box. If you create your simulation with different templates, you may have different unit sets to choose from. You can also make your own in the Unit Sets section by clicking the Unit Sets icon (the rulers). This appears in both the Properties and Simulation modes.		
↔ Connect	Connect to Simulation	ASW (in Excel)	Tutorial 9 Part 3 Page 109
	Connect to an Aspen Plus simulation from within Aspen Simulation Workbook.		
Continuous	Continuous Mode	Batch	Bonus 3 Part 1 Page 207
	Switches the current flowsheet into continuous mode.		
Control Panel F7	Control Panel	Home	Tutorial 2, Part 4 Page 20
	Opens the control panel. The control panel shows useful messages during the simulation.		

	Details	Home (in AEA)	Tutorial 11 Part 2 Page 136

Gets the detailed results of an Aspen Energy Analyzer run.

Energy Available Energy Savings — MW — % of Actual off	Energy Analysis Tab	Home [Above Simulation Page]	Tutorial 11 Part 2 Page 135

Provides an energy analysis of current flowsheet.

	Find Variables	In ASW Organizer	Tutorial 9 Part 3 Page 111

Locate Aspen Plus simulation variables from within Aspen Simulation Workbook.

Find Variables	Find Variables	Elsewhere in Aspen Plus	Bonus 3 Part 1 Page 208

Locate Aspen Plus simulation variables from within Aspen Plus.

Blank Simulation	New Blank Simulation	File	New	Tutorial 1 Part 1 Page 2

Create a new simulation without any templates.

New Strip Chart	New Strip Chart	Batch	Batch Process Plots	Bonus 3 Part 1 Page 208

For a flowsheet in batch mode, creates a new dynamic plot (time on the x-axis) using variables from the current flowsheet. New icons are made next to New Strip Chart for each new plot that you create.

Next F4	Next	Home, Title Bar	Bonus 1 Part 1 Page 163

Moves to the next input form for the flowsheet that requires user input before the simulation can run (usually with a red half-circle). This is useful when the model is incomplete and you are unsure what exactly is missing.

	Organizer	ASW (in Excel)	Tutorial 9 Part 3 Page 109
	Opens the Organizer in Aspen Simulation Workbook, which lets you connect to an Aspen Plus simulation.		
▶ F5	Play	Home, Title Bar ASW (Excel)	Tutorial 1 Part 2 Page 8
	This runs the simulation. When in Properties mode, only the properties portion of the model is executed (such as estimating parameters). In Simulation mode, the entire simulation is run, including Properties if need be. In Aspen Simulation Workbook, this runs the Aspen Plus simulation to which ASW is connected.		
Results Curve Plot	Plot	Special ribbons for each block	Tutorial 3 Part 2 Page 35
	A generic data plotting tool. Use the GUI, or, use CTRL+ALT+X to define x-axis variables, CTRL+ALT+Y to defined y-axis variables, and CTRL+ALT+P to examine the result (there will be no feedback on the +X and +Y keystrokes so you won't know if you did it correctly until you see the plot). The buttons appear when selecting certain blocks or forms when results are available, such as in the Column Design ribbon that appears when selecting a RadFrac block, or the Home ribbon when selecting a Sensitivity block.		
◀	Reinitialize	Home, Title Bar	Tutorial 6 Part 2 Page 70
	Reinitializes the simulation. Resets all stream and block variables to their initial guesses (or default guesses if unreconciled).		
Rename Block CTRL+M	Rename	(When Block or Stream Selected | Right-Click)	Tutorial 1 Part 2 Page 5
	Rename a block or stream.		
Results... CTRL+R	Show Results	(When Block or Stream Selected / Right-Click)	Tutorial 1 Part 2 Page 8 Tutorial 3 Part 1 Page 29
	Shows the results for a stream or block (if results are available).		
Stream Analysis▾	Stream Analysis	Home (when stream selected)	Tutorial 2, Part 4 Page 22
	Perform useful calculations on a stream (once it has been computed), such as generating dew and bubble point curves, and distillation curves.		

Music Index

The feedback from readers of the 1st edition made it clear that the music breaks were quite popular. I hope you enjoy this new selection for the 2nd edition, with curations by Tia Ghantous and yours truly. All the links provided are to legal music streams, where the copyright holder receives a share of ad revenue. Print readers: get this list digitally (clickable links, including a playlist) at:

http://PSEcommunity.org/books/lap24

Artist	Song	Location	Link
Andrew Bayer	Nobody Told Me	Tutorial 2	https://youtu.be/3nuZrkSm9fI
Armin van Buuren	Be In The Moment	Tutorial 12	https://youtu.be/hSPmBKB6XEI
Avicii	Wake Me Up	Tutorial 4	https://youtu.be/IcrbM1I_Bol
Basement Jaxx	Good Luck	Tutorial 5	https://youtu.be/HbVKmSQqELY
Ben Folds	Selfless, Cold and Composed	Tutorial 8	https://youtu.be/aCIzAXc5DpE
Boom Boom Satellites	Shut Up and Explode	Tutorial B3	https://youtu.be/VObVB69daOA
Coldplay	A Sky Full of Stars	Tutorial 11	https://youtu.be/VPRjCeoBqrl
d.notive	Red Light Syndrome	Tutorial B4	https://youtu.be/NjsemNhze7U
Daft Punk	Da Funk	Tutorial 1	https://youtu.be/PwILkY9gRrc
Deadmau5	Closer	Preface	https://youtu.be/52Nla2rMuvA
Fatboy Slim	Rockefeller Skank	Tutorial 5	https://youtu.be/FMrly9zm7QY
Gareth Emery	Long Way Home	Tutorial B2	https://youtu.be/0bj4i-sW44s
Glenn Miller	In the Mood	Tutorial B1	https://youtu.be/_CI-0E_jses
Groove Armada	My Friend	Tutorial 2	https://youtu.be/JxohJX9ElpE

Jan Eggum	Kor e alle helter hen	Tutorial 10	https://youtu.be/IjvfX0HKtRA
Kristin Husøy	Pray for Me	Tutorial B1	https://youtu.be/bHShSfZf7pY
Live	I Alone	Tutorial B2	https://youtu.be/FNrQOUtXYOo
Louis Armstrong	A Kiss to Build a Dream On	Tutorial B3	https://youtu.be/rAJMTd6OO78
Marie-Mai	Trahison sur ma peau	Tutorial 3	https://youtu.be/4A7HWtKLVWA
Miles Davis	Tempus Fugit	Tutorial 9	https://youtu.be/LfefnFfQJrs
Oceanlab	Satellite	Tutorial 3	https://youtu.be/NqPAOslEDcl
Odd Nordstoga	Kveldssong for deg og meg	Tutorial 7	https://youtu.be/Ctepl77QmAU
Queen	Bohemian Rhapsody	Tutorial B4	https://youtu.be/fJ9rUzlMcZQ
Red Hot Chili Peppers	My Friends	Tutorial 7	https://youtu.be/0kT5w27Yxyl
REM	Electrolyte	Tutorial 12	https://youtu.be/1LewYq40Svw
Smashing Pumpkins	Zero	Tutorial 8	https://youtu.be/vDgUTnlyDtQ
Snow Patrol	Chasing Cars (Armin van Buuren Remix)	Tutorial 11	https://youtu.be/wMMyg5e5uTs
Squirrel Nut Zippers	Hell	Tutorial 9	https://youtu.be/iLYB9pvww2M
The Contortionist (Cover)	1979	Tutorial 1	https://youtu.be/MPBvDnQjQJ0
The Crystal Method	Sling the Decks	Tutorial 6	https://youtu.be/2rSyy1riozs
Tony Furtado	Peggy-O	Tutorial 10	https://youtu.be/OzmtDoyhRKA
Tycho	Montana	Tutorial B3	https://youtu.be/QX5XKFn7Ngo
Vance Joy	Riptide	Tutorial 4	https://youtu.be/MsTWpbR_TVE
Vulfpeck	Rango II	Tutorial 6	https://youtu.be/4dpvyZBKg6Q
Weezer	Island in the Sun	Tutorial B2	https://youtu.be/erG5rgNYSdk

Subject Index

Note: Page numbers followed by *f* denote figures; page numbers followed by *n* denote footnotes.